Quantum Mechanics for Nuclear Structure, Volume 2

An intermediate level view

IOP Series in Nuclear Spectroscopy and Nuclear Structure

About the Series:

The IOP Series in Nuclear Spectroscopy and Nuclear Structure provides up-to-date coverage of theoretical and experimental topics in all areas of modern nuclear spectroscopy and structure. Books in the series range from student primers, graduate textbooks, research monographs and practical guides to meet the needs of students and scientists.

All aspects of nuclear spectroscopy and structure research are included, for example:

- Nuclear systematics and data
- Ab-initio models
- Shell model-based descriptions
- Nuclear collective models
- Nuclear symmetries and algebraic descriptions
- Many-body aspects of nuclear structure
- Quantum mechanics for nuclear structure study
- Spectroscopy with gamma-rays, charged particles and neutrons following radioactive decay and reactions
- Spectroscopy of rare isotopes
- Detectors employed for gamma-ray, neutron and charged-particle detection in nuclear spectroscopy and related societal applications.

Quantum Mechanics for Nuclear Structure, Volume 2

An intermediate level view

Kris Heyde
Ghent University, Belgium

John L Wood
School of Physics, Georgia Tech, Atlanta, GA 30332-0430, USA

IOP Publishing, Bristol, UK

ISBN 978-0-7503-2171-6 (ebook)
ISBN 978-0-7503-2169-3 (print)
ISBN 978-0-7503-2172-3 (myPrint)
ISBN 978-0-7503-2170-9 (mobi)

DOI 10.1088/978-0-7503-2171-6

Version: 20200401

IOP ebooks

British Library Cataloguing-in-Publication Data: A catalogue record for this book is available from the British Library.

Published by IOP Publishing, wholly owned by The Institute of Physics, London

IOP Publishing, Temple Circus, Temple Way, Bristol, BS1 6HG, UK

US Office: IOP Publishing, Inc., 190 North Independence Mall West, Suite 601, Philadelphia, PA 19106, USA

Contents

Preface

This book, the second in a two-part work, deals with topics that are essential for a mastery of the quantum mechanics underlying the nuclear many-body problem. These topics represent what can fairly be described as 'an intermediate level view', and so we adopted this subtitle. There are also some more specialised topics that we have selected because it is our opinion that they are important for handling the emerging view of the quantum mechanics needed to understand nuclei.

We begin with a thorough treatment of angular momentum theory, handled in three chapters. First, we present representations of rotations, angular momentum, and spin. This is manifested in all of their popular guises; and in some less familiar ones such as Bargmann representations. Implicit in our treatment is the need to think in terms of representations that have a tensor structure: we note such structures where they appear; but these structures play a much more important role than just their manifestation in the handling of angular momentum and spin. We continue with the details of the coupling of spins and angular momenta. The techniques are fundamental to handling finite many-body quantum systems. The Clebsch–Gordan coefficients that are encountered here are an essential topic that must be mastered by a researcher in nuclear structure, whether theorist or experimentalist. We give a brief introduction to 'recoupling' coefficients, 6-j and 9-j symbols. The manipulation of recoupling is straightforward but takes practice to achieve mastery: that is left for our more advanced reader to take up at a later stage in the series. We complete the angular momentum topics with details of vector and tensor operators. This is also a topic that must be mastered by the nuclear structure researcher, such that they have a clear appreciation of the power of the Wigner–Eckart theorem.

The focus then turns to identical particles and the representation of many-body states and operators. A heavy emphasis is placed on second quantization or the occupation number representation. We pay particular attention to how this language can be used to formulate solvable models of systems with correlations. Such systems exhibit properties that lie completely outside of any classical mechanical concepts. The quasi-spin formalism is a powerful language for the description of many-fermion systems that form Cooper pairs and is developed in detail. We also introduce the Lipkin model: this is a 'toy' model, i.e. it is not realised in nature (it is too simple). But it is exactly solvable and so can be used to test many-body approximation methods. Many-body approximations such as Hartree–Fock theory will be handled later in the series.

We then turn to the role of group theory and of algebraic structures in quantum mechanics. These two topics are closely related in quantum mechanics because of the close relationship between Lie groups and Lie algebras. We give a basic introduction to these topics, using what has been learned via angular momentum theory. We particularly emphasize the role that groups and algebras play in quantum mechanics, both as a way to a deeper understanding of the subject and as a set of tools for formulating models. We provide an introduction to Young diagrams and their manipulation. We take a few steps into the Cartan theory of Lie algebras, sufficient to acquire a deeper appreciation for the mathematics behind their application to quantum systems, especially ladder operators and spectrum generating algebras.

We complete the volume with standard treatments of perturbation theory, the variational method, and a brief handling of the quantization of the electromagnetic field and its interactions with matter.

We have aimed to focus on the quantum mechanics needed for taking up research into the nuclear many-body problem, without going into the details of nuclear modelling and approximation methods. These require familiarity with nuclear data and transformation processes, which adds another 'dimension' to the path to mastery of the subject. These steps will be taken later in the series.

Author biographies

Kris Heyde

Kris Heyde is Professor Emeritus in the Department of Physics and Astronomy at the University of Gent. He continues to work on joint research projects with both experimental and theoretical nuclear physicists. His research interests are on one side directed to learn how collectivity emerges starting from a microscopic point of view. He has a long-standing deep interest in the presence of shape coexistence in atomic nuclei, in particular studying the conditions needed to be realised throughout the nuclear mass table.

John L Wood

John L Wood is a Professor Emeritus in the School of Physics at Georgia Institute of Technology. He continues to collaborate on research projects in both experimental and theoretical nuclear physics. Special research interests include nuclear shapes and systematics of nuclear structure.

IOP Publishing

Quantum Mechanics for Nuclear Structure, Volume 2
An intermediate level view
Kris Heyde and John L Wood

Chapter 1

Representation of rotations, angular momentum and spin

The various representations of rotations in physical space, (3, $\mathbb{R}$) and Hilbert space (n, $\mathbb{C}$) are developed in detail. This leads to an in-depth treatment of the representation of states of well-defined spin and angular momentum in quantum systems. The peculiarities of the physics of spin-$\frac{1}{2}$ systems (spinors) are outlined. The tensorial character of representations is implicit in the treatment. The Schwinger and Bargmann representations are introduced in some detail; and this leads to $SU(2)$ coherent states (which are important for more advanced group representation theory).

Concepts: Euler angles; matrix representations; Pauli spin matrices; ket rotations; $SU(2)$ and $SO(3)$ tensor representations; Schwinger representation; spherical harmonics as Cartesian tensors; spin-$\frac{1}{2}$ neutron interferometry; Bargmann space; measure of a space; $SU(2)$ coherent states; non-unitary representations.

Angular momentum and spin are dynamical variables that are fundamental to finite systems in quantum mechanics, i.e. for molecules, atoms, nuclei and hadrons. To fully handle the quantum mechanics of these systems, the mathematical *representation* of rotations is fundamental. Some elements of these issues in quantum mechanics are introduced in Volume 1. Namely, the concept of a group, the use of matrices, the distinction between rotations in physical space, (3, $\mathbb{R}$), and Hilbert space is presented in chapter 10; and the basic quantization of spin and angular momentum, using algebraic methods, is presented in chapter 11. Further, the facility with which these methods *reduce* the solution of central force problems in quantum mechanics to simple algebraic problems in terms of a single (radial) degree of freedom is presented in chapter 12.

The mathematical representation of rotations is a rich paradigm for the whole of quantum mechanics. In this chapter, a wide range of mathematical *tools* is introduced. Matrix algebra and the algebra of polynomials in real and complex

doi:10.1088/978-0-7503-2171-6ch1

variables feature prominently. The peculiar physics of spin-$\frac{1}{2}$ particles and spinors is presented. But, the primary aim is to initiate a *language* that is suitable for the theoretical formulation of finite many-body quantum systems. Group theory and Lie algebras are implicit in the material presented in this chapter: the groups $SO(2)$, $SO(3)$ and $SU(2)$ feature prominently in their behind-the-scene role. The road into many-body systems necessitates more complicated groups such as $SU(3)$: some of the material in this chapter is intended to 'pave' this road.

1.1 Rotations in (3, $\mathbb{R}$)

One way to describe a rotation in (3, $\mathbb{R}$) is in terms of rotation in a plane through a specified angle[1]. This is defined in terms of an axis of rotation $\hat{n}$ and an angle ϕ. The axis $\hat{n}$ is perpendicular to the plane defined by the initial and final orientations of the vectors $\vec{V}$, $\vec{V}'$: $R(\phi)\vec{V} = \vec{V}'$. The difficulty lies in ascertaining the direction of $\hat{n}$. Although this *'axis-angle'* parameterisation or *Darboux* parameterisation is simple in principle, it is difficult to use in practice.

The most widely used practical parameterisation of rotations in (3, $\mathbb{R}$) is in terms of *Euler* rotations. Consider a space-fixed coordinate frame $Oxyz$ and a body-fixed coordinate frame $O\bar{x}\bar{y}\bar{z}$. The orientation of an object can be specified by the rotation R that rotates the $O\bar{x}\bar{y}\bar{z}$ frame into the $Oxyz$ frame. This can be done in three steps as illustrated in figure 1.1.

Figure 1.1 depicts the following:

$$R(\alpha, \beta, \gamma) = R_z(\gamma)\, R_Y(\beta)\, R_{\bar{z}}(\alpha). \tag{1.1}$$

Note the order of the three rotations. The problem is that these three rotations are about axes belonging to three different frames of reference. The three rotations on the right-hand side of equation (1.1) can be restated in terms of a single frame of reference using *similarity transformations*, specifically

$$R_z(\gamma) = R_Y(\beta)\, R_{\bar{z}}(\gamma)\, R_Y^{-1}(\beta) \tag{1.2}$$

and

$$R_Y(\beta) = R_{\bar{z}}(\alpha)\, R_{\bar{y}}(\beta)\, R_{\bar{z}}^{-1}(\alpha). \tag{1.3}$$

Thus,

$$R(\alpha,\beta,\gamma) = R_Y(\beta)\, R_{\bar{z}}(\gamma)\, \underbrace{R_Y^{-1}(\beta) R_Y(\beta)}_{I}\, R_{\bar{z}}(\alpha), \tag{1.4}$$

$$\therefore\ R(\alpha, \beta, \gamma) = R_Y(\beta)\, R_{\bar{z}}(\gamma)\, R_{\bar{z}}(\alpha); \tag{1.5}$$

and, since $R_{\bar{z}}(\gamma)$ and $R_{\bar{z}}(\alpha)$ commute,

[1] It should be noted that rotations in (3, $\mathbb{R}$) can be elegantly represented using *quaternions*. Use of quaternions avoids gimbal lock; they are used for programming robots and computer games.

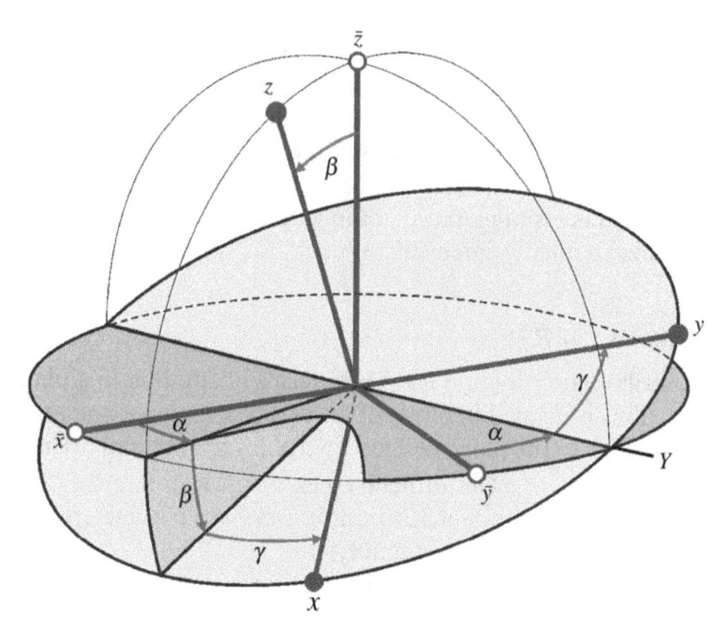

Figure 1.1. The Euler angles (α, β, γ) defined in terms of a three-step sequence of rotations that take an intrinsic or body-fixed frame $O\bar{x}\bar{y}\bar{z}$ into a space-fixed frame $Oxyz$. Note that the axes of rotation are: $O\bar{z}$; the line of intersection of the $O\bar{x}\bar{y}$ and Oxy planes, OY; and Oz. In Step I, $\bar{y}$ rotates to Y and $\bar{z}$ remains fixed; in Step II, $\bar{z}$ rotates to z and Y remains fixed; in Step III, Y rotates to y and z remains fixed. Further, note that the ranges of the angles are: $0 \leqslant \alpha < 2\pi$, $0 \leqslant \beta < \pi$, $0 \leqslant \gamma < 2\pi$. This results in an ambiguity for the rotation $\beta = 0$, $(\alpha, 0, \gamma) \equiv (\alpha', 0, \gamma')$ if $\alpha + \gamma = \alpha' + \gamma'$: this is referred to as '*gimbal lock*' (where 'gimbal' refers to the rotation device or mechanical operator). This figure is adapted from that found on the Easyspin website.

$$\therefore R(\alpha, \beta, \gamma) = R_Y(\beta)\, R_{\bar{z}}(\alpha)\, R_{\bar{z}}(\gamma), \tag{1.6}$$

$$\therefore R(\alpha,\beta,\gamma) = R_{\bar{z}}(\alpha)\, R_{\bar{y}}(\beta)\, \underbrace{R_{\bar{z}}^{-1}(\alpha) R_{\bar{z}}(\alpha)}_{I}\, R_{\bar{z}}(\gamma), \tag{1.7}$$

$$\therefore R(\alpha, \beta, \gamma) = R_{\bar{z}}(\alpha)\, R_{\bar{y}}(\beta)\, R_{\bar{z}}(\gamma). \tag{1.8}$$

Note the new order of the three rotations (cf. equation (1.1)).

1.2 Matrix representations of spin and angular momentum operators

The matrix elements of $\hat{J}_z$, $\hat{J}_\pm$ in the $\{|jm\rangle; j = 0, \frac{1}{2}, 1, \frac{3}{2}, \ldots; m = +j, +j-1, \ldots, -j\}$ basis are (cf. Volume 1, chapter 11):

$$\langle j'm'|\hat{J}_z|jm\rangle = m\hbar\delta_{j'j}\delta_{m'm}, \tag{1.9}$$

$$\langle j'm'|\hat{J}_\pm|jm\rangle = \sqrt{(j \mp m)(j \pm m + 1)}\,\hbar\delta_{j'j}\delta_{m'm\pm1}. \tag{1.10}$$

Matrix elements of $\hat{J}_x$ and $\hat{J}_y$ follow from:

$$\hat{J}_x = \frac{1}{2}(\hat{J}_+ + \hat{J}_-), \tag{1.11}$$

$$\hat{J}_y = \frac{1}{2i}(\hat{J}_+ - \hat{J}_-), \tag{1.12}$$

where, recall $\hat{J}_\pm := \hat{J}_x \pm i\hat{J}_y$. Thus, the matrix representations of $\hat{J}_x$, $\hat{J}_y$, and $\hat{J}_z$ in a $\{|jm\rangle\}$ basis are:

$$\hat{J}_X \leftrightarrow \frac{\hbar}{2}\left(\begin{array}{c|c|c|c|c}
0 & 0 & 0 & 0 & \\ \hline
0 & \begin{matrix}0 & 1\\ 1 & 0\end{matrix} & 0 & & \\ \hline
0 & 0 & \begin{matrix}0 & \sqrt{2} & 0\\ \sqrt{2} & 0 & \sqrt{2}\\ 0 & \sqrt{2} & 0\end{matrix} & 0 & \\ \hline
0 & 0 & 0 & \begin{matrix}0 & \sqrt{3} & 0 & 0\\ \sqrt{3} & 0 & 2 & 0\\ 0 & 2 & 0 & \sqrt{3}\\ 0 & 0 & \sqrt{3} & 0\end{matrix} & \\ \hline
 & & & & \ddots
\end{array}\right), \tag{1.13}$$

$$\hat{J}_Y \leftrightarrow \frac{\hbar}{2i}\left(\begin{array}{c|c|c|c|c}
0 & 0 & 0 & 0 & \\ \hline
0 & \begin{matrix}0 & 1\\ -1 & 0\end{matrix} & 0 & & \\ \hline
0 & 0 & \begin{matrix}0 & \sqrt{2} & 0\\ -\sqrt{2} & 0 & \sqrt{2}\\ 0 & -\sqrt{2} & 0\end{matrix} & 0 & \\ \hline
0 & 0 & 0 & \begin{matrix}0 & \sqrt{3} & 0 & 0\\ -\sqrt{3} & 0 & 2 & 0\\ 0 & -2 & 0 & \sqrt{3}\\ 0 & 0 & -\sqrt{3} & 0\end{matrix} & \\ \hline
 & & & & \ddots
\end{array}\right), \tag{1.14}$$

$$
\hat{J}_Z \leftrightarrow \frac{\hbar}{2}\left(\begin{array}{c|cc|ccc|cccc|c}
0 & 0 & 0 & 0 & 0 & 0 & 0 & 0 & 0 & 0 & \\
\hline
0 & 1 & 0 & & 0 & & & & & & \\
 & 0 & -1 & & & & & & & & \\
\hline
 & & & 2 & 0 & 0 & & & & & \\
0 & 0 & & 0 & 0 & 0 & & 0 & & & \\
 & & & 0 & 0 & -2 & & & & & \\
\hline
 & & & & & & 3 & 0 & 0 & 0 & \\
0 & 0 & & & 0 & & 0 & 1 & 0 & 0 & \\
 & & & & & & 0 & 0 & -1 & 0 & \\
 & & & & & & 0 & 0 & 0 & -3 & \\
\hline
 & & & & & & & & & & \ddots
\end{array}\right). \tag{1.15}
$$

Note the 'block-diagonal' form of $\hat{J}_x$ and $\hat{J}_y$. These blocks correspond to $j = 0, \frac{1}{2}, 1, \frac{3}{2}, \ldots$. The matrix representation of $\hat{J}_z$ is diagonal with eigenvalues $0; \frac{1}{2}\hbar, -\frac{1}{2}\hbar; \hbar, 0, -\hbar; \frac{3}{2}\hbar, \frac{1}{2}\hbar, -\frac{1}{2}\hbar, -\frac{3}{2}\hbar; \ldots$. It is normal practice to *reduce* these (infinite) matrices by breaking apart the blocks to give finite dimensional matrices. Thus, e.g.

$j = \frac{1}{2}$:

$$
\hat{J}_x^{\left(\frac{1}{2}\right)} \leftrightarrow \frac{\hbar}{2}\begin{pmatrix} 0 & 1 \\ 1 & 0 \end{pmatrix}, \qquad \hat{J}_y^{\left(\frac{1}{2}\right)} \leftrightarrow \frac{\hbar}{2}\begin{pmatrix} 0 & -i \\ i & 0 \end{pmatrix}, \qquad \hat{J}_z^{\left(\frac{1}{2}\right)} \leftrightarrow \frac{\hbar}{2}\begin{pmatrix} 1 & 0 \\ 0 & -1 \end{pmatrix}; \tag{1.16}
$$

$j = 1$:

$$
\begin{aligned}
\hat{J}_x^{(1)} &\leftrightarrow \frac{\hbar}{2}\begin{pmatrix} 0 & \sqrt{2} & 0 \\ \sqrt{2} & 0 & \sqrt{2} \\ 0 & \sqrt{2} & 0 \end{pmatrix}, \\
\hat{J}_y^{(1)} &\leftrightarrow \frac{\hbar}{2}\begin{pmatrix} 0 & -\sqrt{2}i & 0 \\ \sqrt{2}i & 0 & -\sqrt{2}i \\ 0 & \sqrt{2}i & 0 \end{pmatrix}, \\
\hat{J}_z^{(1)} &\leftrightarrow \frac{\hbar}{2}\begin{pmatrix} 2 & 0 & 0 \\ 0 & 0 & 0 \\ 0 & 0 & -2 \end{pmatrix}.
\end{aligned} \tag{1.17}
$$

The terminology:

$$
\hat{J}_x^{\left(\frac{1}{2}\right)} := \hat{S}_x, \qquad \hat{J}_y^{\left(\frac{1}{2}\right)} := \hat{S}_y, \qquad \hat{J}_z^{\left(\frac{1}{2}\right)} := \hat{S}_z, \tag{1.18}
$$

$$\sigma_x := \begin{pmatrix} 0 & 1 \\ 1 & 0 \end{pmatrix}, \qquad \sigma_y := \begin{pmatrix} 0 & -i \\ i & 0 \end{pmatrix}, \qquad \sigma_z := \begin{pmatrix} 1 & 0 \\ 0 & -1 \end{pmatrix}, \tag{1.19}$$

(cf. equation (1.16)), where σ_x, σ_y, σ_z are the *Pauli spin matrices*, is in common use.

1.3 The Pauli spin matrices

The Pauli spin matrices, σ_x, σ_y, σ_z, possess a number of useful properties. We redefine them by σ_j, σ_k, σ_l, $(j, k, l) = (x, y, z)$. Then

$$\sigma_j^2 = \sigma_k^2 = \sigma_l^2 = \hat{I}, \tag{1.20}$$

$$\sigma_j\sigma_k + \sigma_k\sigma_j = 0, \qquad \text{for } j \neq k, \tag{1.21}$$

i.e.

$$\{\sigma_j, \sigma_k\} = 2\delta_{jk}\hat{I}, \tag{1.22}$$

where '{, }' is an anticommutator bracket (also written ' $[,]_+$'). Further,

$$[\sigma_j, \sigma_k] = 2i\varepsilon_{jkl}\sigma_l, \tag{1.23}$$

where

$$\varepsilon_{jkl} \equiv \varepsilon_{klj} \equiv \varepsilon_{ljk} \equiv 1; \qquad \varepsilon_{kjl} \equiv \varepsilon_{jlk} \equiv \varepsilon_{lkj} \equiv -1. \tag{1.24}$$

From equations (1.22) and (1.23)

$$\sigma_j\sigma_k = -\sigma_k\sigma_j = i\sigma_l. \tag{1.25}$$

Also,

$$\sigma_j^\dagger = \sigma_j, \tag{1.26}$$

$$\det(\sigma_j) = -1, \tag{1.27}$$

$$\operatorname{tr}(\sigma_j) = 0. \tag{1.28}$$

For the three-dimensional Cartesian vector $\vec{a}$, $\vec{\sigma} \cdot \vec{a}$ is a 2×2 matrix[2]:

$$\vec{\sigma} \cdot \vec{a} := \sigma_x a_x + \sigma_y a_y + \sigma_z a_z, \tag{1.29}$$

$$\therefore \vec{\sigma} \cdot \vec{a} = \begin{pmatrix} a_z & a_x - ia_y \\ a_x + ia_y & -a_z \end{pmatrix}. \tag{1.30}$$

[2] The notation $\vec{\sigma}$, i.e. $(\sigma_x, \sigma_y, \sigma_z)$, viewed as a vector when the components are matrices, needs to be adopted as a powerful language. This will lead to the concept of *vector operators* in chapter 3; therein, the concept is formally developed. (See also Volume 1, equation (7.72).)

This leads to the important identity:

$$(\vec{\sigma} \cdot \vec{a})(\vec{\sigma} \cdot \vec{b}) = \vec{a} \cdot \vec{b}\hat{I} + i\vec{\sigma} \cdot (\vec{a} \times \vec{b}). \tag{1.31}$$

This can be obtained from equations (1.22) and (1.23):

$$(\vec{\sigma} \cdot \vec{a})(\vec{\sigma} \cdot \vec{b}) = \sum_j \sigma_j a_j \sum_k \sigma_k b_k. \tag{1.32}$$

$$\therefore (\vec{\sigma} \cdot \vec{a})(\vec{\sigma} \cdot \vec{b}) = \sum_j \sum_k \left(\frac{1}{2}\{\sigma_j, \sigma_k\} + \frac{1}{2}[\sigma_j, \sigma_k] \right) a_j b_k, \tag{1.33}$$

$$\therefore (\vec{\sigma} \cdot \vec{a})(\vec{\sigma} \cdot \vec{b}) = \sum_j \sum_k (\delta_{jk} + i\varepsilon_{jkl}\sigma_l) a_j b_k, \tag{1.34}$$

$$\therefore (\vec{\sigma} \cdot \vec{a})(\vec{\sigma} \cdot \vec{b}) = \vec{a} \cdot \vec{b}\hat{I} + i\vec{\sigma} \cdot (\vec{a} \times \vec{b}). \tag{1.35}$$

If the components of $\vec{a}$ are real then

$$(\vec{\sigma} \cdot \vec{a})^2 = |\vec{a}|^2 \hat{I}, \tag{1.36}$$

where $|\vec{a}|$ is the magnitude of the vector $\vec{a}$.

1.4 Matrix representations of rotations in ket space

We are now in a position to obtain matrix representations of rotation operators in ket space. For a rotation about an axis $\hat{n}$ through an angle ϕ (cf. Volume 1, chapter 10),

$$\mathcal{D}(R) = \mathcal{D}(\hat{n}, \phi) = \exp\left\{-i\frac{\vec{J} \cdot \hat{n}}{\hbar}\phi\right\}, \tag{1.37}$$

the matrix elements of $\mathcal{D}(\hat{n}, \phi)$ are

$$\langle jm'|\exp\left\{-i\frac{\vec{J} \cdot \hat{n}}{\hbar}\phi\right\}|jm\rangle := \mathcal{D}^{(j)}_{m'm}(\hat{n}, \phi), \tag{1.38}$$

where $j' = j$ is explicitly incorporated: this is because

$$\hat{J}^2 \mathcal{D}(R)|jm\rangle = \mathcal{D}(R)\hat{J}^2|jm\rangle, \tag{1.39}$$

$$\therefore \hat{J}^2\{\mathcal{D}(R)|jm\rangle\} = j(j+1)\hbar^2\{\mathcal{D}(R)|jm\rangle\}, \tag{1.40}$$

which follows from the general relationship $[\hat{A}, \exp\{\hat{A}\}] = 0$. This is sensible because rotations cannot change the length of a vector. Thus, the matrix representation of $\mathcal{D}(R)$ has the form:

$$\mathcal{D}(R) \leftrightarrow \left(\begin{array}{c|c|c|c|c} \mathcal{D}^{(0)} & 0 & 0 & 0 & \\ \hline 0 & \mathcal{D}^{\left(\frac{1}{2}\right)} & 0 & 0 & \\ \hline 0 & 0 & \mathcal{D}^{(1)} & 0 & \\ \hline 0 & 0 & 0 & \mathcal{D}^{\left(\frac{3}{2}\right)} & \\ \hline & & & & \ddots \end{array}\right) \tag{1.41}$$

and we can discuss the $\mathcal{D}^{(j)}$ individually.

The Euler angle parameterisation leads to a simplification when one considers a matrix representation:

$$\mathcal{D}^{(j)}_{m'm}(\alpha, \beta, \gamma) = \langle jm'|e^{\frac{-i\hat{J}_z\alpha}{\hbar}} e^{\frac{-i\hat{J}_y\beta}{\hbar}} e^{\frac{-i\hat{J}_z\gamma}{\hbar}}|jm\rangle; \tag{1.42}$$

but,

$$\langle jm'|e^{\frac{-i\hat{J}_z\alpha}{\hbar}} = \langle jm'|e^{-im'\alpha}, \tag{1.43}$$

$$\therefore \mathcal{D}^{(j)}_{m'm}(\alpha, \beta, \gamma) = e^{-i(m'\alpha+m\gamma)}\langle jm'|e^{\frac{-i\hat{J}_y\beta}{\hbar}}|jm\rangle, \tag{1.44}$$

i.e. only the '$\hat{J}_y$' rotation is non-trivial. We define

$$d^{(j)}_{m'm}(\beta) := \langle jm'|e^{\frac{-i\hat{J}_y\beta}{\hbar}}|jm\rangle. \tag{1.45}$$

The $\mathcal{D}^{(j)}_{m'm}(R)$'s ($R = \hat{n}, \phi$ or α, β, γ) are called *Wigner functions.* They tell us how much of $|jm\rangle$ rotates into $|jm'\rangle$ under the action of R:

$$\mathcal{D}(R)|jm\rangle = \sum_{m'}|jm'\rangle\langle jm'|\mathcal{D}(R)|jm\rangle, \tag{1.46}$$

where the completeness relation has been used.

We are now in a position to obtain explicit matrix representations of $\mathcal{D}(R)$, the so-called *Wigner matrices*:

$\mathcal{D}^{(0)}$: This is trivial. It is the 1×1 matrix (1).

$\mathcal{D}^{(\frac{1}{2})}$: This is a 2×2 matrix and can be evaluated from the properties of the Pauli spin matrices. Consider

$$\mathcal{D}^{\left(\frac{1}{2}\right)}(\hat{n}, \phi) = \exp\left\{-i\frac{\vec{J}^{\left(\frac{1}{2}\right)} \cdot \hat{n}}{\hbar}\phi\right\} = \exp\left\{-i\frac{\vec{\sigma} \cdot \hat{n}\phi}{2}\right\}. \tag{1.47}$$

Then, expanding the exponential:

$$\mathcal{D}^{\left(\frac{1}{2}\right)}(\hat{n}, \phi) = \hat{I} - i\frac{\phi}{2}\vec{\sigma} \cdot \hat{n} - \frac{1}{2!}\left(\frac{\phi}{2}\right)^2(\vec{\sigma} \cdot \hat{n})^2 + \frac{i}{3!}\left(\frac{\phi}{2}\right)^3(\vec{\sigma} \cdot \hat{n})^3 + \cdots. \tag{1.48}$$

But, from equation (1.36),

$$(\vec{\sigma} \cdot \hat{n})^m = \hat{I}, \qquad m \text{ even}, \tag{1.49}$$

$$(\vec{\sigma} \cdot \hat{n})^m = (\vec{\sigma} \cdot \hat{n}), \qquad m \text{ odd}, \tag{1.50}$$

$$\therefore \mathcal{D}^{\left(\frac{1}{2}\right)}(\hat{n}, \phi) = \hat{I}\left\{1 - \frac{1}{2!}\left(\frac{\phi}{2}\right)^2 + \cdots\right\} - i\vec{\sigma} \cdot \hat{n}\left\{\frac{\phi}{2} - \frac{1}{3!}\left(\frac{\phi}{2}\right)^3 + \cdots\right\}, \tag{1.51}$$

$$\therefore \mathcal{D}^{\left(\frac{1}{2}\right)}(\hat{n}, \phi) = \hat{I} \cos\frac{\phi}{2} - i\vec{\sigma} \cdot \hat{n} \sin\frac{\phi}{2}. \tag{1.52}$$

Explicitly,

$$\therefore \mathcal{D}^{\left(\frac{1}{2}\right)}(\hat{n}, \phi) = \begin{pmatrix} \cos\frac{\phi}{2} - in_z \sin\frac{\phi}{2} & (-in_x - n_y)\sin\frac{\phi}{2} \\ (-in_x + n_y)\sin\frac{\phi}{2} & \cos\frac{\phi}{2} + in_z \sin\frac{\phi}{2} \end{pmatrix} \tag{1.53}$$

for an axis-angle parameterisation.
For an Euler angle parameterisation

$$\mathcal{D}^{\left(\frac{1}{2}\right)}(\alpha, \beta, \gamma) = \mathcal{D}_z^{\left(\frac{1}{2}\right)}(\alpha)\mathcal{D}_y^{\left(\frac{1}{2}\right)}(\beta)\mathcal{D}_z^{\left(\frac{1}{2}\right)}(\gamma), \tag{1.54}$$

then using equation (1.52):

$$\mathcal{D}^{\left(\frac{1}{2}\right)}(\alpha, \beta, \gamma) = \begin{pmatrix} e^{-\frac{i\alpha}{2}} & 0 \\ 0 & e^{\frac{i\alpha}{2}} \end{pmatrix} \begin{pmatrix} \cos\frac{\beta}{2} & -\sin\frac{\beta}{2} \\ \sin\frac{\beta}{2} & \cos\frac{\beta}{2} \end{pmatrix} \begin{pmatrix} e^{-\frac{i\gamma}{2}} & 0 \\ 0 & e^{\frac{i\gamma}{2}} \end{pmatrix}, \tag{1.55}$$

$$\therefore \mathcal{D}^{\left(\frac{1}{2}\right)}(\alpha, \beta, \gamma) = \begin{pmatrix} e^{-\frac{i(\alpha+\gamma)}{2}} \cos\frac{\beta}{2} & -e^{-\frac{i(\alpha-\gamma)}{2}} \sin\frac{\beta}{2} \\ e^{\frac{i(\alpha-\gamma)}{2}} \sin\frac{\beta}{2} & e^{\frac{i(\alpha+\gamma)}{2}} \cos\frac{\beta}{2} \end{pmatrix}. \tag{1.56}$$

Note that $\mathcal{D}^{\left(\frac{1}{2}\right)}(\hat{n}, \phi)$ and $\mathcal{D}^{\left(\frac{1}{2}\right)}(\alpha, \beta, \gamma)$ fulfil the unitary unimodular or special unitary form $\begin{pmatrix} a & b \\ -b^* & a^* \end{pmatrix}$, cf. Volume 1, equation (10.42), and herein section 5.10.2.

$\mathcal{D}^{(1)}$: This is a 3×3 matrix. It can be evaluated using a series expansion if we use its Euler angle parameterisation. From

$$\mathcal{D}^{(1)}_{m'm}(\alpha, \beta, \gamma) = e^{-i(m'\alpha+m\gamma)} d^{(1)}_{m'm}(\beta), \tag{1.57}$$

expanding the exponential in $d^{(1)}$:

$$e^{-\frac{i\hat{J}_y^{(1)}\beta}{\hbar}} = \hat{I} - \frac{i\beta}{\hbar}\hat{J}_y^{(1)} - \frac{1}{2!}\frac{\beta^2}{\hbar^2}\left(\hat{J}_y^{(1)}\right)^2 + \frac{i}{3!}\frac{\beta^3}{\hbar^3}\left(\hat{J}_y^{(1)}\right)^3 + \cdots. \tag{1.58}$$

This is greatly simplified by the following identity

$$\begin{aligned}\frac{\left(\hat{J}_y^{(1)}\right)^3}{\hbar^3} = \frac{1}{8}&\begin{pmatrix} 0 & -\sqrt{2}i & 0 \\ \sqrt{2}i & 0 & -\sqrt{2}i \\ 0 & \sqrt{2}i & 0 \end{pmatrix}\begin{pmatrix} 0 & -\sqrt{2}i & 0 \\ \sqrt{2}i & 0 & -\sqrt{2}i \\ 0 & \sqrt{2}i & 0 \end{pmatrix} \\ &\times\begin{pmatrix} 0 & -\sqrt{2}i & 0 \\ \sqrt{2}i & 0 & -\sqrt{2}i \\ 0 & \sqrt{2}i & 0 \end{pmatrix},\end{aligned} \tag{1.59}$$

$$\therefore \frac{\left(\hat{J}_y^{(1)}\right)^3}{\hbar^3} = \frac{1}{8}\begin{pmatrix} 0 & -\sqrt{2}i & 0 \\ \sqrt{2}i & 0 & -\sqrt{2}i \\ 0 & \sqrt{2}i & 0 \end{pmatrix}\begin{pmatrix} 2 & 0 & -2 \\ 0 & 4 & 0 \\ -2 & 0 & 2 \end{pmatrix}, \tag{1.60}$$

$$\therefore \frac{\left(\hat{J}_y^{(1)}\right)^3}{\hbar^3} = \frac{1}{8}\begin{pmatrix} 0 & -4\sqrt{2}i & 0 \\ 4\sqrt{2}i & 0 & -4\sqrt{2}i \\ 0 & 4\sqrt{2}i & 0 \end{pmatrix}, \tag{1.61}$$

$$\therefore \frac{\left(\hat{J}_y^{(1)}\right)^3}{\hbar^3} = \frac{\hat{J}_y^{(1)}}{\hbar}. \tag{1.62}$$

Then, equation (1.58) reduces to

$$e^{-\frac{i\hat{J}_y^{(1)}\beta}{\hbar}} = \hat{I} + \frac{\hat{J}_y^{(1)}}{\hbar}\left\{-i\beta + \frac{i\beta^3}{3!} + \cdots\right\} + \frac{\left(\hat{J}_y^{(1)}\right)^2}{\hbar^2}\left\{\frac{-\beta^2}{2!} + \cdots\right\}, \tag{1.63}$$

$$\therefore e^{-\frac{i\hat{J}_y^{(1)}\beta}{\hbar}} = \hat{I} - \frac{i\hat{J}_y^{(1)}}{\hbar}\sin\beta + \frac{\left(\hat{J}_y^{(1)}\right)^2}{\hbar^2}(\cos\beta - 1). \tag{1.64}$$

Thus,

$$d^{(1)}(\beta) = \begin{pmatrix} \frac{1}{2}(1+\cos\beta) & \frac{-1}{\sqrt{2}}\sin\beta & \frac{1}{2}(1-\cos\beta) \\ \frac{1}{\sqrt{2}}\sin\beta & \cos\beta & \frac{-1}{\sqrt{2}}\sin\beta \\ \frac{1}{2}(1-\cos\beta) & \frac{1}{\sqrt{2}}\sin\beta & \frac{1}{2}(1+\cos\beta) \end{pmatrix}. \tag{1.65}$$

To evaluate $\mathcal{D}^{(1)}(\hat{n}, \phi)$ and $\mathcal{D}^{(j)}(\hat{n}, \phi)$ or $\mathcal{D}^{(j)}(\alpha, \beta, \gamma)$ with $j > 1$, we must develop the theory of tensor bases of representation in ket space.

1.5 Tensor representations for $SU(2)$

Consider the general $SU(2)$ transformation (cf. Volume 1, chapter 10)

$$\begin{pmatrix} a & b \\ -b^* & a^* \end{pmatrix}\begin{pmatrix} u_1 \\ u_2 \end{pmatrix} = \begin{pmatrix} u_1' \\ u_2' \end{pmatrix}, \tag{1.66}$$

where the 2×2 matrix may, for example, have the form given by equation (1.53) or equation (1.56). Then, defining

$$q_1 := u_1^2, \qquad q_2 := \sqrt{2}\,u_1u_2, \qquad q_3 := u_3^2, \tag{1.67}$$

under the transformation, equation (1.66), we obtain

$$q_1' = (u_1')^2 = (au_1 + bu_2)^2 = a^2u_1^2 + 2abu_1u_2 + b^2u_2^2, \tag{1.68}$$

$$\therefore q_1' = a^2q_1 + \sqrt{2}\,abq_2 + b^2q_3; \tag{1.69}$$

and similarly,

$$q_2' = -\sqrt{2}\,ab^*q_1 + (aa^* - bb^*)q_2 + \sqrt{2}\,ba^*q_3, \tag{1.70}$$

$$q_3' = (b^*)^2q_1 - \sqrt{2}\,a^*b^*q_2 + (a^*)^2q_3; \tag{1.71}$$

whence

$$\begin{pmatrix} a^2 & \sqrt{2}\,ab & b^2 \\ -\sqrt{2}\,ab^* & (aa^* - bb^*) & \sqrt{2}\,ba^* \\ (b^*)^2 & -\sqrt{2}\,a^*b^* & (a^*)^2 \end{pmatrix}\begin{pmatrix} q_1 \\ q_2 \\ q_3 \end{pmatrix} = \begin{pmatrix} q_1' \\ q_2' \\ q_3' \end{pmatrix}. \tag{1.72}$$

This is still a representation of an $SU(2)$ transformation: there are no new parameters. However, it is a 3×3 matrix representation of $SU(2)$. From the Euler angle parameterisation, equation (1.56),

$$a = \exp\left\{-\frac{i(\alpha + \gamma)}{2}\right\}\cos\frac{\beta}{2}, \qquad b = -\exp\left\{-\frac{i(\alpha - \gamma)}{2}\right\}\sin\frac{\beta}{2}; \tag{1.73}$$

and substitution of these values of a and b into the matrix in equation (1.72) will yield $\mathcal{D}^{(1)}(\alpha, \beta, \gamma)$, the β-dependent part of which is given by equation (1.65).

The process can be iterated by defining

$$p_1 := u_1^3, \qquad p_2 := \sqrt{3}\,u_1^2u_2, \qquad p_3 := \sqrt{3}\,u_1u_2^2, \qquad p_4 := u_2^3, \tag{1.74}$$

which will yield a 4×4 matrix representation of $SU(2)$, i.e. an expression for $\mathcal{D}^{(\frac{3}{2})}(R)$.

Expressions for $\mathcal{D}^{(j)}(\alpha, \beta, \gamma)$ can be obtained by this process, for any j, together with the values of a and b given in equation (1.73). Likewise, $\mathcal{D}^{(j)}(\hat{n}, \phi)$ can be obtained using (cf. equation (1.53))

$$a = \cos\frac{\phi}{2} - in_z \sin\frac{\phi}{2}, \qquad b = (-in_x - n_y)\sin\frac{\phi}{2}, \tag{1.75}$$

where, recall, the constraint $n_x^2 + n_y^2 + n_z^2 = 1$ ensures that (n_x, n_y, n_z, ϕ) corresponds to three free parameters. To reiterate: $SU(2)$ is a three-parameter group.

The representation associated with the two-component spinor (u_1, u_2) is called the *fundamental representation*. The representation associated with the three-component entity (q_1, q_2, q_3) is a *rank-2 $SU(2)$ tensor*, i.e. it is constituted from quadratic combinations of the fundamental representation. In turn, (p_1, p_2, p_3, p_4) is a rank-3 $SU(2)$ tensor.

1.6 Tensor representations for *SO*(3)

Consider the general $SO(3)$ transformation of the vector $\vec{V}$

$$V_{i'} = \sum_i R_{i'i} V_i, \tag{1.76}$$

where the V_i are the Cartesian components of the vector and the $R_{i'i}$ are the elements of the 3×3 matrix R that effects the orthogonal transformation. This can be generalised to

$$T_{i'j'k'\cdots} = \sum_i \sum_j \sum_{k\cdots} R_{i'i} R_{j'j} R_{k'k} \cdots T_{ijk\cdots}, \tag{1.77}$$

where the $T_{ijk\cdots}$ are the Cartesian components of a tensor, the rank of which is equal to the number of indices and the $R_{i'i}$ are, as before, elements of the 3×3 matrix R. Details are clarified in the following.

The simplest Cartesian tensor is of rank 2 and is often called a *dyad* or *dyadic*. It is formed from two Cartesian vectors, e.g.

$$\vec{U} = (U_1, U_2, U_3), \qquad \vec{V} = (V_1, V_2, V_3), \tag{1.78}$$

$$T_{ij} = U_i V_j, \tag{1.79}$$

of which there are nine components. Thus,

$$T_{i'j'} = \sum_i \sum_j R_{i'i} R_{j'j} T_{ij}, \tag{1.80}$$

where the T_{ij} are the nine components of the dyadic and the $R_{i'i}R_{j'j}$ are the elements of the 9×9 matrix that effects the underlying transformation. We note that there are still only three parameters involved in describing this ($SO(3)$) transformation.

The nine components of **T** are *reducible*, i.e. they can be expressed as linear combinations that form subsets which transform among themselves under rotations. They are:

$$U_1V_1 + U_2V_2 + U_3V_3 = \vec{U} \cdot \vec{V} := T, \tag{1.81}$$

$$\frac{1}{2}(U_2V_3 - U_3V_2) := A_1, \tag{1.82}$$

$$\frac{1}{2}(U_3V_1 - U_1V_3) := A_2, \tag{1.83}$$

$$\frac{1}{2}(U_1V_2 - U_2V_1) := A_3, \tag{1.84}$$

i.e.

$$\frac{1}{2}(U_iV_j - U_jV_i) := A_k\varepsilon_{ijk}. \tag{1.85}$$

Further,

$$\frac{1}{2}(U_1V_2 + U_2V_1) := S_{12}, \tag{1.86}$$

$$\frac{1}{2}(U_2V_3 + U_3V_2) := S_{23}, \tag{1.87}$$

$$\frac{1}{2}(U_3V_1 + U_1V_3) := S_{31}, \tag{1.88}$$

$$U_1V_1 - \frac{1}{3}T := S_{11}, \tag{1.89}$$

$$U_2V_2 - \frac{1}{3}T := S_{22}, \tag{1.90}$$

i.e.

$$\frac{1}{2}(U_iV_j + U_jV_i) - \frac{1}{3}T\delta_{ij} = S_{ij}. \tag{1.91}$$

The combination $U_3V_3 - \frac{1}{3}T$ is excluded because

$$U_3V_3 - \frac{1}{3}T = -\left\{\left(U_1V_1 - \frac{1}{3}T\right) + \left(U_2V_2 - \frac{1}{3}T\right)\right\}. \tag{1.92}$$

T is a scalar product which is invariant under rotations. The A_k are the three independent components of an antisymmetric tensor: by antisymmetric we mean that they change sign under exchange of the indices. The S_{ij} are the five independent components of a traceless second-rank tensor. The trace, $\sum_{i=1}^{3} S_{ii}$ is evidently zero from equation (1.92). Note that

$$U_i V_j = \frac{\vec{U} \cdot \vec{V}}{3} \delta_{ij} + \frac{(U_i V_j - U_j V_i)}{2} + \left(\frac{U_i V_j + U_j V_i}{2} - \frac{\vec{U} \cdot \vec{V}}{3} \delta_{ij} \right), \tag{1.93}$$

i.e.

$$U_i V_j = \frac{T}{3} \delta_{ij} + A_k \varepsilon_{ijk} + \delta_{ij}. \tag{1.94}$$

Consider a rotation around the z-axis through an angle γ:

$$\begin{pmatrix} \cos\gamma & -\sin\gamma & 0 \\ \sin\gamma & \cos\gamma & 0 \\ 0 & 0 & 1 \end{pmatrix} \begin{pmatrix} U_1 \\ U_2 \\ U_3 \end{pmatrix} = \begin{pmatrix} U_1' \\ U_2' \\ U_3' \end{pmatrix}, \tag{1.95}$$

whence

$$U_1' = U_1 \cos\gamma - U_2 \sin\gamma, \qquad U_2' = U_1 \sin\gamma + U_2 \cos\gamma, \qquad U_3' = U_3; \tag{1.96}$$

and similarly

$$V_1' = V_1 \cos\gamma - V_2 \sin\gamma, \qquad V_2' = V_1 \sin\gamma + V_2 \cos\gamma, \qquad V_3' = V_3. \tag{1.97}$$

Then for the A_k,

$$\begin{aligned} A_1' &= \frac{1}{2}(U_2' V_3' - U_3' V_2') \\ &= \frac{1}{2}\{(U_1 \sin\gamma + U_2 \cos\gamma) V_3 - U_3(V_1 \sin\gamma + V_2 \cos\gamma)\} \\ &= A_1 \cos\gamma - A_2 \sin\gamma; \end{aligned} \tag{1.98}$$

similarly

$$A_2' = A_1 \sin\gamma + A_2 \cos\gamma \tag{1.99}$$

and

$$A_3' = A_3, \tag{1.100}$$

i.e.

$$\begin{pmatrix} \cos\gamma & -\sin\gamma & 0 \\ \sin\gamma & \cos\gamma & 0 \\ 0 & 0 & 1 \end{pmatrix} \begin{pmatrix} A_1 \\ A_2 \\ A_3 \end{pmatrix} = \begin{pmatrix} A_1' \\ A_2' \\ A_3' \end{pmatrix}. \tag{1.101}$$

Evidently, we can write

$$\vec{A} = \frac{1}{2}\vec{U} \times \vec{V}. \tag{1.102}$$

Further, for the S_{ij},

$$\begin{aligned} S'_{12} &= \frac{1}{2}\{(U_1 \cos\gamma - U_2 \sin\gamma)(V_1 \sin\gamma + V_2 \cos\gamma) \\ &\quad + (U_1 \sin\gamma + U_2 \cos\gamma)(V_1 \cos\gamma - V_2 \sin\gamma)\} \\ &= (U_1V_1 - U_2V_2)\sin\gamma\cos\gamma + \frac{1}{2}(U_1V_2 + U_2V_1)(\cos^2\gamma - \sin^2\gamma) \\ &= (S_{11} - S_{22})\sin\gamma\cos\gamma + S_{12}(\cos^2\gamma - \sin^2\gamma) \\ &= \frac{1}{2}(S_{11} - S_{22})\sin 2\gamma + S_{12}\cos 2\gamma; \end{aligned} \tag{1.103}$$

similarly,

$$S'_{23} = S_{23}\cos\gamma + S_{31}\sin\gamma, \tag{1.104}$$

$$S'_{31} = -S_{23}\sin\gamma + S_{31}\cos\gamma, \tag{1.105}$$

$$S'_{11} = -S_{12}\sin 2\gamma + \frac{1}{2}S_{11}(1 + \cos 2\gamma) + \frac{1}{2}S_{22}(1 - \cos 2\gamma), \tag{1.106}$$

and

$$S'_{22} = S_{12}\sin 2\gamma + \frac{1}{2}S_{11}(1 - \cos 2\gamma) + \frac{1}{2}S_{22}(1 + \cos 2\gamma), \tag{1.107}$$

i.e.

$$\begin{pmatrix} \cos 2\gamma & 0 & 0 & \frac{1}{2}\sin 2\gamma & -\frac{1}{2}\sin 2\gamma \\ 0 & \cos\gamma & \sin\gamma & 0 & 0 \\ 0 & -\sin\gamma & \cos\gamma & 0 & 0 \\ -\sin 2\gamma & 0 & 0 & \frac{1}{2}(1 + \cos 2\gamma) & \frac{1}{2}(1 - \cos 2\gamma) \\ \sin 2\gamma & 0 & 0 & \frac{1}{2}(1 - \cos 2\gamma) & \frac{1}{2}(1 + \cos 2\gamma) \end{pmatrix} \begin{pmatrix} S_{12} \\ S_{23} \\ S_{31} \\ S_{11} \\ S_{22} \end{pmatrix} = \begin{pmatrix} S'_{12} \\ S'_{23} \\ S'_{31} \\ S'_{11} \\ S'_{22} \end{pmatrix}. \tag{1.108}$$

Similar equations can be obtained for $\{S_{12}, S_{23}, S_{31}, S_{11}, S_{22}\}$ for rotations about the x and y axes.

The $\{A_k\}$, T, and the $\{S_{ij}\}$ transform separately under rotations. Dyadics are said to possess a *reducible* structure with respect to rotations.

1.7 The Schwinger representations for *SU*(2)

Representations of $SU(2)$ can be constructed using a method due to Schwinger. Consider

$$\left| j = \frac{1}{2}, m = \frac{1}{2} \right\rangle := a_+^\dagger|0\rangle, \qquad \left| j = \frac{1}{2}, m = -\frac{1}{2} \right\rangle := a_-^\dagger|0\rangle; \tag{1.109}$$

i.e. $a_+^\dagger$ creates a state (particle in a state) of spin-$\frac{1}{2}$ up and $a_-^\dagger$ creates a state of spin-$\frac{1}{2}$ down by action on the 'vacuum' $|0\rangle$ (which has no particles), where

$$\left[a_i, a_j^\dagger\right] = \delta_{ij}, \qquad \{i, j\} = \{+, -\}. \tag{1.110}$$

Then, defining

$$\hat{J}_+ := a_+^\dagger a_-, \qquad \hat{J}_- := a_-^\dagger a_+, \qquad \hat{J}_0 := \frac{(a_+^\dagger a_+ - a_-^\dagger a_-)}{2}, \tag{1.111}$$

it follows that

$$[\hat{J}_0, \hat{J}_\pm] = \pm\hat{J}_\pm, \tag{1.112}$$

$$[\hat{J}_+, \hat{J}_-] = 2\hat{J}_0, \tag{1.113}$$

which define the structure developed for angular momentum and spin (here, $\hbar \equiv 1$).

Although the elementary building blocks in the Schwinger representation have spin-$\frac{1}{2}$, they should not be regarded as fermions. These spin-$\frac{1}{2}$ 'objects' are designed to be combined to produce any desired value of total spin: the number of spin-$\frac{1}{2}$'s needed to produce a total spin of j will be $2j$. These building blocks can be regarded as bosons. They can be visualised in terms of a two-dimensional harmonic oscillator:

$$[a_+, a_+^\dagger] = 1, \qquad [a_-, a_-^\dagger] = 1, \tag{1.114}$$

$$\hat{N}_+ = a_+^\dagger a_+, \qquad \hat{N}_- = a_-^\dagger a_-, \tag{1.115}$$

$$|n_+\rangle = \frac{(a_+^\dagger)^{n_+}}{\sqrt{n_+!}}|0\rangle, \qquad |n_-\rangle = \frac{(a_-^\dagger)^{n_-}}{\sqrt{n_-!}}|0\rangle, \tag{1.116}$$

$$\hat{N}_+|n_+\rangle = n_+|n_+\rangle, \qquad \hat{N}_-|n_-\rangle = n_-|n_-\rangle, \tag{1.117}$$

$$a_+^\dagger|n_+\rangle = \sqrt{n_+ + 1}\,|n_+ + 1\rangle, \qquad a_-^\dagger|n_-\rangle = \sqrt{n_- + 1}\,|n_- + 1\rangle, \tag{1.118}$$

$$a_+|n_+\rangle = \sqrt{n_+}\,|n_+ - 1\rangle, \qquad a_-|n_-\rangle = \sqrt{n_-}\,|n_- - 1\rangle, \tag{1.119}$$

$$a_+|0\rangle = 0, \qquad a_-|0\rangle = 0. \tag{1.120}$$

The states $|n_+\rangle$, $|n_-\rangle$ can be written in the combined form $|n_+ n_-\rangle$ which, from

$$[a_-, a_+^\dagger] = [a_-, a_+] = [a_-^\dagger, a_+^\dagger] = [a_-^\dagger, a_+] = 0, \tag{1.121}$$

obey

$$|n_+n_-\rangle = \frac{(a_+^\dagger)^{n_+}(a_-^\dagger)^{n_-}}{\sqrt{n_+!n_-!}}|00\rangle, \tag{1.122}$$

$$\hat{N}_+|n_+n_-\rangle = n_+|n_+n_-\rangle, \qquad \hat{N}_-|n_+n_-\rangle = n_-|n_+n_-\rangle, \tag{1.123}$$

$$a_+^\dagger|n_+n_-\rangle = \sqrt{n_+ + 1}\,|n_+ + 1, n_-\rangle, \qquad a_-^\dagger|n_+n_-\rangle = \sqrt{n_- + 1}\,|n_+, n_- + 1\rangle, \tag{1.124}$$

$$a_+|n_+n_-\rangle = \sqrt{n_+}\,|n_+ - 1, n_-\rangle, \qquad a_-|n_+n_-\rangle = \sqrt{n_-}\,|n_+, n_- - 1\rangle, \tag{1.125}$$

$$a_+|00\rangle = 0, \qquad a_-|00\rangle = 0. \tag{1.126}$$

Then, from equations (1.111) and (1.115), i.e.

$$\hat{J}_+ = a_+^\dagger a_-, \qquad \hat{J}_- = a_-^\dagger a_+, \qquad \hat{J}_0 = \frac{1}{2}(a_+^\dagger a_+ - a_-^\dagger a_-) = \frac{1}{2}(\hat{N}_+ - \hat{N}_-), \tag{1.127}$$

together with

$$\hat{N} := \hat{N}_+ + \hat{N}_- = a_+^\dagger a_+ + a_-^\dagger a_- \tag{1.128}$$

and

$$\hat{J}^2 := \hat{J}_0^2 + \frac{1}{2}(\hat{J}_+\hat{J}_- + \hat{J}_-\hat{J}_+), \tag{1.129}$$

we obtain:

$$\hat{J}_+|n_+n_-\rangle = \sqrt{n_-(n_+ + 1)}\,|n_+ + 1, n_- - 1\rangle, \tag{1.130}$$

$$\hat{J}_-|n_+n_-\rangle = \sqrt{n_+(n_- + 1)}\,|n_+ - 1, n_- + 1\rangle, \tag{1.131}$$

$$\hat{J}_0|n_+n_-\rangle = \frac{1}{2}(n_+ - n_-)|n_+n_-\rangle, \tag{1.132}$$

$$\hat{N}|n_+n_-\rangle = (n_+ + n_-)|n_+n_-\rangle, \tag{1.133}$$

and

$$\begin{aligned}
\hat{J}^2|n_+n_-\rangle &= \hat{J}_0^2|n_+n_-\rangle + \frac{1}{2}\hat{J}_+\hat{J}_-|n_+n_-\rangle + \frac{1}{2}\hat{J}_-\hat{J}_+|n_+n_-\rangle \\
&= \frac{1}{4}(n_+ - n_-)^2|n_+n_-\rangle + \frac{1}{2}n_+(n_- + 1)|n_+n_-\rangle \\
&\quad + \frac{1}{2}n_-(n_+ + 1)|n_+n_-\rangle \\
&= \left(\frac{n_+ + n_-}{2}\right)\left(\frac{n_+ + n_-}{2} + 1\right)|n_+n_-\rangle \\
&= \frac{n}{2}\left(\frac{n}{2} + 1\right)|n_+n_-\rangle,
\end{aligned} \tag{1.134}$$

where

$$n = n_+ + n_-. \tag{1.135}$$

Evidently, by making the associations

$$n_+ \leftrightarrow j + m, \qquad n_- \leftrightarrow j - m, \tag{1.136}$$

we obtain

$$n = 2j; \tag{1.137}$$

and from equations (1.130)–(1.132) and (1.134)

$$\hat{J}_+|n_+n_-\rangle = \sqrt{(j - m)(j + m + 1)}\,|n_+ + 1, n_- - 1\rangle, \tag{1.138}$$

$$\hat{J}_-|n_+n_-\rangle = \sqrt{(j + m)(j - m + 1)}\,|n_+ - 1, n_- + 1\rangle, \tag{1.139}$$

$$\hat{J}_0|n_+n_-\rangle = m|n_+n_-\rangle, \tag{1.140}$$

$$\hat{J}^2|n_+n_-\rangle = j(j + 1)|n_+n_-\rangle, \tag{1.141}$$

respectively. Thus, by comparison with

$$\hat{J}_+\lfloor jm\rangle = \sqrt{(j - m)(j + m + 1)}\,\lfloor j\ m + 1\rangle, \tag{1.142}$$

$$\hat{J}_-\lfloor jm\rangle = \sqrt{(j + m)(j - m + 1)}\,\lfloor j\ m - 1\rangle, \tag{1.143}$$

$$\hat{J}_0\lfloor jm\rangle = m\lfloor jm\rangle, \tag{1.144}$$

$$\hat{J}^2\lfloor jm\rangle = j(j + 1)\lfloor jm\rangle, \tag{1.145}$$

we can assert that

$$|n_+n_-\rangle \equiv \lfloor jm\rangle, \tag{1.146}$$

and from equation (1.122)

$$\lfloor jm\rangle \equiv \frac{(a_+^\dagger)^{j+m}(a_-^\dagger)^{j-m}}{\sqrt{(j + m)!(j - m)!}}|00\rangle. \tag{1.147}$$

Two special cases of note are: $m = +j$, i.e.

$$\lfloor jj\rangle \equiv \frac{(a_+^\dagger)^{2j}}{\sqrt{(2j)!}}|00\rangle; \tag{1.148}$$

and $m = -j$, i.e.

$$|j, -j\rangle \equiv \frac{(a_-^\dagger)^{2j}}{\sqrt{(2j)!}}|00\rangle. \tag{1.149}$$

Consider then the rotation of the states $|j = \frac{1}{2}, m = \frac{1}{2}\rangle \equiv |\frac{1}{2}, \frac{1}{2}\rangle$ and $|j = \frac{1}{2}, m = -\frac{1}{2}\rangle \equiv |\frac{1}{2}, -\frac{1}{2}\rangle$:

$$\mathcal{D}_y(\beta)\left|\frac{1}{2}, \frac{1}{2}\right\rangle \leftrightarrow \begin{pmatrix} \cos\frac{\beta}{2} & -\sin\frac{\beta}{2} \\ \sin\frac{\beta}{2} & \cos\frac{\beta}{2} \end{pmatrix}\begin{pmatrix} 1 \\ 0 \end{pmatrix} = \begin{pmatrix} \cos\frac{\beta}{2} \\ \sin\frac{\beta}{2} \end{pmatrix}, \tag{1.150}$$

$$\mathcal{D}_y(\beta)\left|\frac{1}{2}, -\frac{1}{2}\right\rangle \leftrightarrow \begin{pmatrix} \cos\frac{\beta}{2} & -\sin\frac{\beta}{2} \\ \sin\frac{\beta}{2} & \cos\frac{\beta}{2} \end{pmatrix}\begin{pmatrix} 0 \\ 1 \end{pmatrix} = \begin{pmatrix} -\sin\frac{\beta}{2} \\ \cos\frac{\beta}{2} \end{pmatrix}; \tag{1.151}$$

i.e.

$$\mathcal{D}_y(\beta)\left|\frac{1}{2}, -\frac{1}{2}\right\rangle = \cos\frac{\beta}{2}\left|\frac{1}{2}, \frac{1}{2}\right\rangle + \sin\frac{\beta}{2}\left|\frac{1}{2}, -\frac{1}{2}\right\rangle, \tag{1.152}$$

$$\mathcal{D}_y(\beta)\left(\frac{1}{2}, -\frac{1}{2}\right) = -\sin\frac{\beta}{2}\left|\frac{1}{2}, \frac{1}{2}\right\rangle + \cos\frac{\beta}{2}\left|\frac{1}{2}, -\frac{1}{2}\right\rangle. \tag{1.153}$$

Then from

$$\left|\frac{1}{2}, \frac{1}{2}\right\rangle = a_+^\dagger|0\rangle, \qquad \left|\frac{1}{2}, -\frac{1}{2}\right\rangle = a_-^\dagger|0\rangle, \tag{1.154}$$

we have

$$\mathcal{D}_y(\beta)\left|\frac{1}{2}, \frac{1}{2}\right\rangle = \mathcal{D}_y(\beta)a_+^\dagger\mathcal{D}_y^{-1}(\beta)\mathcal{D}_y(\beta)|0\rangle, \tag{1.155}$$

$$\mathcal{D}_y(\beta)\left|\frac{1}{2}, -\frac{1}{2}\right\rangle = \mathcal{D}_y(\beta)a_-^\dagger\mathcal{D}_y^{-1}(\beta)\mathcal{D}_y(\beta)|0\rangle; \tag{1.156}$$

whence

$$\mathcal{D}_y(\beta)a_+^\dagger\mathcal{D}_y^{-1}(\beta) \equiv a_+^{\dagger'} = \cos\frac{\beta}{2}a_+^\dagger + \sin\frac{\beta}{2}a_-^\dagger, \tag{1.157}$$

$$\mathcal{D}_y(\beta)a_-^\dagger\mathcal{D}_y^{-1}(\beta) \equiv a_-^{\dagger'} = -\sin\frac{\beta}{2}a_+^\dagger + \cos\frac{\beta}{2}a_-^\dagger. \tag{1.158}$$

Thus,

$$\mathcal{D}_y(\beta)|jm\rangle := \frac{(a_+^{\dagger'})^{j+m}(a_-^{\dagger'})^{j-m}}{\sqrt{(j+m)!(j-m)!}}|00\rangle. \tag{1.159}$$

$$\therefore\ \mathcal{D}_y(\beta)|jm\rangle = \frac{\left(\cos\frac{\beta}{2}a_+^\dagger + \sin\frac{\beta}{2}a_-^\dagger\right)^{j+m}\left(-\sin\frac{\beta}{2}a_+^\dagger + \cos\frac{\beta}{2}a_-^\dagger\right)^{j-m}}{\sqrt{(j+m)!(j-m)!}}|00\rangle. \tag{1.160}$$

The right-hand side of equation (1.160) can be expanded using the binomial theorem:

$$\begin{aligned}\mathcal{D}_y(\beta)|jm\rangle = &\frac{1}{\sqrt{(j+m)!(j-m)!}}\sum_l \frac{(j+m)!}{l!(j+m-l)!}\left(a_+^\dagger\cos\frac{\beta}{2}\right)^l\left(a_-^\dagger\sin\frac{\beta}{2}\right)^{j+m-l}\\ &\times\sum_k\frac{(j-m)!}{k!(j-m-k)!}(-a_+^\dagger\sin\frac{\beta}{2})^k\left(a_-^\dagger\cos\frac{\beta}{2}\right)^{j-m-k}|00\rangle.\end{aligned} \tag{1.161}$$

$$\begin{aligned}\therefore\ \mathcal{D}_y(\beta)|jm\rangle = &\sqrt{(j+m)!(j-m)!}\sum_{l,k}(-1)^k\frac{\left(\cos\frac{\beta}{2}\right)^{j-m+l-k}\left(\sin\frac{\beta}{2}\right)^{j+m-l+k}}{l!(j+m-l)!k!(j-m-k)!}\\ &\times(a_+^\dagger)^{l+k}(a_-^\dagger)^{2j-l-k}|00\rangle,\end{aligned} \tag{1.162}$$

and comparing with

$$\mathcal{D}_y(\beta)|jm\rangle = \sum_{m'}|jm'\rangle d_{m'm}^{(j)}(\beta), \tag{1.163}$$

i.e.

$$\mathcal{D}_y(\beta)|jm\rangle = \sum_{m'}d_{m'm}^{(j)}(\beta)\frac{(a_+^\dagger)^{j+m'}(a_-^\dagger)^{j-m'}}{\sqrt{(j+m')!(j-m')!}}|00\rangle, \tag{1.164}$$

equating coefficients of powers of $a_+^\dagger$ in equations (1.162) and (1.164),

$$l+k=j+m'. \tag{1.165}$$

Then, for a particular choice of m',

$$l=j+m'-k \tag{1.166}$$

and

$$\begin{aligned}d_{m'm}^{(j)}(\beta) = &\sum_{\substack{k\\ \text{(no negative factorials)}}}(-1)^k\frac{\sqrt{(j+m)!(j-m)!(j+m')!(j-m')!}}{(j+m'-k)!(m-m'+k)!k!(j-m-k)!}\\ &\times\left(\cos\frac{\beta}{2}\right)^{2j-2k+m'-m}\times\left(\sin\frac{\beta}{2}\right)^{2k+m-m'}.\end{aligned} \tag{1.167}$$

1.8 A spinor function basis for $SU(2)$

The Schwinger representation and its associated basis leads directly to a *spinor function basis* for $SU(2)$:

$$\left| j = \frac{1}{2}, m = \frac{1}{2} \right\rangle \leftrightarrow u, \qquad \left| j = \frac{1}{2}, m = -\frac{1}{2} \right\rangle \leftrightarrow v, \tag{1.168}$$

where u and v are independent functions. We require u and v to satisfy (again, $\hbar \equiv 1$)

$$\hat{J}_0 u = \frac{1}{2} u, \; \hat{J}_0 v = -\frac{1}{2} v, \tag{1.169}$$

$$\hat{J}_+ u = 0, \quad \hat{J}_+ v = u, \tag{1.170}$$

$$\hat{J}_- u = v, \quad \hat{J}_- v = 0. \tag{1.171}$$

Thus, we deduce the *realisation*

$$\hat{J}_0 \leftrightarrow \frac{1}{2}\left(u\frac{\partial}{\partial u} - v\frac{\partial}{\partial v} \right), \tag{1.172}$$

$$\hat{J}_+ \leftrightarrow u\frac{\partial}{\partial v}, \tag{1.173}$$

$$\hat{J}_- \leftrightarrow v\frac{\partial}{\partial u}. \tag{1.174}$$

It follows from equations (1.148), (1.149) and (1.147) that

$$|jj\rangle := \frac{u^{2j}}{\sqrt{(2j)!}}, \qquad |j, -j\rangle := \frac{v^{2j}}{\sqrt{(2j)!}}, \tag{1.175}$$

and

$$|jm\rangle := \frac{u^{j+m} v^{j-m}}{\sqrt{(j+m)!(j-m)!}}. \tag{1.176}$$

It should be noted that $\{u, v\}$ are elements of a complex function space which is formally developed in section 1.17 under the title of the *Bargmann representation*, i.e. this function space is known as *Bargmann space*. These bases are *irreducible* (unlike Cartesian tensors).

1.9 A spherical harmonic basis for $SO(3)$

The use of spinor functions as a basis for $SU(2)$ and the relations for $\hat{J}_0$, $\hat{J}_\pm$ given in equations (1.172)–(1.174) lead to the consideration of a functional representation of the $|lm\rangle$ for ($\hbar \equiv 1$)

$$\hat{L}_0 = \hat{L}_z = \hat{x}\hat{p}_y - \hat{y}\hat{p}_x \leftrightarrow -ix\frac{\partial}{\partial y} + iy\frac{\partial}{\partial x}; \tag{1.177}$$

$$\hat{L}_+ = \hat{L}_x + i\hat{L}_y = \hat{y}\hat{p}_z - \hat{z}\hat{p}_y + i\hat{z}\hat{p}_x - i\hat{x}\hat{p}_z, \tag{1.178}$$

$$\therefore \hat{L}_+ \leftrightarrow -iy\frac{\partial}{\partial z} + iz\frac{\partial}{\partial y} + z\frac{\partial}{\partial x} - x\frac{\partial}{\partial z}; \tag{1.179}$$

$$\hat{L}_- \leftrightarrow -iy\frac{\partial}{\partial z} + iz\frac{\partial}{\partial y} - z\frac{\partial}{\partial x} + x\frac{\partial}{\partial z}; \tag{1.180}$$

where the postion representation has been used. Evidently, $\hat{L}_0$, $\hat{L}_\pm$ in the form given by equations (1.177), (1.179) and (1.180) leave the degree of a polynomial in x, y and z unchanged. Therefore, we consider the space of homogeneous polynomials in x, y and z, i.e.

$$f(x, y, z) = (ax + by + cz)^l, \tag{1.181}$$

where a, b, and c are complex numbers.

We start with the homogeneous polynomials $\phi_{lm=-l}(\vec{r})$, $\vec{r} := (x, y, z)$, that satisfy the so-called '*lowest weight*' conditions:

$$\hat{L}_0\phi_{l,-l}(\vec{r}) = -l\phi_{l,-l}(\vec{r}) \tag{1.182}$$

and

$$\hat{L}_-\phi_{l,-l}(\vec{r}) = 0. \tag{1.183}$$

Then, consider

$$\begin{aligned} \hat{L}_0(ax + by + cz)^l &= \left(-ix\frac{\partial}{\partial y} + iy\frac{\partial}{\partial x}\right)(ax + by + cz)^l \\ &= -ixl(ax + by + cz)^{l-1}b + iyl(ax + by + cz)^{l-1}a \\ &= l(ax + by + cz)^{l-1}(-ibx + iay), \end{aligned} \tag{1.184}$$

and the right-hand side fulfils equation (1.182), i.e.

$$\hat{L}_0(ax + by + cz)^l = -l(ax + by + cz)^l, \tag{1.185}$$

provided $a = 1$, $b = -i$, $c = 0$. Thus,

$$\phi_{l,-l}(\vec{r}) = (x - iy)^l. \tag{1.186}$$

Evidently,

$$\begin{aligned} \hat{L}_-\phi_{l,-l}(\vec{r}) &= \left(-iy\frac{\partial}{\partial z} + iz\frac{\partial}{\partial y} - z\frac{\partial}{\partial x} + x\frac{\partial}{\partial z}\right)(x - iy)^l \\ &= izl(-i)(x - iy)^{l-1} - zl(x - iy)^{l-1} \\ &= 0. \end{aligned} \tag{1.187}$$

We can construct the $\phi_{lm}(\vec{r})$ using ($\hbar \equiv 1$)

$$\hat{L}_{+}\phi_{lm}(\vec{r}) = \sqrt{(l - m)(l + m + 1)}\,\phi_{l,m+1}(\vec{r}). \tag{1.188}$$

For $l = 1$: from

$$\phi_{1,-1}(\vec{r}) = x - iy, \tag{1.189}$$

$$\begin{aligned} \hat{L}_{+}\phi_{1,-1}(\vec{r}) &= \left(-iy\frac{\partial}{\partial z} + iz\frac{\partial}{\partial y} + z\frac{\partial}{\partial x} - x\frac{\partial}{\partial z}\right)(x - iy) \\ &= iz(-i) + z \\ &= 2z \\ &:= \sqrt{2}\,\phi_{1,0}(\vec{r}); \end{aligned} \tag{1.190}$$

$$\therefore\ \phi_{1,0}(\vec{r}) = \sqrt{2}\,z. \tag{1.191}$$

Then,

$$\begin{aligned} \hat{L}_{+}\phi_{1,0}(\vec{r}) &= \sqrt{2}\left(-iy\frac{\partial}{\partial z} + iz\frac{\partial}{\partial y} + z\frac{\partial}{\partial x} - x\frac{\partial}{\partial z}\right)z \\ &= \sqrt{2}(-iy - x) \\ &= -\sqrt{2}(x + iy) \\ &:= \sqrt{2}\,\phi_{1,1}(\vec{r}); \end{aligned} \tag{1.192}$$

$$\therefore\ \phi_{1,1}(\vec{r}) = -(x + iy). \tag{1.193}$$

For $l = 2$: from

$$\phi_{2,-2}(\vec{r}) = (x - iy)^2, \tag{1.194}$$

$$\begin{aligned} \hat{L}_{+}\phi_{2,-2}(\vec{r}) &= \left(-iy\frac{\partial}{\partial z} + iz\frac{\partial}{\partial y} + z\frac{\partial}{\partial x} - x\frac{\partial}{\partial z}\right)(x - iy)^2 \\ &= iz2(x - iy)(-i) + z2(x - iy) \\ &= 4z(x - iy) \\ &:= 2\phi_{2,-1}(\vec{r}); \end{aligned} \tag{1.195}$$

$$\therefore\ \phi_{2,-1}(\vec{r}) = 2z(x - iy). \tag{1.196}$$

Then,

$$
\begin{aligned}
\hat{L}_{+}\phi_{2,-1}(\vec{r}) &= \left(-iy\frac{\partial}{\partial z} + iz\frac{\partial}{\partial y} + z\frac{\partial}{\partial x} - x\frac{\partial}{\partial z}\right)2z(x - iy) \\
&= -iy2(x - iy) + iz2z(-i) + z2z - x2(x - iy) \\
&= -2(x - iy)(x + iy) + 4z^2 \\
&= -2(x^2 + y^2) + 4z^2 \\
&:= \sqrt{6}\phi_{2,0}(\vec{r});
\end{aligned} \tag{1.197}
$$

$$
\therefore \phi_{2,0}(\vec{r}) = \sqrt{\frac{2}{3}}(-x^2 - y^2 + 2z^2). \tag{1.198}
$$

Similarly,

$$
\phi_{2,1}(\vec{r}) = -2z(x + iy), \tag{1.199}
$$

$$
\phi_{2,2}(\vec{r}) = (x + iy)^2. \tag{1.200}
$$

The functions $\phi_{lm}(\vec{r})$ are proportional to the spherical harmonics, $Y_{lm}(\theta, \phi)$ (see table 1.1). This follows from the relationship between Cartesian coordinates and spherical polar coordinates:

$$
x = r\sin\theta\cos\phi, \qquad y = r\sin\theta\sin\phi, \qquad z = r\cos\theta, \tag{1.201}
$$

whence

$$
\begin{aligned}
\phi_{1,\pm1} &= \mp(x \pm iy) \\
&= \mp r\sin\theta e^{\pm i\phi} \\
&= r\sqrt{\frac{8\pi}{3}}\, Y_{1,\pm1}(\theta, \phi).
\end{aligned} \tag{1.202}
$$

Similarly,

$$
\begin{aligned}
\phi_{1,0}(\vec{r}) &= \sqrt{2}r\cos\theta \\
&= r\sqrt{\frac{8\pi}{3}}\, Y_{1,0}(\theta, \phi).
\end{aligned} \tag{1.203}
$$

$$
\begin{aligned}
\phi_{2,\pm2}(\vec{r}) &= r^2\sin^2\theta e^{\pm 2i\phi} \\
&= r^2\sqrt{\frac{32\pi}{15}}\, Y_{2,\pm2}(\theta, \phi).
\end{aligned} \tag{1.204}
$$

$$
\begin{aligned}
\phi_{2,\pm1}(\vec{r}) &= \mp 2r^2\cos\theta\sin\theta e^{\pm i\phi} \\
&= r^2\sqrt{\frac{32\pi}{15}}\, Y_{2,\pm1}(\theta, \phi).
\end{aligned} \tag{1.205}
$$

Table 1.1. The spherical harmonics, $Y_{lm}(\theta, \phi)$, $m = l, l-1, l-2, \ldots, 1, 0, -1, \ldots, -l+1, -l$, for $l = 0, 1, 2$, and 3. They are normalized for $0 \leqslant \phi \leqslant 2\pi$, $0 \leqslant \theta \leqslant \pi$.

l	m	$Y_{lm}(\theta, \phi)$
0	0	$\frac{1}{\sqrt{4\pi}}$
1	0	$\sqrt{\frac{3}{4\pi}}\cos\theta$
1	± 1	$\mp\sqrt{\frac{3}{8\pi}}e^{\pm i\phi}\sin\theta$
2	0	$\sqrt{\frac{5}{16\pi}}(3\cos^2\theta - 1)$
2	± 1	$\mp\sqrt{\frac{15}{8\pi}}e^{\pm i\phi}\cos\theta\sin\theta$
2	± 2	$\sqrt{\frac{15}{32\pi}}e^{\pm 2i\phi}\sin^2\theta$
3	0	$\sqrt{\frac{63}{16\pi}}\left(\frac{5}{3}\cos^3\theta - \cos\theta\right)$
3	± 1	$\mp\sqrt{\frac{21}{64\pi}}e^{\pm i\phi}(5\cos^2\theta - 1)\sin\theta$
3	± 2	$\sqrt{\frac{105}{32\pi}}e^{\pm 2i\phi}\sin^2\theta\cos\theta$
3	± 3	$\mp\sqrt{\frac{35}{64\pi}}e^{\pm 3i\phi}\sin^3\theta$

$$\begin{aligned}\phi_{2,0}(\vec{r}) &= r^2\sqrt{\frac{2}{3}}(3\cos^2\theta - 1)\\ &= r^2\sqrt{\frac{32\pi}{15}}\,Y_{2,0}(\theta, \phi).\end{aligned} \tag{1.206}$$

In general,

$$\phi_{l,\pm l}(\vec{r}) = (\mp 1)^l(r\sin\theta\cos\phi \pm ir\sin\theta\sin\phi)^l, \tag{1.207}$$

i.e.

$$\phi_{l,\pm l}(\vec{r}) = (\mp 1)^l r^l \sin^l\theta e^{\pm il\phi}. \tag{1.208}$$

The spherical harmonics $Y_{l,m=\pm l}(\theta, \phi)$ are:

$$Y_{l,\pm l}(\theta, \phi) = \sqrt{\frac{(2l+1)!!}{4\pi(2l)!!}}e^{\pm il\phi}\sin^l\theta, \tag{1.209}$$

where $(2l)!! := (2l)(2l-2)(2l-4)\ldots 2$ or 1 and

$$\int_0^{2\pi} d\phi \int_0^{\pi} \sin\theta d\theta |Y_{l,\pm l}(\theta, \phi)|^2 = 1. \tag{1.210}$$

Thus,

$$\phi_{l,\pm l}(\vec{r}) = r^l \sqrt{\frac{4\pi(2l)!!}{(2l+1)!!}} Y_{l,\pm l}(\theta, \phi). \tag{1.211}$$

It then follows from

$$\phi_{lm}(\vec{r}) = \sqrt{\frac{(l-m)!}{(2l)!(l+m)!}} (\hat{L}_+)^{l+m}(x - iy)^l, \tag{1.212}$$

which is obtained by repeated application of equations (1.186)–(1.188), that a general spherical harmonic is given by

$$Y_{lm}(\theta, \phi) = \frac{1}{2^l l!} \sqrt{\frac{(2l+1)(l-m)!}{4\pi(l+m)!}} \frac{1}{r^l} (\hat{L}_+)^{l+m}(x - iy)^l. \tag{1.213}$$

This leads to the general expression for spherical harmonics:

$$Y_{lm}(\theta, \phi) = \frac{1}{2^l l!} \sqrt{\frac{(2l+1)(l-m)!}{4\pi(l+m)!}} e^{im\phi}(-\sin\theta)^m \left[\frac{d}{d(\cos\theta)}\right]^{l+m} (\cos^2\theta - 1)^l. \tag{1.214}$$

The spherical harmonics are related to the Legendre polynomials, P_l by:

$$P_l(\cos\theta) = \sqrt{\frac{4\pi}{2l+1}} Y_{l,m=0}(\theta, \phi). \tag{1.215}$$

1.10 Spherical harmonics and wave functions

Spherical harmonics naturally arise when using three-dimensional position wave functions in quantum mechanics. Thus, for the position eigenkets $|\vec{r}\rangle$:

$$|\alpha\rangle = \int d\vec{r} |\vec{r}\rangle\langle\vec{r}|\alpha\rangle, \tag{1.216}$$

the position wave function $\Psi_\alpha(\vec{r})$ is the amplitude $\langle\vec{r}|\alpha\rangle$ and $\Psi_\alpha(\vec{r})$ is often expressed in spherical polar coordinates:

$$\Psi_\alpha(\vec{r}) = R_\alpha(r)\Omega_\alpha(\theta, \phi). \tag{1.217}$$

The functions $\Omega_\alpha(\theta, \phi)$ are then expanded in terms of spherical harmonics

$$\Omega_\alpha(\theta, \phi) = \sum_{lm} c_{\alpha lm} Y_{lm}(\theta, \phi). \tag{1.218}$$

Within the above framework, we can define direction eigenkets $|\hat{n}\rangle$, $\hat{n} = \frac{\vec{r}}{r}$:

$$|\alpha\rangle = \int d\hat{n}|\hat{n}\rangle\langle\hat{n}|\alpha\rangle; \tag{1.219}$$

and for

$$|lm\rangle = \int d\hat{n}|\hat{n}\rangle\langle\hat{n}|lm\rangle, \tag{1.220}$$

$$\langle\hat{n}|lm\rangle = Y_{lm}(\theta, \phi) = Y_{lm}(\hat{n}), \tag{1.221}$$

i.e. $Y_{lm}(\theta, \phi)$ is the amplitude for the state $|lm\rangle$ to be found in the direction $\hat{n}$ specified by θ and ϕ.

1.11 Spherical harmonics and rotation matrices

Spherical harmonics can be related to (the elements of) rotation matrices because of their connection to direction eigenkets:

$$|\hat{n}\rangle = \sum_{lm}|lm\rangle\langle lm|\hat{n}\rangle = \sum_{lm} Y^*_{lm}(\theta, \phi)|lm\rangle. \tag{1.222}$$

To see this, consider

$$|\hat{n}\rangle = \mathcal{D}(R)|\hat{z}\rangle, \tag{1.223}$$

i.e. $|\hat{n}\rangle$ is obtained by the rotation of $|\hat{z}\rangle$. Evidently,

$$\mathcal{D}(R) = \mathcal{D}(\alpha = \phi, \beta = \theta, \gamma = 0) \tag{1.224}$$

will do the job. Then for equation (1.223), from the completeness relation:

$$|\hat{n}\rangle = \sum_{lm}\mathcal{D}(R)|lm\rangle\langle lm|\hat{z}\rangle, \tag{1.225}$$

$$\begin{aligned} \therefore\ \langle l'm'|\hat{n}\rangle &= \sum_{lm}\langle l'm'|\mathcal{D}(R)|lm\rangle\langle lm|\hat{z}\rangle \\ &= \mathcal{D}^{(l')}_{m'm}(\alpha = 0, \beta = \theta, \gamma = 0)\langle l'm|\hat{z}\rangle. \end{aligned} \tag{1.226}$$

But, $\langle l'm|\hat{z}\rangle$ is just $Y^*_{l'm}(\theta = 0, \phi)$ and $Y_{l'm}(\theta = 0, \phi) = 0$ for $m \neq 0$: this is seen by inspection of table 1.1. Thus,

$$\begin{aligned} \langle l'm|\hat{z}\rangle &= Y^*_{l'm}(\theta = 0, \phi)\delta_{m0} \\ &= \sqrt{\frac{2l' + 1}{4\pi}} P_{l'}(\cos\theta)|_{\theta=0}\delta_{m0} \\ &= \sqrt{\frac{2l' + 1}{4\pi}}\delta_{m0}, \end{aligned} \tag{1.227}$$

where the $P_{l'}(\cos\theta)$ are the Legendre polynomials given by equation (1.215). Hence, from equations (1.226), (1.223) and (1.227):

$$Y^*_{l'm'}(\theta, \phi) = \mathcal{D}^{(l')}_{m'0}(\alpha = \phi, \beta = \theta, \gamma = 0)\sqrt{\frac{2l'+1}{4\pi}}, \tag{1.228}$$

or

$$\mathcal{D}^{(l)}_{m0}(\alpha, \beta, \gamma = 0) = \sqrt{\frac{4\pi}{2l+1}}\, Y^*_{lm}(\theta, \phi)|_{\theta=\beta, \phi=\alpha}; \tag{1.229}$$

and for $m = 0$

$$\mathcal{D}^{(l)}_{00}(\alpha, \beta, \gamma) = d^{(l)}_{00}(\beta), \tag{1.230}$$

and

$$\therefore\ d^{(l)}_{00}(\beta) = P_l(\cos\theta)|_{\theta=\beta}. \tag{1.231}$$

Theorem 1.11.1. *The addition theorem for spherical harmonics,*

$$P_l(\cos\theta) = \sum_m \frac{4\pi}{2l+1} Y_{lm}(\theta_2, \phi_2) Y^*_{lm}(\theta_1, \phi_1), \tag{1.232}$$

where θ is defined by

$$\cos\theta := \cos\theta_1 \cos\theta_2 + \sin\theta_1 \sin\theta_2 \cos(\phi_1 - \phi_2). \tag{1.233}$$

Proof. Consider

$$\langle l0|\mathcal{D}(\phi, \theta, 0)|l0\rangle = \langle l0|\mathcal{D}(\phi_2, \theta_2, 0)\mathcal{D}(\phi_1, \theta_1, 0)|l0\rangle, \tag{1.234}$$

where the group properties of rotations in ket space have been used. Then, from the completeness relation

$$\langle l0|\mathcal{D}(\phi, \theta, 0)|l0\rangle = \sum_m \langle l0|\mathcal{D}(\phi_2, \theta_2, 0)|lm\rangle\langle lm|\mathcal{D}(\phi_1, \theta_1, 0)|l0\rangle, \tag{1.235}$$

$$\therefore\ \mathcal{D}^{(l)}_{00}(\phi, \theta, 0) = \sum_m \mathcal{D}^{(l)}_{0m}(\phi_2, \theta_2, 0)\mathcal{D}^{(l)}_{m0}(\phi_1, \theta_1, 0), \tag{1.236}$$

and from equations (1.229) and (1.231),

$$P_l(\cos\theta) = \sum_m \frac{4\pi}{2l+1} Y_{lm}(\theta_2, \phi_2) Y^*_{lm}(\theta_1, \phi_1). \tag{1.237}$$

□

1.12 Properties of the rotation matrices

The rotation matrices $\mathcal{D}^{(j)}(\alpha, \beta, \gamma)$ are unitary. Thus, their matrix elements $\mathcal{D}^{(j)}_{mm'}(\alpha, \beta, \gamma)$ obey:

$$\mathcal{D}^{(j)}_{m'm}(-\gamma, -\beta, -\alpha) = \mathcal{D}^{(j)*}_{mm'}(\alpha, \beta, \gamma), \tag{1.238}$$

$$\sum_m \mathcal{D}^{(j)*}_{mm'}(\alpha, \beta, \gamma)\mathcal{D}^{(j)}_{mm''}(\alpha, \beta, \gamma) = \delta_{m'm''}, \tag{1.239}$$

$$\sum_m \mathcal{D}^{(j)*}_{m'm}(\alpha, \beta, \gamma)\mathcal{D}^{(j)}_{m''m}(\alpha, \beta, \gamma) = \delta_{m'm''}. \tag{1.240}$$

The reduced rotation matrices $d^{(j)}(\beta)$ are real. Thus, their matrix elements, $d^{(j)}_{mm'}(\beta)$, from equation (1.238), obey:

$$d^{(j)}_{m'm}(-\beta) = d^{(j)}_{mm'}(\beta). \tag{1.241}$$

From the general expression for the matrix elements of $d^{(j)}_{mm'}(\beta)$, equation (1.167), it follows that

$$(-1)^{m'-m}d^{(j)}_{-m',-m}(\beta) = d^{(j)}_{m'm}(\beta) = (-1)^{m'-m}d^{(j)}_{mm'}(\beta), \tag{1.242}$$

and hence

$$\mathcal{D}^{(j)}_{m'm}(\alpha, \beta, \gamma) = (-1)^{m'-m}\mathcal{D}^{(j)*}_{-m',-m}(\alpha, \beta, \gamma). \tag{1.243}$$

1.13 The rotation of $\langle jm|$

The rotation of $\langle jm|$ involves an important phase factor. From the rotation of $|jm\rangle$ by $\mathcal{D}^{(j)}(\alpha, \beta, \gamma)$:

$$\mathcal{D}(\alpha, \beta, \gamma)|jm\rangle = \sum_{m'} \mathcal{D}^{(j)}_{m'm}(\alpha, \beta, \gamma)|jm'\rangle, \tag{1.244}$$

$$\therefore \langle jm|\mathcal{D}^{\dagger}(\alpha, \beta, \gamma) = \sum_{m'} \mathcal{D}^{(j)*}_{m'm}(\alpha, \beta, \gamma)\langle jm'|. \tag{1.245}$$

Then, from the complex conjugate of equation (1.243):

$$\langle jm|\mathcal{D}^{\dagger}(\alpha, \beta, \gamma) = \sum_{m'} (-1)^{m'-m}\mathcal{D}^{(j)}_{-m',-m}(\alpha, \beta, \gamma)\langle jm'|, \tag{1.246}$$

and replacing $-m' \leftrightarrow m'$, $-m \leftrightarrow m$, and noting that the sum is over $m' = -j, -j+1, \dots, j-1, j$ and so is unaffected,

$$\therefore \langle j, -m|\mathcal{D}^{\dagger}(\alpha, \beta, \gamma) = \sum_{m'} (-1)^{-m'+m}\mathcal{D}^{(j)}_{m'm}(\alpha, \beta, \gamma)\langle j, -m'|, \tag{1.247}$$

$$\therefore (-1)^{-m}\langle j, -m|\mathcal{D}^{\dagger}(\alpha, \beta, \gamma) = \sum_{m'}(-1)^{-m'}\mathcal{D}^{(j)}_{m'm}(\alpha, \beta, \gamma)\langle j, -m'|, \tag{1.248}$$

i.e. $(-1)^{-m}\langle j, -m|$ transforms like $|jm\rangle$. It is conventional to multiply both sides of equation (1.248) by $(-1)^{j}$ and then $(-1)^{j-m}\langle j, -m|$ transforms like $|jm\rangle$, and the phase is real.

1.14 The rotation of the $Y_{lm}(\theta, \phi)$

The transformations of the $Y_{lm}(\theta, \phi)$ under rotation follow from equation (1.221)

$$\langle \hat{n}|lm\rangle = Y_{lm}(\theta, \phi) = Y_{lm}(\hat{n}), \tag{1.249}$$

and

$$\mathcal{D}(R)|\hat{n}\rangle = |\hat{n}'\rangle; \tag{1.250}$$

whence from

$$\mathcal{D}(R^{-1})|lm\rangle = \sum_{m'}|lm'\rangle\langle lm'|\mathcal{D}(R^{-1})|lm\rangle, \tag{1.251}$$

i.e.

$$\mathcal{D}(R^{-1})|lm\rangle = \sum_{m'}|lm'\rangle\mathcal{D}^{(l)}_{m'm}(R^{-1}), \tag{1.252}$$

then

$$\langle \hat{n}|\mathcal{D}(R^{-1})|lm\rangle = \sum_{m'}\langle \hat{n}|lm'\rangle\mathcal{D}^{(l)}_{m'm}(R^{-1}). \tag{1.253}$$

But

$$\langle \hat{n}|\mathcal{D}(R^{-1}) = \langle \hat{n}'|, \tag{1.254}$$

$$\therefore\ Y_{lm}(\hat{n}') = \sum_{m'} Y_{lm'}(\hat{n})\mathcal{D}^{(l)*}_{mm'}(R), \tag{1.255}$$

or

$$Y_{lm}(\theta_R, \phi_R) = \sum_{m'} Y_{lm'}(\theta, \phi)\mathcal{D}^{(l)*}_{mm'}(R). \tag{1.256}$$

Similarly, from equation (1.248)

$$(-1)^{-m}Y_{l,-m}(\theta_R, \phi_R) = \sum_{m'}(-1)^{-m'}Y_{l,-m'}(\theta, \phi)\mathcal{D}^{(l)*}_{mm'}(R). \tag{1.257}$$

1.15 Exercises

1.1. Explore the commutator properties of

$$T_1 = \begin{pmatrix} 0 & 0 & 0 \\ 0 & 0 & -i \\ 0 & i & 0 \end{pmatrix}, \qquad T_2 = \begin{pmatrix} 0 & 0 & i \\ 0 & 0 & 0 \\ -i & 0 & 0 \end{pmatrix}, \qquad T_3 = \begin{pmatrix} 0 & -i & 0 \\ i & 0 & 0 \\ 0 & 0 & 0 \end{pmatrix}, \tag{1.258}$$

in comparison with $SO(3)$ and $SU(2)$, $(3, \mathbb{R})$ and $(2, \mathbb{C})$.

1.2. Show that

$$d^{\left(\frac{3}{2}\right)}(\beta) = \begin{pmatrix} c^3 & -\sqrt{3}\,c^2 s & \sqrt{3}\,cs^2 & -s^3 \\ \sqrt{3}\,c^2 s & c^3 - 2cs^2 & s^3 - 2c^2 s & \sqrt{3}\,cs^2 \\ \sqrt{3}\,cs^2 & -s^3 + 2c^2 s & c^3 - 2cs^2 & -\sqrt{3}\,c^2 s \\ s^3 & \sqrt{3}\,cs^2 & \sqrt{3}\,c^2 s & c^3 \end{pmatrix}, \tag{1.259}$$

where $c := \cos\frac{\beta}{2}$, $s := \sin\frac{\beta}{2}$.

1.3. Show that the results of equation (1.167) agree with equation (1.229) for $\phi = \alpha = 0$ and $j = l = 1$, 2, and 3.

1.4. Show that

$$d^{(j)}_{m'm}(\beta) = (-1)^{m'-m} d^{(j)}_{mm'}(\beta). \tag{1.260}$$

[Hint: in the binomial expansion of equation (1.160), which results in equation (1.162) and eventually equation (1.167), reverse the positions of $\cos\frac{\beta}{2}a_+^\dagger$, $\sin\frac{\beta}{2}a_-^\dagger$ and $-\sin\frac{\beta}{2}a_+^\dagger$, $\cos\frac{\beta}{2}a_-^\dagger$, i.e. express the expansion so that it contains $\left(a_+^\dagger \cos\frac{\beta}{2}\right)^{j+m-l}$, etc.]

1.5. Show that for $R = (0, \beta, 0)$ the $Y_{1\mu}(\theta, \phi)$, $\mu = 0, \pm 1$ obey equation (1.256). [Hint: express the $Y_{1\mu}(\theta, \phi)$ in terms of x, y and z (cf. equations (1.188), (1.191), (1.193), (1.202) and (1.203)), obtain $(x, y, z)_R$ using $R_y(\beta)$, and show that $d^{(1)}(\beta)$ (equation (1.65)) transforms the $Y_{1\mu}(\theta, \phi)$ into the $Y_{1\mu}(\theta_R, \phi_R)$.]

1.16 Spin-$\frac{1}{2}$ particles; neutron interferometry

The constituents of matter—electrons, protons, and neutrons—all have intrinsic spin of $\frac{1}{2}\hbar$. 'Intrinsic' is the term coined to convey the fact that the dynamics of spin does not occur in physical space. 'Spin space' is not accessible to the physicist in the sense that the spin of a particle cannot be changed: it is intrinsic to the particle. In fact, it is not known what spin is. It is only known what spin does, namely 'couple' to other spins and angular momenta such that it behaves as a $j = \frac{1}{2}$ representation of $SU(2)$.

In the absence of other particles and when its own angular momentum is zero, the quantum mechanics of a spin-$\frac{1}{2}$ particle is almost trivial. It can exist in two possible states: 'spin up' and 'spin down'. These are directional components of the spin vector and are usually defined by

$$\hat{s}_z \left| s = \frac{1}{2}, m_s = \pm\frac{1}{2} \right\rangle = \pm\frac{1}{2}\hbar \left| s = \frac{1}{2}, m_s = \pm\frac{1}{2} \right\rangle, \tag{1.261}$$

where the direction is defined to be the z-axis in $(3, \mathbb{R})$. However, there is one extraordinary property of spin-$\frac{1}{2}$ particles: a rotation through 2π does not leave their state kets unchanged! This is seen immediately from equation (1.52) for $\phi = 2\pi$, whence (using $|sm_s\rangle \leftrightarrow \chi_\pm$)

$$\mathcal{D}^{\left(\frac{1}{2}\right)}(\hat{n}, 2\pi)\chi_\pm = (\hat{I}\cos\pi - i\vec{\sigma}\cdot\hat{n}\sin\pi)\chi_\pm, \tag{1.262}$$

$$\therefore \mathcal{D}^{\left(\frac{1}{2}\right)}(\hat{n}, 2\pi)\chi_\pm = -\chi_\pm. \tag{1.263}$$

This property is not observable where expectation values are involved; but it has a dramatic effect on the interferometry of beams of spin-$\frac{1}{2}$ particles.

The interferometry (diffractive splitting and recombination) of particle beams is a well-established property of quantum mechanical particles. It is most elegantly illustrated using beams of neutrons. Neutrons, being electrically neutral, are not subject to stray electric fields which can obscure the interferometric properties of electrically charged particles. However, neutrons have magnetic moments and through the use of suitable magnetic fields it is possible to effect the rotation of the state of a neutron. This has been done using the experimental arrangement shown schematically in figure 1.2. A picture of the silicon crystal, which is the essential component of the interferometer is shown in figure 1.3. The neutron beam is divided and recombined in such a way that one part passes through a magnetic field B which causes the neutron state ket to undergo a phase change. The recombined beam exhibits an interference pattern which can be varied by changing B. Some results are shown in figure 1.4. (Note: by 'divided' it is meant that for each individual neutron, it is not certain which path it takes. It is not a situation where some neutrons take one path and the other neutrons take the other path.)

The phase change produced by the magnetic field is $e^{\frac{i\omega T}{2}}$, where T is the time spent by the neutrons in the magnetic field, ω is the spin-precession frequency,

$$\omega = \frac{2\mu_n B}{\hbar}, \tag{1.264}$$

μ_n is the magnetic moment of the neutron, and the magnetic field is assumed to be of uniform constant strength B. The phase change is the standard result for a magnetic field B acting for a time T on a magnetic moment μ_n, causing the spin to precess. The connection between precession and rotation is seen to follow directly from the Hamiltonian for a neutron in the magnetic field (chosen to be in the z direction)

$$\hat{H} = \omega\hat{S}_z, \tag{1.265}$$

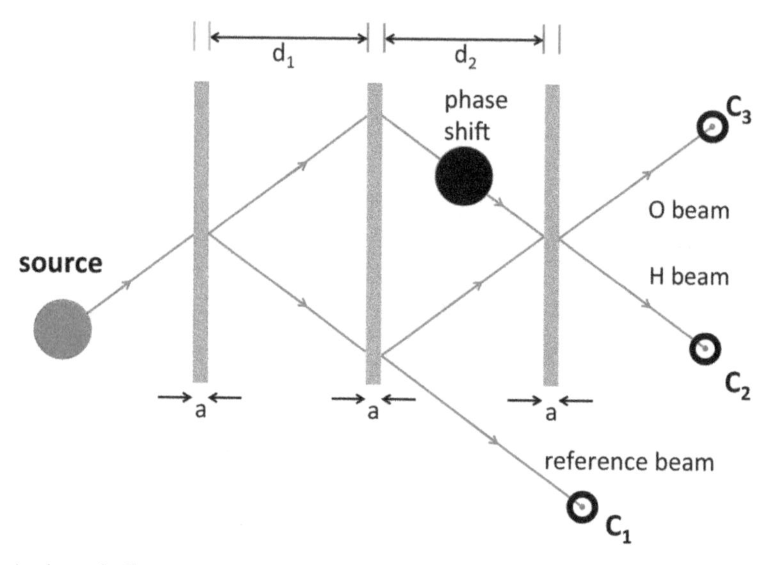

Figure 1.2. A schematic diagram of the paths of a beam of neutrons through the neutron interferometer shown in figure 1.3. The lattice planes are continuous from slab to slab and the distances a, d_1, and d_2 are machined to optical precision. The phase shift (state ket rotation) is effected in the darkened region using a magnetic field. The distances d_1 and d_2 are typically 3 cm and a is typically 0.5 cm.

Figure 1.3. The essential component of the neutron interferometer in use at the University of Missouri. It consists of three silicon slabs machined from a single crystal of high-purity silicon to ensure alignment of crystal planes from slab to slab. (Reproduced from [1], with the permission of the American Institute of Physics.)

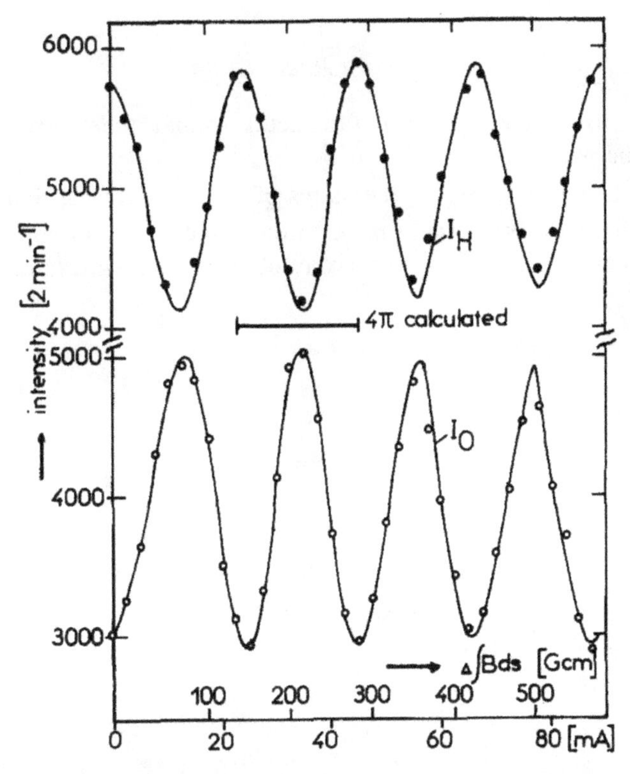

Figure 1.4. Observed neutron intensities in counts/2 min in the O beam and H beam, i.e. in counters C_3 and C_2, respectively, in figure 1.2. This is effected by changing the magnetic field action (given in Gauss cm) on the neutron beam by varying the magnet current (given in milliamps). One oscillation corresponds to a rotation of 4π not 2π. (Reprinted from [2], Copyright (1975), with permission from Elsevier.)

the time evolution operator for the system

$$U(t, 0) = \exp\left\{-\frac{i\hat{H}t}{\hbar}\right\} = \exp\left\{-\frac{i\hat{S}_z\omega t}{\hbar}\right\}, \tag{1.266}$$

and a comparison with the rotation operator about the z-axis

$$\mathcal{D}_z(\phi) = \exp\left\{-\frac{i\hat{S}_z\phi}{\hbar}\right\}, \tag{1.267}$$

i.e.

$$\phi = \omega t. \tag{1.268}$$

For a monoenergetic beam of neutrons, T is fixed. To produce the results shown in figure 1.4, B is varied (by varying the current to the magnet). The change in B necessary to yield successive maxima is given by

$$\Delta B = \frac{\hbar^2}{\mu_n \lambda m d}, \tag{1.269}$$

where λ is the de Broglie wavelength of the neutrons, m is the neutron mass, and d is the length of the path for which $B \neq 0$.

The extraordinary property of the states of spin-$\frac{1}{2}$ particles, that they must be rotated through 4π to 'bring them back to their unchanged orientation', does not parallel our experience of rotating everyday objects. Such states are called *spinors*.

1.17 The Bargmann representation

The functions

$$\chi_n(z) := \frac{z^n}{\sqrt{n!}} \tag{1.270}$$

provide an orthonormal basis for expanding functions realised on z-space (the complex plane), with scalar products defined in terms of z-space integrals with *Bargmann measure*[3], $\frac{e^{-|z|^2}}{\pi}$ [3]. This space is called *Bargmann space*.

The relevance of these functions to coherent states is implicit in the normalized coherent state form $|z\rangle_I$, i.e. (cf. Volume 1, section 5.5)

$$|z\rangle_I := e^{\frac{-|z|^2}{2}} \sum_{n=0}^{\infty} \frac{(z^*)^n}{\sqrt{n!}} |n\rangle. \tag{1.271}$$

Whence, consider

$$K := \int\int \mathrm{d}z |z\rangle_{I\,I}\langle z| = \int\int \mathrm{d}z e^{-|z|^2} \sum_n \frac{(z^*)^n}{\sqrt{n!}} |n\rangle \sum_m \frac{z^m}{\sqrt{m!}} \langle m|; \tag{1.272}$$

which, for $z = re^{i\phi}$, gives

$$K = \int_0^{\infty} r\mathrm{d}r \int_0^{2\pi} \mathrm{d}\phi e^{-r^2} \sum_{n,m} e^{i(m-n)\phi} \frac{r^{n+m}}{\sqrt{n!m!}} |n\rangle\langle m|. \tag{1.273}$$

Now,

$$\int_0^{2\pi} \mathrm{d}\phi e^{i(m-n)\phi} = 2\pi\delta_{mn}, \tag{1.274}$$

[3] The measure of a space appears in the infinitesimal volumes under integrals. For example, polar coordinates in three dimensions possess an infinitesimal volume expressed as $r^2\mathrm{d}r \sin\theta \mathrm{d}\theta \mathrm{d}\phi$ and the measure is $r^2 \sin\theta$. Cartesian coordinates possess a trivial measure because the infinitesimal volume under an integral is $\mathrm{d}x\mathrm{d}y\mathrm{d}z$ (this space could be said to be 'flat').

$$\begin{aligned}\therefore K &= \sum_n \int_0^\infty \mathrm{d}r\, r^{2n+1} e^{-r^2} \frac{2\pi}{n!} |n\rangle\langle n| \\ &= \sum_n \frac{\Gamma(n+1)}{2} \frac{2\pi}{n!} |n\rangle\langle n| \\ &= \pi \sum_n |n\rangle\langle n| \\ &= \pi \mathbf{I}.\end{aligned} \tag{1.275}$$

Thus, the resolution of the identity on Bargmann space is:

$$\mathbf{I} = \int\int \frac{\mathrm{d}z}{\pi} |z\rangle_{I\,I}\langle z| := \int\int \mathrm{d}z \frac{e^{-|z|^2}}{\pi} |z\rangle_{II\,II}\langle z|, \tag{1.276}$$

where $|z\rangle_{II} \leftrightarrow \chi_n(z)$, cf. equation (1.270). Then,

$$\begin{aligned}\langle \Psi_1 | \Psi_2 \rangle &= \int\int \mathrm{d}z \frac{e^{-|z|^2}}{\pi} \langle \Psi_1 | z \rangle_{II\,II}\langle z | \Psi_2 \rangle \\ &= \int\int \mathrm{d}z \frac{e^{-|z|^2}}{\pi} \Psi_1^*(z) \Psi_2(z) \\ &= \int\int \mathrm{d}\mu(z) \Psi_1^*(z) \Psi_2(z),\end{aligned} \tag{1.277}$$

where

$$\Psi(z) := {}_{II}\langle z | \Psi \rangle, \tag{1.278}$$

$$\mathrm{d}\mu(z) := \frac{e^{-|z|^2}}{\pi} \mathrm{d}z. \tag{1.279}$$

Bargmann representations of functions are transformed into position representations of functions by the *Bargmann transformation*,

$$\Psi(x) = \int\int \mathrm{d}\mu(z) A(x, z^*) \Psi(z), \tag{1.280}$$

where

$$A(x, z^*) := \frac{1}{\pi^{\frac{1}{4}}} \exp\left\{ -\frac{1}{2} x^2 + \sqrt{2}\, x z^* - \frac{1}{2} (z^*)^2 \right\} \tag{1.281}$$

is the *Bargmann kernel function*.

Comments:

1. The orthogonality of the $\chi_n(z)$ is evident in a polar coordinate representation which gives $(z^*)^n z^m \to e^{i(m-n)\phi}$ and $\int_0^{2\pi} \mathrm{d}\phi e^{i(m-n)\phi} = 2\pi\delta_{mn}$.
2. The normalizability of the $\chi_n(z)$ is evident from the Gaussian form of Bargmann measure which 'quenches' the scalar products for large $|z|$. (Indeed, the scalar products involve 'camouflaged' Hermite polynomials.)

3. The functions $\chi_n(z)$ are trivially generalised to tensor product functions,

$$\chi_{n_1}(z_1) \otimes \chi_{n_2}(z_2) \otimes \cdots$$

which yields functions

$$\sum_{n_1,n_2,\ldots} \alpha_{n_1,n_2,\ldots} \frac{z_1^{n_1}}{\sqrt{n_1!}} \frac{z_2^{n_2}}{\sqrt{n_2!}} \cdots$$

(cf. equations (1.147) and (1.176)).

1.17.1 Representation of operators

Consider the operator $\mathcal{O}$ and its representation, $\mathcal{O} \leftrightarrow \Gamma(\mathcal{O})$ in terms of z and $\frac{\partial}{\partial z}$, $\mathcal{O}(z, \frac{\partial}{\partial z})$ acting on z-space wave functions, $\Psi(z)$. This is similar to the procedure presented in Volume 1, chapter 8, where, e.g. the operator p_x (momentum in the x direction) was shown to have a 'position' representation $p_x \leftrightarrow -i\hbar\frac{\partial}{\partial x}$ when acting on Cartesian-space wave functions $\Psi(x, y, z)$. The key there is to define a position eigenket basis $\{|x\rangle\}$ and arrive at statements such as $\langle x|\hat{p}_x|\Psi\rangle = -i\hbar\frac{\partial}{\partial x}\langle x|\Psi\rangle = -i\hbar\frac{\partial\Psi(x)}{\partial x}$. Thus, we proceed with the $|z\rangle_{II}$ basis, ($|z\rangle_{II} := e^{z^* a^\dagger}|0\rangle$, cf. Volume 1, section 5.5, equation (5.118))

$$\begin{aligned} \mathcal{O}|\Psi\rangle \Rightarrow \Gamma\left(\mathcal{O}\left(z, \frac{\partial}{\partial z}\right)\right)\Psi(z) &= {}_{II}\langle z|\mathcal{O}|\Psi\rangle = \langle 0|e^{za}\mathcal{O}|\Psi\rangle \\ &= \langle 0|(e^{za}\mathcal{O}e^{-za})e^{za}|\Psi\rangle \\ &= \langle 0|\left(\mathcal{O} + [za, \mathcal{O}] + \frac{1}{2}[za, [za, \mathcal{O}]] + \cdots\right)e^{za}|\Psi\rangle, \end{aligned} \tag{1.282}$$

where the Baker–Campbell–Hausdorff lemma is used (cf. Volume 1, chapter 5, equation (5.110)). Essentially all operators of relevance can be expressed in terms of a and $a^\dagger$, whence: for $\mathcal{O} = a$

$$\mathcal{O}|\Psi\rangle = \langle 0|\left(a + \cancelto{0}{[za, a]} + \cdots\right)|e\rangle^{za}|\Psi\rangle \tag{1.283}$$

and from $\frac{\partial}{\partial z}(e^{za}) = ae^{za}$

$$\Rightarrow \Gamma(a) = \frac{\partial}{\partial z}. \tag{1.284}$$

For $\mathcal{O} = a^\dagger$

$$\begin{aligned} \mathcal{O}|\Psi\rangle &= \langle 0|\left(a^\dagger + \underbrace{[za, a^\dagger]}_{z} + \frac{1}{2}\cancelto{0}{[za, \underbrace{[za, a^\dagger]}_{z}]} + \cdots\right)e^{za}|\Psi\rangle \\ &= \langle 0|ze^{za}|\Psi\rangle \Rightarrow \Gamma(a^\dagger) = z. \end{aligned} \tag{1.285}$$

Note:

1.

$$\left[\frac{\partial}{\partial z}, z\right] = 1, \text{ cf. } [a, a^\dagger] = 1. \tag{1.286}$$

2. z and $\frac{\partial}{\partial z}$ are Hermitian adjoints for scalar products defined on Bargmann measure:

$$\text{e. g. for } \Psi_a = \sum_n a_n z^n, \;\; \Psi_b = \sum_n b_n z^n, \tag{1.287}$$

$$\int\int dz \frac{e^{-|z|^2}}{\pi} \Psi_a^* \frac{\partial}{\partial z} \Psi_b = \sum_n a_n^* b_{n+1}(n+1)! = \int\int dz \frac{e^{-|z|^2}}{\pi} \Psi_b (z\Psi_a)^*. \tag{1.288}$$

1.18 Coherent states for *SU*(2)

The generalisation of the coherent state concept from the one-dimensional harmonic oscillator (Volume 1, section 5.5) to angular momentum is effected through their respective algebras: the *Heisenberg–Weyl algebra* in one dimension, *hw*(1) and *su*(2).

	hw(1)	*su*(2)
Generators	$a^\dagger$	J_+
	a	J_-
	I	J_0
Commutator relations	$[a, a^\dagger] = I$	$[J_-, J_+] = -2J_0$
	$[I, a^\dagger] = 0$	$[J_0, J_+] = +J_+$
	$[I, a] = 0$	$[J_0, J_-] = -J_-$
Lowest-weight state	$\vert 0\rangle$	$\vert j, -j\rangle := \vert -j\rangle$
	$a\vert 0\rangle = 0$	$J_-\vert -j\rangle = 0$

Generalising the type-I coherent state from $HW(1)$ to $SU(2)$

$$|\zeta_I\rangle := \exp\{\zeta^* J_+ - \zeta J_-\} | -j\rangle, \tag{1.289}$$

for $\zeta := \frac{1}{2}\theta e^{i\phi}$,

$$e^{\zeta^* J_+ - \zeta J_-} = e^{-i\theta(J_x \sin\phi - J_y \cos\phi)} = e^{-i\theta(\vec{J}\cdot\hat{n})}, \tag{1.290}$$

where $\hat{n}$ is a unit vector in the x, y plane making an angle ϕ with the negative y-axis. This is illustrated in figure 1.5. All physically significant rotations are accommodated by this formalism (the apparent exclusion of rotations about the z-axis only excludes changes in phase, which could be introduced using $e^{-i\chi J_0}$).

The state $|\zeta\rangle_I$, $\zeta = \zeta(\theta, \phi)$, can be expressed:

$$\begin{aligned}|\zeta\rangle_I = |\theta, \phi\rangle_I &= e^{-i\theta(\vec{J}\cdot\hat{n})}|j, -j\rangle \\ &= \sum_m |jm\rangle\langle jm|e^{-i\theta(\vec{J}\cdot\hat{n})}|j, -j\rangle \\ &= \sum_m |jm\rangle\mathcal{D}^{(j)}_{m,-j}(\phi, \theta, 0)^*.\end{aligned} \tag{1.291}$$

From the orthonormality of the $\mathcal{D}$ functions, sections 1.11 and 1.12,

$$\mathbf{I} = \frac{(2j+1)}{4\pi}\int \mathrm{d}\Omega|\theta, \phi\rangle_{I\,I}\langle\theta, \phi|, \qquad \mathrm{d}\Omega = \sin\theta\,\mathrm{d}\theta\,\mathrm{d}\phi. \tag{1.292}$$

The states $\exp\{\zeta^* J_+ - \zeta J_-\}|j, -j\rangle$ are sometimes called 'atomic coherent' or 'Bloch' states (see, e.g. [4]).

The type-II coherent states of $HW(1)$ can be generalised to $SU(2)$:

$$|z\rangle_{II} := \exp\{(z^* J_+)\}|j, -j\rangle. \tag{1.293}$$

($|\zeta\rangle_I$ and $|z\rangle_{II}$ are no longer trivially related, hence the use of z and ζ.)

The $SU(2)$ states can be expressed in terms of the $\{|z\rangle_{II}\}$:

$$|\Psi\rangle \to \Psi(z) = {}_{II}\langle z|\Psi\rangle = \langle -j|e^{zJ_-}|\Psi\rangle := \Psi_j(z). \tag{1.294}$$

Operators are mapped into z-space realisations, $\Gamma(\mathcal{O})$ by

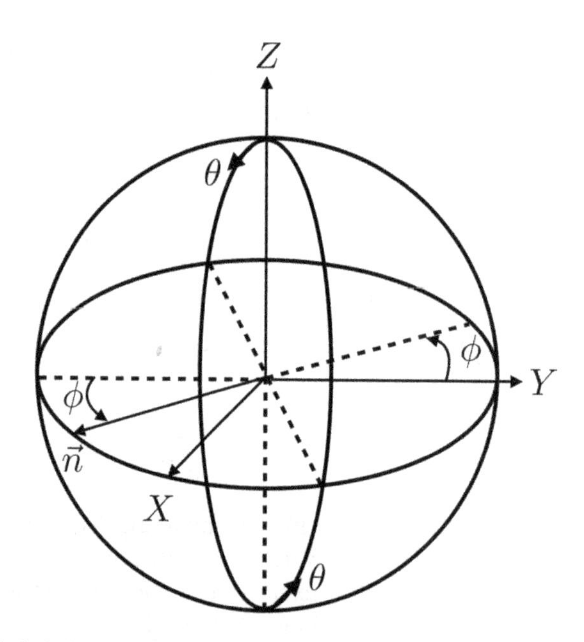

Figure 1.5. A depiction of the parameters ϕ and θ that define a type-I $SU(2)$ coherent state.

$$\begin{aligned}\mathcal{O}|\Psi\rangle \to \Gamma(\mathcal{O})\Psi_J(z) &= \langle z|\mathcal{O}|\Psi\rangle = \langle -j|e^{zJ_-}\mathcal{O}|\Psi\rangle \\ &= \langle -j|(e^{zJ_-}\mathcal{O}e^{-zJ_-})e^{zJ_-}|\Psi\rangle \\ &= \langle -j|(\mathcal{O} + [zJ_-, \mathcal{O}] + \frac{1}{2}[zJ_-, [zJ_-, \mathcal{O}]] + \cdots)e^{zJ_-}|\Psi\rangle.\end{aligned} \tag{1.295}$$

Essentially all operators of relevance can be expressed in terms of J_-, J_0, and J_+, whence: for $\mathcal{O} = J_-$

$$\Gamma(J_-) = \langle -j|(J_- + \underbrace{[zJ_-, J_-]}_{\to 0} + \cdots)e^{z}J_-|\Psi\rangle, \tag{1.296}$$

and from $\frac{\partial}{\partial z}(e^{zJ_-}) = J_-e^{zJ_-}$

$$\Rightarrow \Gamma(J_-) = \frac{\partial}{\partial z}. \tag{1.297}$$

For $\mathcal{O} = J_0$

$$\Gamma(J_0) = \langle -j|\left(\underbrace{J_0}_{-j} + \underbrace{[zJ_-, J_0]}_{zJ_-} + \underbrace{[zJ_-, [zJ_-, J_0]]}_{zJ_-} {}^{\to 0} + \cdots\right)e^{zJ_-}|\Psi\rangle, \tag{1.298}$$

and from $z\frac{\partial}{\partial z}(e^{zJ_-}) = zJ_-e^{zJ_-}$

$$\Rightarrow \Gamma(J_0) = -j + z\frac{\partial}{\partial z}. \tag{1.299}$$

For $\mathcal{O} = J_+$

$$\Gamma(J_+) = \langle -j|\left(J_+^{\to 0} + \underbrace{[zJ_-, J_+]}_{-2zJ_0} + \frac{1}{2}[zJ_-, \underbrace{[zJ_-, J_+]}_{-2zJ_0}] + \cdots\right)e^{zJ_-}|\Psi\rangle, \tag{1.300}$$

$$\Rightarrow \Gamma(J_+) = 2jz - z^2\frac{\partial}{\partial z}. \tag{1.301}$$

Then

$$\Gamma(J_0)\frac{z^n}{\sqrt{n!}} = (-j + n)\frac{z^n}{\sqrt{n!}}, \qquad n = 0, 1, 2, \ldots, \tag{1.302}$$

$$\Gamma(J_+)\frac{z^n}{\sqrt{n!}} = (2j - n)\frac{z^{n+1}}{\sqrt{n!}} = (2j - n)\sqrt{n + 1}\frac{z^{n+1}}{\sqrt{(n + 1)!}}, \tag{1.303}$$

$$\Gamma(J_-)\frac{z^n}{\sqrt{n!}} = n\frac{z^{n-1}}{\sqrt{n!}} = \sqrt{n}\frac{z^{n-1}}{\sqrt{(n - 1)!}}. \tag{1.304}$$

Comments

1. Starting from the state $|j, -j\rangle$, the raising action of $\Gamma(J_+)$ terminates at $n = 2j$ (as it should for $SU(2)$).
2. The representation is non-unitary, i.e. $\Gamma(J_+)^\dagger \neq \Gamma(J_-)$ for scalar products defined on Bargmann measure (for which $(\frac{\partial}{\partial z})^\dagger = z$).
3. This type of non-unitary representation is an example of a *Dyson representation* [5].
4. The $|z\rangle_{II}$ basis is defined on an $SU(2)$ irrep labelled by j, i.e. it is a linear combination of the states $|jm\rangle$, $m = j, j-1, \ldots, -j$.

1.19 Properties of $SU(2)$ from coherent states

To obtain the properties of a quantum system which possesses an algebraic structure requires that unitary representations of the operators be found. This can be done in different ways. One way (which will not be developed here) is to change the *measure* of the z-space to enforce orthonormality. Thus, for the atomic coherent states:

$$\mathbf{I} = \int\int \mathrm{d}z \frac{(2j+1)}{\pi(1+|z|^2)^{2j+2}} |z\rangle_{I\,I}\langle z| \tag{1.305}$$

provides resolution into orthonormal (unitary) representations in z-space. A second way (which is developed here) uses a similarity transformation. This choice is made because for higher symmetry algebras changing the measure of the z-space (even if it can be found!) involves very complicated integrals.

The non-unitarity of the z-space realisation of J_-, J_0, J_+ —$\Gamma(J_-)$, $\Gamma(J_0)$, $\Gamma(J_+)$; $(\Gamma(J_+)^\dagger \neq \Gamma(J_-))$—can be converted to a unitary realisation by a similarity transformation with an operator K:

$$\gamma(J_-) = K^{-1}\Gamma(J_-)K, \tag{1.306}$$

$$\gamma(J_+) = K^{-1}\Gamma(J_+)K, \tag{1.307}$$

$$\gamma(J_0) = K^{-1}\Gamma(J_0)K, \tag{1.308}$$

where it is required that

$$\gamma(J_+) = (\gamma(J_-))^\dagger. \tag{1.309}$$

Now, from the form of $\Gamma(J_0)(=-j + z\frac{\partial}{\partial z})$, it is already Hermitian ($z^\dagger = \frac{\partial}{\partial z}$). Thus, we have the condition that K commutes with $\Gamma(J_0)$ and so K is diagonal in m. Therefore, it is sufficient for K to simply normalize each ladder step in m, $m = -j, -j+1, \ldots, +j$. Hence, for $SU(2)$ it is sufficient for K to be a diagonal matrix. (For higher symmetry groups K will be more complicated.) The '*K-matrix method*' is being *introduced* in the context of $SU(2)$, where standard methods are simpler, to 'see how it works'.

Then, from $\gamma(J_+) = K^{-1}\Gamma(J_+)K$, multiplying from the left with K and from the right with $K^\dagger$

$$K\gamma(J_+)K^\dagger = KK^{-1}\Gamma(J_+)KK^\dagger. \tag{1.310}$$

But

$$\begin{aligned}\gamma(J_+) = (\gamma(J_-))^\dagger &= (K^{-1}\Gamma(J_-)K)^\dagger = \left(K^{-1}\frac{\partial}{\partial z}K\right)^\dagger \\ &= K^\dagger z(K^{-1})^\dagger,\end{aligned} \tag{1.311}$$

$$\therefore\ K\gamma(J_+)K^\dagger = KK^\dagger z(K^{-1})^\dagger K^\dagger = KK^\dagger z(KK^{-1})^\dagger, \tag{1.312}$$

$$\therefore\ \Gamma(J_+)KK^\dagger = KK^\dagger z. \tag{1.313}$$

Thus, from the above condition that for $SU(2)$, K is diagonal with real matrix elements,

$$\therefore\ \Gamma(J_+)K^2 = K^2 z. \tag{1.314}$$

The matrix elements of $KK^\dagger$ (a real diagonal matrix for $SU(2)$, $KK^\dagger = K^2$) can be obtained by proceeding in either of two ways:

1. Take matrix elements, with Bargmann measure, between the states n, $n + 1$, viz.

$$\langle\chi_{n+1}|\Gamma(J_+)K_n^2|\chi_n\rangle = \langle\chi_{n+1}|K_{n+1}^2 z|\chi_n\rangle. \tag{1.315}$$

2. Introduce the auxiliary operator, Λ_{op} with the property

$$[\Lambda_{op}, z] = \Gamma(J_+), \tag{1.316}$$

whence

$$(\Lambda_{op}z - z\Lambda_{op})K^2 = K^2 z, \tag{1.317}$$

and then take matrix elements, with Bargmann measure, between states n, $n + 1$, see the following.

The second way is easier to use for higher symmetry algebras and is the one developed here. (In particular, the second way solves for the representation of an algebra by obtaining the ratios of matrix elements of K, which are all that is ever needed.)

The motivation for defining Λ_{op} can be seen from the analogy between $[\Lambda_{op}, z] = \Gamma(J_+)$ and $[J_0, J_+] = J_+$, recalling that for the $hw(1)$ algebra, $z = \Gamma(a^\dagger)$.

From

$$[\Lambda_{op}, z] = \Gamma(J_+) = 2jz - z^2\frac{\partial}{\partial z}, \tag{1.318}$$

the right-hand side suggests that Λ_{op} must have terms of the form $z\frac{\partial}{\partial z}$ and $z^2\frac{\partial}{\partial z^2}$, viz.

$$\left(z\frac{\partial}{\partial z}\right)(z) = z + z^2\frac{\partial}{\partial z}, \tag{1.319}$$

$$\begin{aligned}\left(z^2\frac{\partial}{\partial z^2}\right)(z) &= \left(z^2\frac{\partial}{\partial z}\right)\left(I + z\frac{\partial}{\partial z}\right) = z^2\frac{\partial}{\partial z} + \left(z^2\frac{\partial}{\partial z}\right)\left(z\frac{\partial}{\partial z}\right)\\ &= z^2\frac{\partial}{\partial z} + z^2\frac{\partial}{\partial z} + z^3\frac{\partial}{\partial z^2},\end{aligned} \tag{1.320}$$

whence

$$\Lambda_{op} = 2jz\frac{\partial}{\partial z} - \frac{1}{2}z^2\frac{\partial}{\partial z^2} = \frac{1}{2}\left(4j - z\frac{\partial}{\partial z} + 1\right)z\frac{\partial}{\partial z}. \tag{1.321}$$

Now

$$z\frac{\partial}{\partial z}\frac{z^n}{\sqrt{n!}} = n\frac{z^n}{\sqrt{n!}}, \tag{1.322}$$

hence

$$\Lambda_{\text{eigenvalue}} := \Lambda_n = \frac{1}{2}(4j - n + 1)n. \tag{1.323}$$

Then, from $\Gamma(J_+)K^2 = K^2 z$ and $\Gamma(J_+) = [\Lambda_{op}, z]$, taking matrix elements,

$$\langle\chi_{n+1}|(\Lambda_{op}z - z\Lambda_{op})K^2|\chi_n\rangle = \langle\chi_{n+1}|K^2 z|\chi_n\rangle, \tag{1.324}$$

$$\therefore (\Lambda_{n+1} - \Lambda_n)\langle\chi_{n+1}|z|\chi_n\rangle K_n^2 = K_{n+1}^2\langle\chi_{n+1}|z|\chi_n\rangle, \tag{1.325}$$

and so

$$\frac{K_{n+1}^2}{K_n^2} = 2j - n. \tag{1.326}$$

Starting with a normalized $|-j\rangle$, $K_0^2 = 1$ and we obtain on iteration

$$K_n^2 = \frac{(2j)!}{(2j-n)!}. \tag{1.327}$$

The matrix elements of J_0, J_+, and J_- are then straightforwardly deduced:

$$\begin{aligned}\langle m|J_0|n\rangle &= \langle \chi_m|\gamma(J_0)|\chi_n\rangle \\ &= \int\int \mathrm{d}z \frac{e^{-|z|^2}}{\pi}\chi_m^*(z)(K^{-1}\Gamma(J_0)K)\chi_n(z) \\ &= \int\int \mathrm{d}z \frac{e^{-|z|^2}}{\pi}\frac{(z^*)^m}{\sqrt{m!}}K_m^{-1}\left(-j + z\frac{\partial}{\partial z}\right)K_n\frac{z^n}{\sqrt{n!}} \\ &= \int\int \mathrm{d}z \frac{e^{-|z|^2}}{\pi}\frac{(z^*)^m}{\sqrt{m!}}\frac{K_n}{K_m}(-j+n)\frac{z^n}{\sqrt{n!}} \\ &= \delta_{m,n}\frac{K_n}{K_m}(-j+n), \qquad n = 0, 1, 2, \ldots;\end{aligned} \tag{1.328}$$

$$\begin{aligned}\langle m|J_+|n\rangle &= \langle \chi_m|\gamma(J_+)|\chi_n\rangle \\ &= \int\int \mathrm{d}z \frac{e^{-|z|^2}}{\pi}\chi_m^*(z)(KzK^{-1})\chi_n(z) \\ &= \int\int \mathrm{d}z \frac{e^{-|z|^2}}{\pi}\frac{(z^*)^m}{\sqrt{m!}}K_m z K_n^{-1}\frac{z^n}{\sqrt{n!}} \\ &= \int\int \mathrm{d}z \frac{e^{-|z|^2}}{\pi}\frac{(z^*)^m}{\sqrt{m!}}\frac{K_m}{K_n}\frac{z^{n+1}}{\sqrt{(n+1)!}}\sqrt{n+1} \\ &= \delta_{m,n+1}\frac{K_m}{K_n}\sqrt{n+1}, \qquad n = 0, 1, 2, \ldots;\end{aligned} \tag{1.329}$$

and, similarly,

$$\langle m|J_-|n\rangle = \delta_{m,n-1}\frac{K_m}{K_n}\sqrt{n}, \qquad n = 0, 1, 2, \ldots. \tag{1.330}$$

Specifically, the matrix elements of J_0, J_+, and J_- are:

$$\begin{aligned}\langle n|J_0|n\rangle &= (-j+n), \qquad n = 0, 1, 2, \ldots \\ &= -j, -j+1, -j+2, \ldots;\end{aligned} \tag{1.331}$$

$$\begin{aligned}\langle n+1|J_+|n\rangle &= \frac{K_{n+1}}{K_n}\sqrt{n+1} = \sqrt{\frac{(2j)!}{(2j-(n+1))!}\frac{(2j-n)!}{(2j)!}}\sqrt{n+1} \\ &= \sqrt{(2j-n)(n+1)},\end{aligned} \tag{1.332}$$

and, from $m = -j + n$, i.e. $n = m + j$

$$\langle n+1|J_+|n\rangle = \sqrt{(j-m)(j+m+1)}; \tag{1.333}$$

and, similarly,

$$\langle n-1|J_-|n\rangle = \sqrt{(2j-n+1)n}, \tag{1.334}$$

$$\therefore \langle n-1|J_-|n\rangle = \sqrt{(j-m+1)(j+m)}. \tag{1.335}$$

Note: for $n = 2j$, $\langle n+1|J_+|n\rangle = 0$, i.e. $n_{max} = 2j$ (as it should for $SU(2)$); and it follows also, therefore, that $2j$ = integer.

1. The associations $\frac{\partial}{\partial z} \leftrightarrow a$, $z \leftrightarrow a^\dagger$, $n \leftrightarrow a^\dagger a$ reveal that:

$$\Gamma(J_-) \leftrightarrow a, \tag{1.336}$$

$$\Gamma(J_0) \leftrightarrow -j + a^\dagger a, \tag{1.337}$$

$$\Gamma(J_+) \leftrightarrow a^\dagger(2j - a^\dagger a); \tag{1.338}$$

   and that $\Gamma(J_-)$, $\Gamma(J_+)$ can be made into an adjoint pair via

$$\Gamma(J_-) \to \gamma(J_-) := \sqrt{2j - a^\dagger a}\; a, \tag{1.339}$$

$$\Gamma(J_+) \to \gamma(J_+) := a^\dagger\sqrt{2j - a^\dagger a}. \tag{1.340}$$

   This type of representation is called a *Holstein–Primakoff representation* [6].
2. There is an extensive literature on boson realisations of Lie algebras (see, e.g. [7]).

The material in this section relies on a treatment adopted by K T Hecht [8].

1.20 Exercises

1.6.
   (a) Derive the result, equation (1.269).
   (b) For $d = 1.00$ cm and neutrons of de Broglie wavelength $\lambda = 1.82$ Å, show that a 4π rotation is produced by $\Delta Bd = 149$ Gauss $\cdot$cm. ($\mu_n = 9.65 \times 10^{-24}$ erg $\cdot$Gauss^{-1}, $m = 1.67 \times 10^{-27}$ kg.)

1.7. Show that substituting $\Psi(z) = \chi_n(z)$ (cf. equation (1.170)) into equation (1.280) gives

$$\Psi(x) = \frac{H_n(x)e^{\frac{-x^2}{2}}}{\sqrt{2^n n!\sqrt{\pi}}}. \tag{1.341}$$

   Use the generating function for Hermite polynomials

$$e^{-s^2+2sx} = \sum_n H_n(x)\frac{s^n}{n!}, \text{ with } s = \frac{z^*}{\sqrt{2}}. \tag{1.342}$$

1.8. Show that $\Psi(z) = \sum_n c_n \frac{z^n}{\sqrt{n!}}$, where the c_n are the expansion coefficients for $\Psi(x)$ in the (orthonormal) basis defined by the one-dimensional harmonic oscillator energy eigenfunctions.

1.9. Show that

$$[\Gamma(J_-), \Gamma(J_+)] = -2\Gamma(J_0),$$
$$[\Gamma(J_0), \Gamma(J_+)] = +\Gamma(J_+),$$
$$[\Gamma(J_0), \Gamma(J_-)] = -\Gamma(J_-).$$

1.10. Show that:

$$\Gamma(J^2) := \frac{1}{2}\{\Gamma(J_-)\Gamma(J_+) + \Gamma(J_+)\Gamma(J_-)\} + \Gamma(J_0)^2 = j(j+1).$$

References

[1] Werner S A 1980 *Phys. Today* **33** 24
[2] Rauch H *et al* 1975 *Phys. Lett.* **54A** 425
[3] Bargmann V 1962 *Mod. Phys. Rev.* **34** 829
[4] Arecchi F T *et al* 1972 *Phys. Rev.* A **6** 2211
[5] Dyson F J 1956 *Phys. Rev.* **102** 1217
[6] Holstein T and Primakoff H 1940 *Phys. Rev.* **58** 1098
[7] Klein A and Marshalek E R 1991 *Mod. Phys. Rev.* **63** 375
[8] Hecht K T 1987 *The Vector Coherent State Method and Its Application to Problems of Higher Symmetries* (Lecture Notes in Physics vol 290) (Berlin: Springer)

IOP Publishing

Quantum Mechanics for Nuclear Structure, Volume 2

An intermediate level view

Kris Heyde and John L Wood

Chapter 2

Addition of angular momenta and spins

The coupling of spins and angular momenta is introduced at the simplest possible level: the coupling of two spin-$\frac{1}{2}$ particles. The concepts of reducible and irreducible representations are clarified. A thorough introduction to Clebsch–Gordan coefficients (CG coeffs) and the related Wigner or 3-*j* coefficients (3-*j* symbols) is given. Spin–orbit coupling is described. The interconnection between CG coeffs. and rotation matrices is presented. A brief introduction to recoupling coefficients—6-*j* and 9-*j* coefficients—is given. Tables of simple CG and 3-*j* coefficients are given in appendix A.

Concepts: direct, tensor, or Kronecker products; reducibility of a representation; irreducible representations (irreps); Clebsch–Gordan (CG) coefficients; Wigner or 3-*j* coefficients; CG series; irrep character; spin–orbit interaction and coupling; vector spherical harmonics; 6-*j* coefficients; 9-*j* coefficients.

The epitome of quantum systems possessing $SO(3)$ and $SU(2)$ symmetry is provided by quantum particles with angular momentum and spin. This naturally leads us to consider the coupling of spins and angular momenta in many-particle quantum systems. This is an essential part of the physics of molecules, atoms, nuclei and hadrons.

2.1 The coupling of two spin-$\frac{1}{2}$ particles

The simplest possible coupling in $SU(2)$ is that of two spin-$\frac{1}{2}$ particles. This is realised, for example, in the ground state of the hydrogen atom where one is concerned with a proton and an electron (both spin-$\frac{1}{2}$ particles) in a state of relative motion with angular momentum zero ($l = 0$). Thus, the possible ground-state configurations of the hydrogen atom are completely described by

$$|\, s = ½,\, m_s\rangle_p \otimes |\, s = ½,\, m_s\rangle_e = |½,\, ½\rangle_p \otimes |½,\, ½\rangle_e, \tag{2.1}$$

or

doi:10.1088/978-0-7503-2171-6ch2

$$=|½, ½\rangle_p \otimes |½, -½\rangle_e, \tag{2.2}$$

or

$$=|½, -½\rangle_p \otimes |½, ½\rangle_e, \tag{2.3}$$

or

$$=|½, -½\rangle_p \otimes |½, -½\rangle_e. \tag{2.4}$$

These basis states, made up of the $s_p = \frac{1}{2}$, $m_p = \pm\frac{1}{2}$ and $s_e = \frac{1}{2}$, $m_e = \pm\frac{1}{2}$ basis states for the proton and electron, respectively, are in a form referred to as a *direct product*, *tensor product*, or *Kronecker product*.

Just as we considered the rotation of spin-$\frac{1}{2}$ particles in chapter 1, we can discuss the separate rotation of the proton and the electron in the four states, equations (2.1)–(2.4)

$$\left|\frac{1}{2}m_s\right\rangle_p^R \otimes \left|\frac{1}{2}m_s\right\rangle_e^R = \mathcal{D}^{(\frac{1}{2})}(\alpha_p, \beta_p, \gamma_p)\left|\frac{1}{2}m_s\right\rangle_p \otimes \mathcal{D}^{(\frac{1}{2})}(\alpha_e, \beta_e, \gamma_e)\left|\frac{1}{2}m_s\right\rangle_e, \tag{2.5}$$

where α_p, β_p, and γ_p and α_e, β_e, and γ_e are the Euler angles describing the rotations of the proton and the electron ket states, respectively. These rotations can be viewed in the form

$$\begin{aligned}\left|\tfrac{1}{2}m_s\right\rangle_p^R \otimes \left|\tfrac{1}{2}m_s\right\rangle_e^R &= \left(\mathcal{D}^{(\frac{1}{2})}(\alpha_p, \beta_p, \gamma_p) \otimes \mathcal{D}^{(\frac{1}{2})}(\alpha_e, \beta_e, \gamma_e)\right)\\ &\quad\times\left(\left|\frac{1}{2}m_s\right\rangle_p \otimes \left|\frac{1}{2}m_s\right\rangle_e\right),\end{aligned} \tag{2.6}$$

i.e. as transformations $\mathcal{D}_p^{(\frac{1}{2})} \otimes \mathcal{D}_e^{(\frac{1}{2})}$ on a four-dimensional space spanned by $\left|\frac{1}{2}m_s\right\rangle_p \otimes \left|\frac{1}{2}m_s\right\rangle_e$. The matrices representing $\mathcal{D}_p^{(\frac{1}{2})} \otimes \mathcal{D}_e^{(\frac{1}{2})}$ are 4×4. For example, if $\alpha_p = 0$, $\gamma_p = 0$, $\alpha_e = 0$, $\gamma_e = 0$, they have the form

$$\begin{aligned}&\mathcal{D}^{(\frac{1}{2})}(0, \beta_p, 0) \otimes \mathcal{D}^{(\frac{1}{2})}(0, \beta_e, 0) := \mathcal{D}^{(\frac{1}{2}, \frac{1}{2})}(0, \beta_p, 0; 0, \beta_e, 0)\\ &:= \begin{pmatrix} \cos\frac{\beta_p}{2}\cos\frac{\beta_e}{2} & -\cos\frac{\beta_p}{2}\sin\frac{\beta_e}{2} & -\sin\frac{\beta_p}{2}\cos\frac{\beta_e}{2} & \sin\frac{\beta_p}{2}\sin\frac{\beta_e}{2} \\ \cos\frac{\beta_p}{2}\sin\frac{\beta_e}{2} & \cos\frac{\beta_p}{2}\cos\frac{\beta_e}{2} & -\sin\frac{\beta_p}{2}\sin\frac{\beta_e}{2} & -\sin\frac{\beta_p}{2}\cos\frac{\beta_e}{2} \\ \sin\frac{\beta_p}{2}\cos\frac{\beta_e}{2} & -\sin\frac{\beta_p}{2}\sin\frac{\beta_e}{2} & \cos\frac{\beta_p}{2}\cos\frac{\beta_e}{2} & -\cos\frac{\beta_p}{2}\sin\frac{\beta_e}{2} \\ \sin\frac{\beta_p}{2}\sin\frac{\beta_e}{2} & \sin\frac{\beta_p}{2}\cos\frac{\beta_e}{2} & \cos\frac{\beta_p}{2}\sin\frac{\beta_e}{2} & \cos\frac{\beta_p}{2}\cos\frac{\beta_e}{2} \end{pmatrix},\end{aligned} \tag{2.7}$$

where the 4×4 matrix is generated from

$$\mathcal{D}^{(\frac{1}{2})}(0, \beta_p, 0) = \begin{pmatrix} \cos\frac{\beta_p}{2} & -\sin\frac{\beta_p}{2} \\ \sin\frac{\beta_p}{2} & \cos\frac{\beta_p}{2} \end{pmatrix} \tag{2.8}$$

and

$$\mathcal{D}^{(\frac{1}{2})}(0, \beta_e, 0) = \begin{pmatrix} \cos\frac{\beta_e}{2} & -\sin\frac{\beta_e}{2} \\ \sin\frac{\beta_e}{2} & \cos\frac{\beta_e}{2} \end{pmatrix} \tag{2.9}$$

in a straightforward way.

The proton and the electron in hydrogen experience a 'spin–spin' interaction through their magnetic moments: this results in a splitting in energy of the states, equations (2.1)–(2.4). Thus, transformations such as that described by equation (2.7) do not leave the energy of the hydrogen atom ground state unchanged. However, 'rigid' rotations with $\alpha_p = \alpha_e$, $\beta_p = \beta_e$ and $\gamma_p = \gamma_e$, which preserve the relative orientation of the two spins, will leave the energy unchanged. In equation (2.7), for example, if $\beta_p = \beta_e \equiv \beta$ then

$$\mathcal{D}^{(\frac{1}{2})}(0, \beta, 0) \otimes \mathcal{D}^{(\frac{1}{2})}(0, \beta, 0)$$
$$= \begin{pmatrix} \cos^2\frac{\beta}{2} & -\sin\frac{\beta}{2}\cos\frac{\beta}{2} & -\sin\frac{\beta}{2}\cos\frac{\beta}{2} & \sin^2\frac{\beta}{2} \\ \sin\frac{\beta}{2}\cos\frac{\beta}{2} & \cos^2\frac{\beta}{2} & -\sin^2\frac{\beta}{2} & -\sin\frac{\beta}{2}\cos\frac{\beta}{2} \\ \sin\frac{\beta}{2}\cos\frac{\beta}{2} & -\sin^2\frac{\beta}{2} & \cos^2\frac{\beta}{2} & -\sin\frac{\beta}{2}\cos\frac{\beta}{2} \\ \sin^2\frac{\beta}{2} & \sin\frac{\beta}{2}\cos\frac{\beta}{2} & \sin\frac{\beta}{2}\cos\frac{\beta}{2} & \cos^2\frac{\beta}{2} \end{pmatrix}. \tag{2.10}$$

This matrix is *reducible*:

$$C^\dagger\mathcal{D}^{(\frac{1}{2})}(0, \beta, 0) \otimes \mathcal{D}^{(\frac{1}{2})}(0, \beta, 0)C$$
$$= \begin{pmatrix} \frac{1}{2}(1+\cos\beta) & -\sin\frac{\beta}{\sqrt{2}} & \frac{1}{2}(1-\cos\beta) & 0 \\ \sin\frac{\beta}{\sqrt{2}} & \cos\beta & -\sin\frac{\beta}{\sqrt{2}} & 0 \\ \frac{1}{2}(1-\cos\beta) & \sin\frac{\beta}{\sqrt{2}} & \frac{1}{2}(1+\cos\beta) & 0 \\ 0 & 0 & 0 & 1 \end{pmatrix}, \tag{2.11}$$

where

$$C = \begin{pmatrix} 1 & 0 & 0 & 0 \\ 0 & \frac{1}{\sqrt{2}} & 0 & \frac{1}{\sqrt{2}} \\ 0 & \frac{1}{\sqrt{2}} & 0 & -\frac{1}{\sqrt{2}} \\ 0 & 0 & 1 & 0 \end{pmatrix}. \tag{2.12}$$

A comparison of the matrix in equation (2.11) with that in equation (1.65) reveals that

$$C^{\dagger}\mathcal{D}^{(\frac{1}{2})}(0, \beta, 0) \otimes \mathcal{D}^{(\frac{1}{2})}(0, \beta, 0)C = \mathcal{D}^{(1)}(0, \beta, 0) \oplus \mathcal{D}^{(0)}(0, \beta, 0), \tag{2.13}$$

where the $\oplus$ symbol describes the combining of the 3×3 matrix, $\mathcal{D}^{(1)}(0, \beta, 0)$ (equation (1.65)) and the 1×1 matrix $\mathcal{D}^{(0)}(0, \beta, 0) = (1)$ to yield the 4×4 matrix in equation (2.11), i.e.

$$\mathcal{D}^{(1)}(0, \beta, 0) \oplus \mathcal{D}^{(0)}(0, \beta, 0) := \left(\begin{array}{c|c} \mathcal{D}^{(1)}(0, \beta, 0) & 0 \\ \hline 0 & \mathcal{D}^{(0)}(0, \beta, 0) \end{array}\right). \tag{2.14}$$

Thus, under rotations the four spin–spin couplings of the hydrogen atom ground state behave as a (total) spin $J = 1$ triplet and a (total) spin $J = 0$ singlet.

The relationship between the basis, equations (2.1)–(2.4) and the basis for the spin-1 and spin-0 couplings is obtained from the form of the matrix C. Recall that the elements of a unitary transformation are ⟨old basis | new basis⟩ (cf. Volume 1, equation (6.85)). Thus, denoting the old basis (equations (2.1)–(2.4)) by $\left|\frac{1}{2}\frac{1}{2}\right\rangle_p \otimes \left|\frac{1}{2}\frac{1}{2}\right\rangle_e \leftrightarrow |\uparrow\uparrow\rangle$, etc., and the new basis by $\{|1\rangle, |2\rangle, |3\rangle, |4\rangle\}$, we have

$$\langle\uparrow\uparrow|1\rangle = 1, \tag{2.15}$$

$$\langle\uparrow\downarrow|2\rangle = \frac{1}{\sqrt{2}}, \quad \langle\uparrow\downarrow|4\rangle = \frac{1}{\sqrt{2}}, \tag{2.16}$$

$$\langle\downarrow\uparrow|2\rangle = \frac{1}{\sqrt{2}}, \quad \langle\downarrow\uparrow|4\rangle = -\frac{1}{\sqrt{2}}, \tag{2.17}$$

$$\langle\downarrow\downarrow|3\rangle = 1, \tag{2.18}$$

whence

$$|1\rangle = |\uparrow\uparrow\rangle, \tag{2.19}$$

$$|2\rangle = \frac{1}{\sqrt{2}}(|\uparrow\downarrow\rangle + |\downarrow\uparrow\rangle), \tag{2.20}$$

$$|3\rangle = |\downarrow\downarrow\rangle, \tag{2.21}$$

and

$$|4\rangle = \frac{1}{\sqrt{2}}(|\uparrow\downarrow\rangle - |\downarrow\uparrow\rangle). \tag{2.22}$$

The kets $|1\rangle$, $|2\rangle$, and $|3\rangle$ form a triplet representing the spin-1 coupling of the proton and the electron. The ket $|4\rangle$ forms a singlet representing the spin-0 coupling of the proton and the electron. The expansion coefficients, e.g. $\langle\uparrow\downarrow|2\rangle = \frac{1}{\sqrt{2}}$, are called *Clebsch–Gordan coefficients*. They are the amplitudes of the uncoupled base kets needed for combinations that have definite total spin ($S = 0$ and $S = 1$ in this case).

These results can also be obtained algebraically. For 'rigid' rotations, $\alpha_p = \alpha_e \equiv \alpha$, $\beta_p = \beta_e \equiv \beta$, $\gamma_p = \gamma_e \equiv \gamma$, and

$$\begin{aligned} \mathcal{D}^{(\frac{1}{2})}(\alpha_p, \beta_p, \gamma_p) \otimes \mathcal{D}^{(\frac{1}{2})}(\alpha_e, \beta_e, \gamma_e) &= e^{-\frac{i\hat{S}_z^{(p)}\alpha_p}{\hbar}} e^{-\frac{i\hat{S}_y^{(p)}\beta_p}{\hbar}} e^{-\frac{i\hat{S}_z^{(p)}\gamma_p}{\hbar}} e^{-\frac{i\hat{S}_z^{(e)}\alpha_e}{\hbar}} e^{-\frac{i\hat{S}_y^{(e)}\beta_e}{\hbar}} e^{-\frac{i\hat{S}_z^{(e)}\gamma_e}{\hbar}} \\ &= e^{-\frac{i\hat{S}_z\alpha}{\hbar}} e^{-\frac{i\hat{S}_y\beta}{\hbar}} e^{-\frac{i\hat{S}_z\gamma}{\hbar}}, \end{aligned} \tag{2.23}$$

where

$$\hat{S}_z := \hat{S}_z^{(p)} + \hat{S}_z^{(e)} = \hat{S}_z^{(p)} \otimes \hat{I}^{(e)} + \hat{I}^{(p)} \otimes \hat{S}_z^{(e)}, \tag{2.24}$$

$$\hat{S}_y := \hat{S}_y^{(p)} + \hat{S}_y^{(e)} = \hat{S}_y^{(p)} \otimes \hat{I}^{(e)} + \hat{I}^{(p)} \otimes \hat{S}_y^{(e)}, \tag{2.25}$$

and similarly,

$$\hat{S}_x := \hat{S}_x^{(p)} + \hat{S}_x^{(e)} = \hat{S}_x^{(p)} \otimes \hat{I}^{(e)} + \hat{I}^{(p)} \otimes \hat{S}_x^{(e)}, \tag{2.26}$$

i.e.

$$\vec{S} := \vec{S}^{(p)} + \vec{S}^{(e)} = \vec{S}^{(p)} \otimes \hat{I}^{(e)} + \hat{I}^{(p)} \otimes \vec{S}^{(e)}. \tag{2.27}$$

Then for

$$\hat{S}_\pm^{(p)} = \hat{S}_x^{(p)} \pm i\hat{S}_y^{(p)}, \tag{2.28}$$

$$\hat{S}_\pm^{(e)} = \hat{S}_x^{(e)} \pm i\hat{S}_y^{(e)}, \tag{2.29}$$

$$\hat{S}_\pm = \hat{S}_\pm^{(p)} + \hat{S}_\pm^{(e)}, \tag{2.30}$$

consider

$$\begin{aligned}\hat{S}_+\left|\frac{1}{2},-\frac{1}{2}\right\rangle_p\otimes\left|\frac{1}{2},-\frac{1}{2}\right\rangle_e &= \hat{S}_+^{(p)}\left|\frac{1}{2},-\frac{1}{2}\right\rangle_p\otimes\left|\frac{1}{2},-\frac{1}{2}\right\rangle_e\\ &\quad+\hat{S}_+^{(e)}\left|\frac{1}{2},-\frac{1}{2}\right\rangle_p\otimes\left|\frac{1}{2},-\frac{1}{2}\right\rangle_e\\ &=\left|\frac{1}{2}\frac{1}{2}\right\rangle_p\otimes\left|\frac{1}{2},-\frac{1}{2}\right\rangle_e\\ &\quad+\left|\frac{1}{2},-\frac{1}{2}\right\rangle_p\otimes\left|\frac{1}{2},\frac{1}{2}\right\rangle_e.\end{aligned} \tag{2.31}$$

Similarly,

$$\begin{aligned}\hat{S}_-\left|\frac{1}{2},\frac{1}{2}\right\rangle_p\otimes\left|\frac{1}{2},\frac{1}{2}\right\rangle_e &=\left|\frac{1}{2},-\frac{1}{2}\right\rangle_p\otimes\left|\frac{1}{2},\frac{1}{2}\right\rangle_e+\left|\frac{1}{2},\frac{1}{2}\right\rangle_p\\ &\quad\otimes\left|\frac{1}{2},-\frac{1}{2}\right\rangle_e,\end{aligned} \tag{2.32}$$

and

$$\begin{aligned}&\hat{S}_\pm\left(\left|\frac{1}{2},\frac{1}{2}\right\rangle_p\otimes\left|\frac{1}{2},-\frac{1}{2}\right\rangle_e+\left|\frac{1}{2},-\frac{1}{2}\right\rangle_p\otimes\left|\frac{1}{2},\frac{1}{2}\right\rangle_e\right)\\ &=2\left|\frac{1}{2},\pm\frac{1}{2}\right\rangle_p\otimes\left|\frac{1}{2},\pm\frac{1}{2}\right\rangle_e.\end{aligned} \tag{2.33}$$

But (cf. Volume 1, equations (11.58) and (11.60)), ($\hbar\equiv 1$)

$$\hat{S}_\pm|SM\rangle=\sqrt{(S\mp M)(S\pm M+1)}\,|SM\pm 1\rangle, \tag{2.34}$$

whence

$$\hat{S}_+|1,-1\rangle=\sqrt{2}\,|1,0\rangle,\qquad \hat{S}_-|1,-1\rangle=0, \tag{2.35}$$

$$\hat{S}_+|1,0\rangle=\sqrt{2}\,|1,1\rangle,\qquad \hat{S}_-|1,0\rangle=\sqrt{2}\,|1,-1\rangle, \tag{2.36}$$

$$\hat{S}_+|1,1\rangle=0,\qquad \hat{S}_-|1,1\rangle=\sqrt{2}\,|1,0\rangle. \tag{2.37}$$

Hence, equations (2.19)–(2.21) follow. Equation (2.22) follows from equation (2.20) by orthogonal construction, and

$$\hat{S}_\pm\left(\left|\frac{1}{2},\frac{1}{2}\right\rangle_p\otimes\left|\frac{1}{2},-\frac{1}{2}\right\rangle_e-\left|\frac{1}{2},-\frac{1}{2}\right\rangle_p\otimes\left|\frac{1}{2},\frac{1}{2}\right\rangle_e\right)=0, \tag{2.38}$$

cf.

$$\hat{S}_{\pm}|0, 0\rangle = 0. \tag{2.39}$$

A spin–spin interaction in the hydrogen atom can be defined as

$$V_{S-S} := k\vec{S}^{(p)} \cdot \vec{S}^{(e)} \text{ (k constant).} \tag{2.40}$$

This can be expressed as

$$V_{S-S} = \frac{k}{2}\left\{\vec{S} \cdot \vec{S} - \vec{S}^{(p)} \cdot \vec{S}^{(p)} - \vec{S}^{(e)} \cdot \vec{S}^{(e)}\right\}, \tag{2.41}$$

$$\therefore V_{S-S} = \frac{k}{2}\left\{\hat{S}^2 - \hat{S}^{(p)^2} - \hat{S}^{(e)^2}\right\}. \tag{2.42}$$

The eigenvalues of V_{S-S} for the states $|S, S^{(p)}, S^{(e)}, M\rangle = \left|1\frac{1}{2}\frac{1}{2}1\right\rangle, \left|1\frac{1}{2}\frac{1}{2}0\right\rangle, \left|1\frac{1}{2}\frac{1}{2}, -1\right\rangle$, and $\left|0\frac{1}{2}\frac{1}{2}0\right\rangle$ follow:

$$\begin{aligned} V_{S-S}\left|1\frac{1}{2}\frac{1}{2}1\right\rangle &= \frac{k}{2}\left\{1(1+1) - \frac{1}{2}\left(\frac{1}{2}+1\right) - \frac{1}{2}\left(\frac{1}{2}+1\right)\right\}\left|1\frac{1}{2}\frac{1}{2}1\right\rangle \\ &= \frac{k}{4}\left|1\frac{1}{2}\frac{1}{2}1\right\rangle, \end{aligned} \tag{2.43}$$

and similarly for $\left|1\frac{1}{2}\frac{1}{2}0\right\rangle$ and $\left|1\frac{1}{2}\frac{1}{2}, -1\right\rangle$; and

$$\begin{aligned} V_{S-S}\left|0\frac{1}{2}\frac{1}{2}0\right\rangle &= \frac{k}{2}\left\{0(0+1) - \frac{1}{2}\left(\frac{1}{2}+1\right) - \frac{1}{2}\left(\frac{1}{2}+1\right)\right\}\left|0\frac{1}{2}\frac{1}{2}0\right\rangle \\ &= -\frac{3k}{4}\left|0\frac{1}{2}\frac{1}{2}0\right\rangle. \end{aligned} \tag{2.44}$$

For $k > 0$, this can be depicted as shown in figure 2.1. Thus, algebraically the coupled basis emerges as the eigenbasis for the ground-state multiplet of the hydrogen atom, as it did by considering rotational invariants for the hydrogen atom ground state under the condition of 'rigid' rotations.

2.2 The general coupling of two particles with spin or angular momentum

The general coupling problem for a pair of $SU(2)$ representations is realised in the coupling of the spins or angular momenta j_1, j_2 of two particles. The result can be stated as a theorem:

Theorem 2.1. *For two $SU(2)$ representations $\mathcal{D}^{(j_1)}$ and $\mathcal{D}^{(j_2)}$, the Kronecker product $\mathcal{D}^{(j_1)} \otimes \mathcal{D}^{(j_2)}$ contains the $SU(2)$ representations $\mathcal{D}^{(j)}$, $|j_1 - j_2| \leqslant j \leqslant j_1 + j_2$, i.e.*

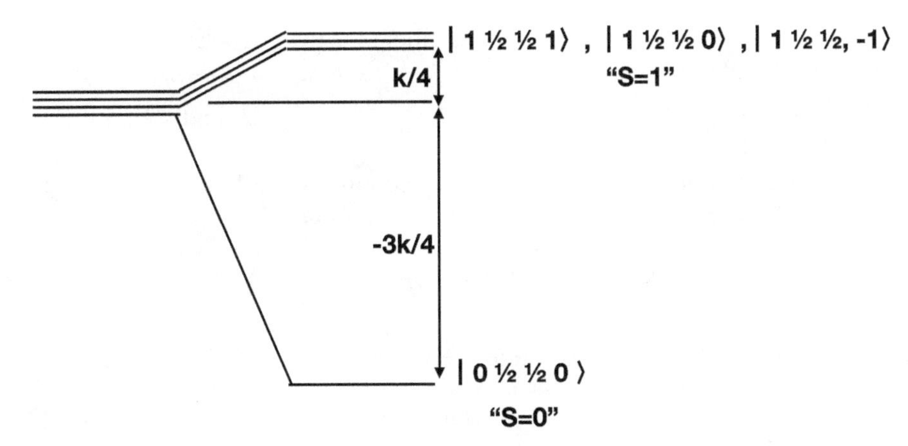

Figure 2.1. Spin–spin splitting of the hydrogen atom ground state. The $S = 1$ to $S = 0$ transition corresponds to a wavelength of 21 cm: it is well known in radioastronomy.

$$\mathcal{D}^{(j_1)} \otimes \mathcal{D}^{(j_2)} = \sum_{j=|j_1-j_2|}^{j_1+j_2} \mathcal{D}^{(j)} = \mathcal{D}^{(|j_1-j_2|)} \oplus \mathcal{D}^{(|j_1-j_2|+1)} \oplus \cdots \mathcal{D}^{(j_1+j_2)}. \tag{2.45}$$

(The expansion on the right-hand side is called the ($SU(2)$) *Clebsch–Gordan series*; the sum $\oplus$ is defined in equation (2.14).)

Proof. The theorem can be proved by the brute-force method of computing the dimensions of the representations on each side of equation (2.45). The left-hand side is the product of two representations of dimension $(2j_1 + 1)$ and $(2j_2 + 1)$, respectively ($m_1 = -j_1, -j_1 + 1, \ldots, j_1$; $m_2 = -j_2, -j_2 + 1, \ldots, j_2$); the right-hand side is the sum of representations of dimension $(2j + 1)$ where $j = |j_1 - j_2|, |j_1 - j_2| + 1, \ldots, j_1 + j_2$.

The theorem is more elegantly proved by considering the traces of the Wigner matrices and the fact that the trace of a matrix is invariant under a similarity transformation (cf. Volume 1, theorem 6.2). Thus,

$$\mathrm{Tr}\,\{\mathcal{D}^{(j)}(\hat{n}, \phi)\} = \mathrm{Tr}\,\{\mathcal{D}^{(j)}(\hat{z}, \phi)\} = \sum_{m=-j}^{+j} e^{im\phi}, \tag{2.46}$$

whence

$$\sum_{m_1=-j_1}^{+j_1} \sum_{m_2=-j_2}^{+j_2} e^{i(m_1+m_2)\phi} = \sum_{j=|j_1-j_2|}^{j_1+j_2} \sum_{m=-j}^{+j} e^{im\phi}. \tag{2.47}$$

□

(The quantity $\sum_{m=-j}^{+j} e^{im\phi} := \chi^{(j)}(\phi)$ is called the *character* of the $SU(2)$ irrep j for angle ϕ.)

Two choices of basis are available for describing the two particles: the coupled basis $|jm\rangle \equiv |jj_1j_2m\rangle$ and the uncoupled basis $|j_1m_1j_2m_2\rangle \equiv |j_1m_1\rangle \otimes |j_2m_2\rangle$. These two bases are simultaneous eigenkets of the operators $\{\hat{J}^2, \hat{J}_1^2, \hat{J}_2^2, \hat{J}_z\}$ and $\{\hat{J}_1^2, \hat{J}_{1z}, \hat{J}_2^2, \hat{J}_{2z}\}$, respectively.

For two particles described by the uncoupled basis states $|j_1m_1\rangle, |j_2m_2\rangle$, they can be alternatively described by the coupled basis states $|jm\rangle$:

$$|jm\rangle = \sum_{m_1(m_2=m-m_1)} |j_1m_1j_2m_2\rangle\langle j_1m_1j_2m_2|jm\rangle, \tag{2.48}$$

where j_1 and j_2 are fixed and j can take the values

$$j = |j_1 - j_2|, |j_1 - j_2| + 1, \cdots, j_1 + j_2, \tag{2.49}$$

(note each j value occurs *only once*, i.e. it is *multiplicity free* with $m = m_1 + m_2$). The expansion coefficients $\langle j_1m_1j_2m_2|jm\rangle$ are called *Clebsch–Gordan* coefficients or *vector-coupling* coefficients or *Wigner SU(2)* coefficients. They are the amplitudes of $|j_1m_1\rangle \otimes |j_2m_2\rangle$ needed for combinations that have total spin j. Stated in another way, they are the amplitudes for combining the elements of the uncoupled basis so as to give an irreducible form. The transformation between the two bases is unitary. The relation, $m = m_1 + m_2$, follows from $\hat{J}_z = \hat{J}_{1z} + \hat{J}_{2z}$. The Clebsch–Gordan coefficients are real. The range of j values looks 'obvious' in that, with the restriction to integer increments, the values are given by a classical vector picture; BUT: the coupling involves cones of indeterminacy, as depicted in figure 2.2.

A powerful method for computing allowed J values for the coupling of two particles is the so-called *m-scheme*:

Ex. j1=1, j2=1 J=2, 1, 0

m_1	m_2	M
+1	+1	+2
+1	0	+1
+1	-1	0
0	+1	+1
0	0	0
0	-1	-1
-1	+1	0
-1	0	-1
-1	-1	-2

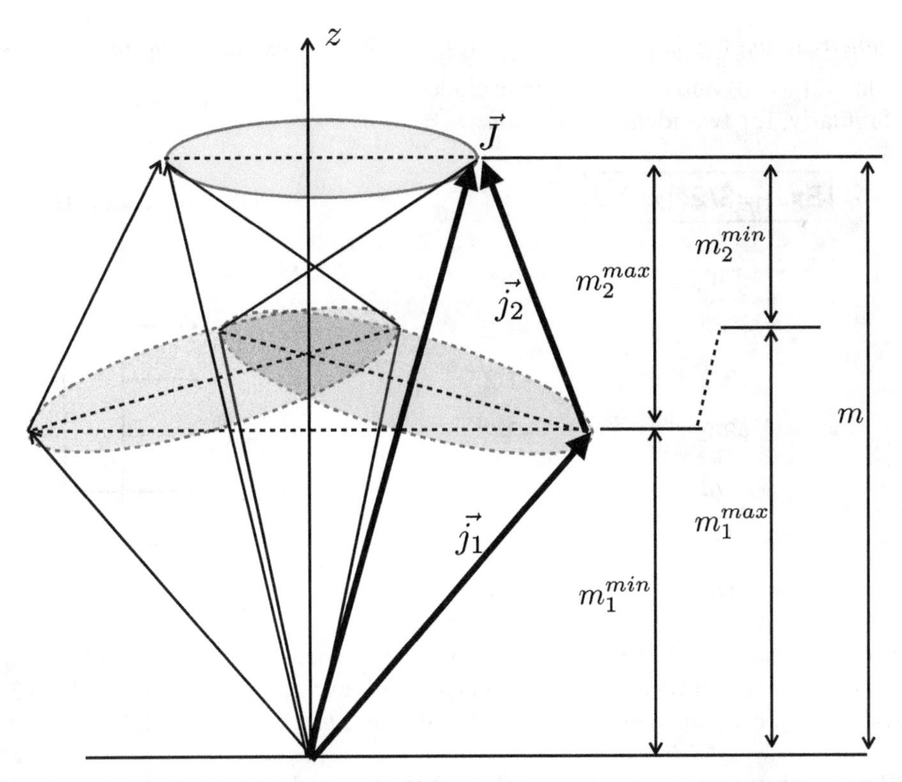

Figure 2.2. Depiction of the combining of the 'cones of uncertainty' for directional components m_1 and m_2 when coupling j_1 and j_2 to definite J and M. The vectors $\vec{j}_1$ and $\vec{j}_2$ can be viewed as 'rotating' around $\vec{J}$ which is in turn rotating around the z-axis in the sense of generating (hyper)cones of uncertainty. However, this descriptive use of the word 'rotating' must not be adopted as a hidden view of such uncertainty.

The method is especially useful when there are *restrictions*, e.g. for two identical *bosons*:

Ex. j_1=1, j_2=1 — **J=2, 0**

m_1	m_2	M
+1	+1	+2
+1	0	+1
+1	-1	0
0	0	0
0	-1	-1
-1	-1	-2

Symmetrization, e.g. $|m_1, m_2\rangle = \frac{1}{\sqrt{2}}\{|+1, 0\rangle + |0, +1\rangle\}$ excludes $(m_1, m_2) = (0, +1)$; similarly, $(-1, 0)$ and $(-1, +1)$ are excluded.

Similarly, for two identical *fermions*:

Ex. j_1=3/2, j_2=3/2 — **J=2, 0**

m_1	m_2	M
+3/2	+1/2	+2
+3/2	-1/2	+1
+3/2	-3/2	0
+1/2	-1/2	0
+1/2	-3/2	-1
-1/2	-3/2	-2

(the j_1 and j_2 states are presumed identical in every respect except for m_1 and m_2, i.e. all other quantum numbers equal). The *Pauli principle* excludes $(m_1, m_2) = (+3/2, +3/2)$, $(+1/2, +1/2)$, etc.; *antisymmetrization* excludes $(m_1, m_2) = (+1/2, +3/2)$, etc.

The m-scheme can be used for any number of particles to ascertain the allowed J values, e.g. for three identical fermions:

Ex. j_1=3/2, j_2=3/2, j_3 = 3/2 — **J=3/2**

m_1	m_2	m_3	M
+3/2	+1/2	-1/2	+3/2
+3/2	+1/2	-3/2	+1/2
+3/2	-1/2	-3/2	-1/2
+1/2	-1/2	-3/2	-3/2

This leads to the common artifice of describing an almost filled fermion 'shell' in terms of *holes*, i.e. $(j = 3/2)^3 \equiv (j = 3/2)^{-1}$.

A dramatisation of the non-classical nature of quantum mechanical angular momentum is provided by:

$$\vec{L}_{op} \times \vec{L}_{op} = \begin{vmatrix} e_x & e_y & e_z \\ \hat{L}_x & \hat{L}_y & \hat{L}_z \\ \hat{L}_x & \hat{L}_y & \hat{L}_z \end{vmatrix} = e_x(\hat{L}_y\hat{L}_z - \hat{L}_z\hat{L}_y)$$

$$+ e_y(\hat{L}_z\hat{L}_x - \hat{L}_x\hat{L}_z) + e_z(\hat{L}_x\hat{L}_y - \hat{L}_y\hat{L}_x)$$

$$= e_x[\hat{L}_y\hat{L}_z] + e_y[\hat{L}_z\hat{L}_x] + e_z[\hat{L}_x\hat{L}_y]$$

$$= i\hbar(e_x\hat{L}_x + e_y\hat{L}_y + e_z\hat{L}_z) = i\hbar\vec{L}_{op}.$$

The Clebsch–Gordan coefficients can be depicted conveniently using 'm_1, m_2' diagrams. For example, figure 2.3 shows the diagrams for the possible couplings of $j_1 = 2$, $j_2 = 2$.

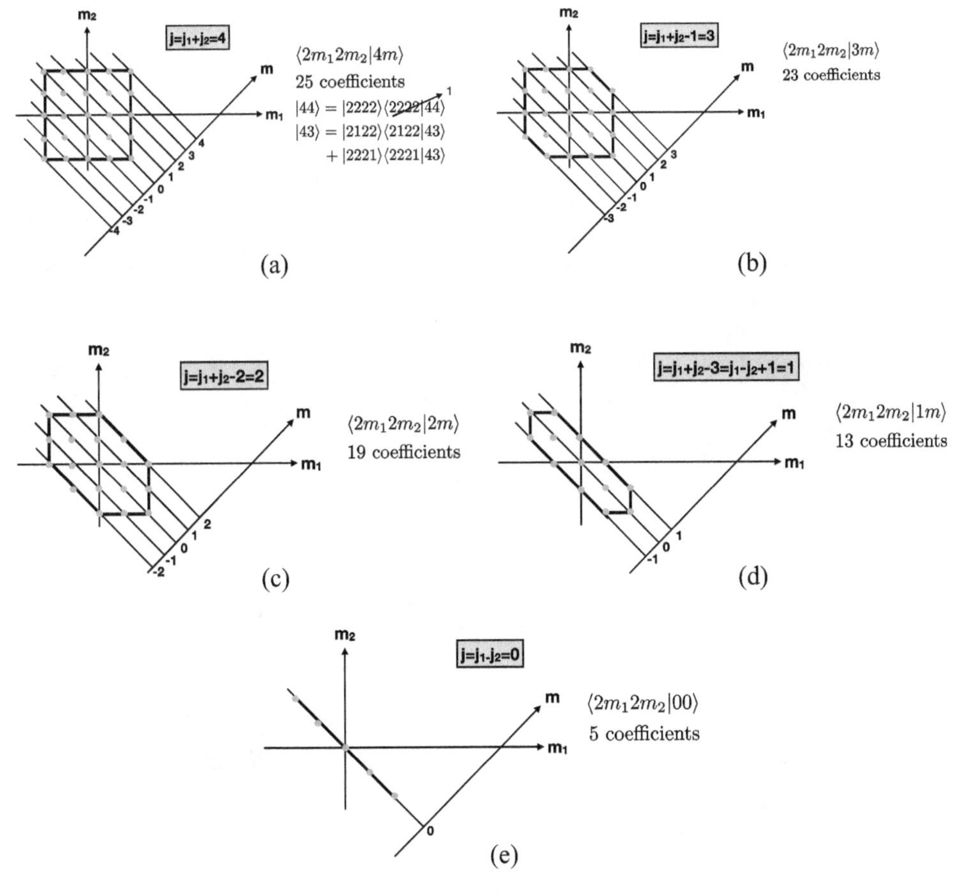

Figure 2.3. Depiction of the Clebsch–Gordan coefficient (a) $\langle 2m_1 2m_2|4\ m\rangle$, (b) $\langle 2m_1 2m_2|3\ m\rangle$, (c) $\langle 2m_1 2m_2|2\ m\rangle$, (d) $\langle 2m_1 2m_2|1\ m\rangle$, and (e) $\langle 2m_1 2m_2|00\rangle$.

It remains to compute the numerical values of the Clebsch–Gordan coefficients: so far we have determined their values only for the coupling $j_1 = \frac{1}{2}, j_2 = \frac{1}{2}$ to $j = 0$ or 1. In principle, one can repeat the process illustrated for $j_1 = \frac{1}{2}, j_2 = \frac{1}{2}$, and block-diagonalize $d^{(j_1)}(\beta) \otimes d^{(j_2)}(\beta)$: the Clebsch–Gordan coefficients are the matrix elements of C in:

$$C^\dagger d^{(j_1)}(\beta) \otimes d^{(j_2)}(\beta) C = d^{(j_1+j_2)}(\beta) \oplus d^{(j_1+j_2-1)}(\beta) \oplus \cdots \oplus d^{(|j_1-j_2|)}(\beta). \tag{2.50}$$

However, as for $j_1 = \frac{1}{2}, j_2 = \frac{1}{2}$, one can obtain Clebsch–Gordan coefficients by algebraic methods. Some examples are given below.

Algebraic methods can be used to generate Clebsch–Gordan coefficients by noting two points:

(i) The raising and lowering operators in the coupled and uncoupled basis are related by

$$\hat{J}_\pm = \hat{J}_{1\pm} + \hat{J}_{2\pm}. \tag{2.51}$$

(ii) The coefficient

$$\langle j_1, m_1 = j_1, j_2, m_2 = j_2 | j = j_1 + j_2, m = j_1 + j_2 \rangle = 1 \tag{2.52}$$

because

$$|j_1, m_1 = j_1\rangle \otimes |j_2, m_2 = j_2\rangle = |j = j_1 + j_2, m = j_1 + j_2\rangle. \tag{2.53}$$

The second point is clear from figure 2.3, where $|22\rangle \otimes |22\rangle = |44\rangle$. Thus, applying $\hat{J}_- = \hat{J}_{1-} + \hat{J}_{2-}$ to the 'maximum weight' state, equation (2.53),

$$\hat{J}_- |j = j_1 + j_2, m = j_1 + j_2\rangle = (\hat{J}_{1-} + \hat{J}_{2-}) |j_1, m_1 = j_1\rangle |j_2, m_2 = j_2\rangle, \tag{2.54}$$

from

$$\hat{J}_- |jm\rangle = \sqrt{(j+m)(j-m+1)}\, |j, m-1\rangle \tag{2.55}$$

(cf. Volume 1, equation (11.60) ($\hbar \equiv 1$)),

$$\begin{aligned} \therefore \sqrt{2(j_1 + j_2)}\, |j_1 + j_2, m = j_1 + j_2 - 1\rangle = &\sqrt{2j_1}\, |j_1, m_1 = j_1 - 1\rangle |j_2, m_2 = j_2\rangle \\ &+ \sqrt{2j_2}\, |j_1, m_1 = j_1\rangle |j_2, m_2 = j_2 - 1\rangle, \end{aligned} \tag{2.56}$$

$$\begin{aligned} \therefore |j_1 + j_2, m = j_1 + j_2 - 1\rangle = &\sqrt{\frac{j_1}{j_1 + j_2}}\, |j_1, m_1 = j_1 - 1\rangle |j_2, m_2 = j_2\rangle \\ &+ \sqrt{\frac{j_2}{j_1 + j_2}}\, |j_1, m_1 = j_1\rangle |j_2, m_2 = j_2 - 1\rangle, \end{aligned} \tag{2.57}$$

$$\therefore \langle j_1, m_1 = j_1 - 1, j_2, m_2 = j_2 | j = j_1 + j_2, m = j_1 + j_2 - 1 \rangle = \sqrt{\frac{j_1}{j_1 + j_2}} \tag{2.58}$$

and

$$\langle j_1, m_1 = j_1, j_2, m_2 = j_2 - 1 | j = j_1 + j_2, m = j_1 + j_2 - 1 \rangle = \sqrt{\frac{j_2}{j_1 + j_2}}. \tag{2.59}$$

For $j_1 = j_2$, these reduce to $\frac{1}{\sqrt{2}}$.

The procedure then divides into two paths. The other coefficients with $j = j_1 + j_2$ (cf. figure 2.3) are generated by successive application of equations (2.51)–(2.57), whence:

$$\begin{aligned} &\langle j_1 m_1 j_2 m_2 | j_1 + j_2, m_1 + m_2 \rangle \\ &= \sqrt{\frac{(j_1 + j_2 - m_1 - m_2)!(j_1 + j_2 + m_1 + m_2)!(2j_1)!(2j_2)!}{(j_1 - m_1)!(j_2 - m_2)!(j_1 + m_1)!(j_2 + m_2)!(2j_1 + 2j_2)!}}. \end{aligned} \tag{2.60}$$

The second path leads to the coefficients with $j = j_1 + j_2 - 1$ (cf. figure 2.3) by constructing $|j = j_1 + j_2 - 1, m = j_1 + j_2 - 1\rangle$ which must be orthogonal to $|j = j_1 + j_2, m = j_1 + j_2 - 1\rangle$ (equation (2.57)), i.e.

$$\begin{aligned} |j_1 + j_2 - 1, m = j_1 + j_2 - 1\rangle = &\sqrt{\frac{j_2}{j_1 + j_2}} |j_1, m_1 = j_1 - 1\rangle |j_2, m_2 = j_2\rangle \\ &- \sqrt{\frac{j_1}{j_1 + j_2}} |j_1, m_1 = j_1\rangle |j_2, m_2 = j_2 - 1\rangle. \end{aligned} \tag{2.61}$$

The other coefficients with $j = j_1 + j_2 - 1$ are generated by successive application of equations (2.51)–(2.61). Coefficients with $j = j_1 + j_2 - 2$, $j = j_1 + j_2 - 3$, etc., are constructed by appropriate orthogonalization and application of equation (2.51).

General expressions for some Clebsch–Gordan coefficients are given in tables A.1–A.4.

Clebsch–Gordan coefficients possess a number of general properties:

(i) They are real: this is by convention.

(ii) $j_1 + j_2 \geqslant j \geqslant |j_1 - j_2|$: this is proved in theorem 2.2.1.

(iii) $m = m_1 + m_2$: this follows from
$\hat{J}_z = \hat{J}_{1z} + \hat{J}_{2z}$, (cf. equation (2.51))
whence

$$\begin{aligned} 0 &= (\hat{J}_z - \hat{J}_{1z} - \hat{J}_{2z})|jm\rangle \\ &= \langle j_1 m_1 j_2 m_2 | (\hat{J}_z - \hat{J}_{1z} - \hat{J}_{2z})|jm\rangle \\ &= \langle j_1 m_1 j_2 m_2 | (m - m_1 - m_2)|jm\rangle, \end{aligned} \tag{2.62}$$

$$\therefore 0 = (m - m_1 - m_2)\langle j_1 m_1 j_2 m_2 | j_m \rangle. \tag{2.63}$$

Thus, either

$$m = m_1 + m_2 \quad \text{or} \quad \langle j_1 m_1 j_2 m_2 | jm \rangle = 0. \tag{2.64}$$

(iv)

$$\langle j_1 m_1 j_2 m_2 | jm \rangle = \langle jm | j_1 m_1 j_2 m_2 \rangle: \tag{2.65}$$

this follows from (i).

(v)

$$\sum_j \sum_m \langle j_1 m_1 j_2 m_2 | jm \rangle \langle j_1 m_1' j_2 m'_2 | jm \rangle = \delta_{m_1 m_1'} \delta_{m_2 m_2'}: \tag{2.66}$$

this follows from the orthogonality of the $|j_1 m_1 j_2 m_2\rangle$ basis

$$\langle j_1 m_1 j_2 m_2 | jm \rangle \langle j_1 m_1' j_2 m_2' | jm \rangle = \delta_{m_1 m_1'} \delta_{m_2 m_2'} \tag{2.67}$$

by using the completeness relation for the $|jm\rangle$ basis,

$$\sum_j \sum_m \langle j_1 m_1 j_2 m_2 | jm \rangle \langle jm | j_1 m_1' j_2 m_2' \rangle = \delta_{m_1 m_1'} \delta_{m_2 m_2'}, \tag{2.68}$$

from which equation (2.66) follows using (iv).

(vi)

$$\sum_{m_1} \sum_{m_2} \langle j_1 m_1 j_2 m_2 | jm \rangle \langle j_1 m_1 j_2 m_2 | j' m' \rangle = \delta_{jj'} \delta_{mm'}: \tag{2.69}$$

this follows in a manner similar to (v).

(vii)

$$\sum_{m_1} \sum_{m_2} |\langle j_1 m_1 j_2 m_2 | jm \rangle|^2 = 1: \tag{2.70}$$

this follows from equation (2.69) for $j' = j$, $m' = m$, and using (iv).

(viii)

$$\begin{aligned} &\sqrt{(j \mp \mu)(j \pm \mu + 1)} \langle j_1 m_1 j_2 m_2 | j_1 \mu \pm 1 \rangle \\ &= \sqrt{(j_1 \mp m_1 + 1)(j_1 \pm m_1)} \langle j_1 m_1 \mp 1, j_2 m_2 | j\mu \rangle \\ &+ \sqrt{(j_2 \mp m_2 + 1)(j_2 \pm m_2)} \langle j_1 m_1, j_2 m_2 \mp 1 | j\mu \rangle: \end{aligned} \tag{2.71}$$

this is a general recursion relationship for Clebsch–Gordan coefficients which follows from

$$\hat{J}_\pm |j\mu\rangle = (\hat{J}_{1\pm} + \hat{J}_{2\pm}) \sum_{m_1'} \sum_{m_2'} | j_1 m_1' j_2 m_2' \rangle \langle j_1 m_1' j_2 m_2' | j\mu \rangle, \tag{2.72}$$

where the completeness relation for the $|j_1 m_1' j_2 m_2'\rangle$ basis has been used; whence

$$\begin{aligned}&\sqrt{(j \mp \mu)(j \pm \mu + 1)}\,|j\mu \pm 1\rangle \\ &= \sum_{m_1'}\sum_{m_2'}\Big\{\sqrt{(j_1 \mp m_1')(j_1 \pm m_1' + 1)}\,|j_1 m_1' \pm 1, j_2 m_2'\rangle \\ &+ \sqrt{(j_2 \mp m_2')(j_2 \pm m_2' + 1)}\,|j_1 m_1' j_2 m_2' \pm 1\rangle\Big\}\langle j_1 m_1' j_2 m_2'|j\mu\rangle,\end{aligned} \tag{2.73}$$

and taking the inner products on both sides of this equation with $\langle j_1 m_1 j_2 m_2|$,

$$\begin{aligned}&\therefore \sqrt{(j \mp \mu)(j \pm \mu + 1)}\,\langle j_1 m_1 j_2 m_2|j\mu \pm 1\rangle \\ &= \sum_{m_1'}\sum_{m_2'}\Big\{\sqrt{(j_1 \mp m_1')(j_1 \pm m_1' + 1)}\,\langle j_1 m_1 j_2 m_2|j_1 m_1' \pm 1, j_2 m_2'\rangle \\ &+ \sqrt{(j_2 \mp m_2')(j_2 \pm m_2' + 1)}\,\langle j_1 m_1 j_2 m_2|j_1 m_1' j_2 m_2' \pm 1\rangle\Big\}\langle j_1 m_1' j_2 m_2'|j\mu\rangle,\end{aligned} \tag{2.74}$$

and equation (2.71) follows from the orthonormality of the $|j_1 m_1 j_2 m_2\rangle$ basis.

$$\text{Note: } m_1 + m_2 = \mu \pm 1. \tag{2.75}$$

The recursion relations embodied in equation (2.71) can be viewed usefully in a graphical way by plotting the (m_1, m_2) values of the coefficients in the recursion relation. This is shown in figure 2.4. Thus, $\hat{J}_+$ connects $(m_1 - 1, m_2)$ and $(m_1, m_2 - 1)$ to (m_1, m_2); and $\hat{J}_-$ connects

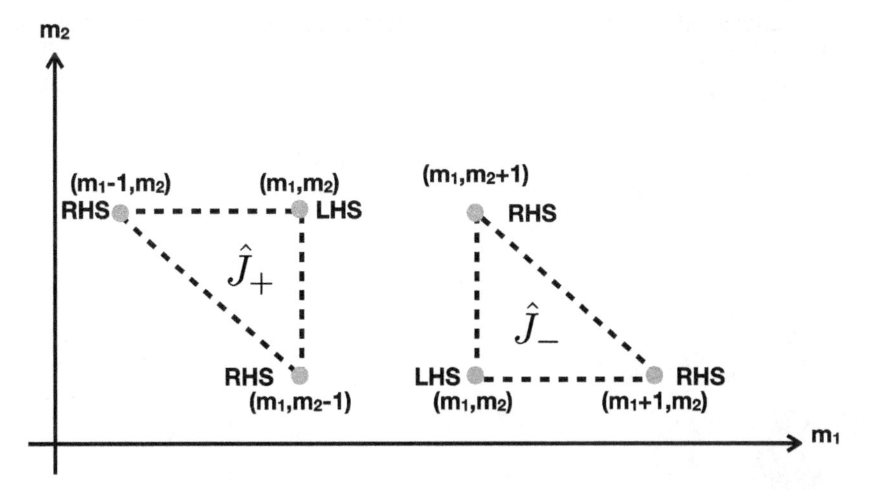

Figure 2.4. The action of $\hat{J}_+$ and $\hat{J}_-$ in the m_1–m_2 plane. LHS and RHS refer to the left-hand side and right-hand side of equation (2.71). Thus, the points in the m_1–m_2 plane corresponding to the RHS of equation (2.71), which are connected to the LHS of equation (2.71) by $\hat{J}_-$, are $(m_1 + 1, m_2)$ and $(m_1, m_2 + 1)$.

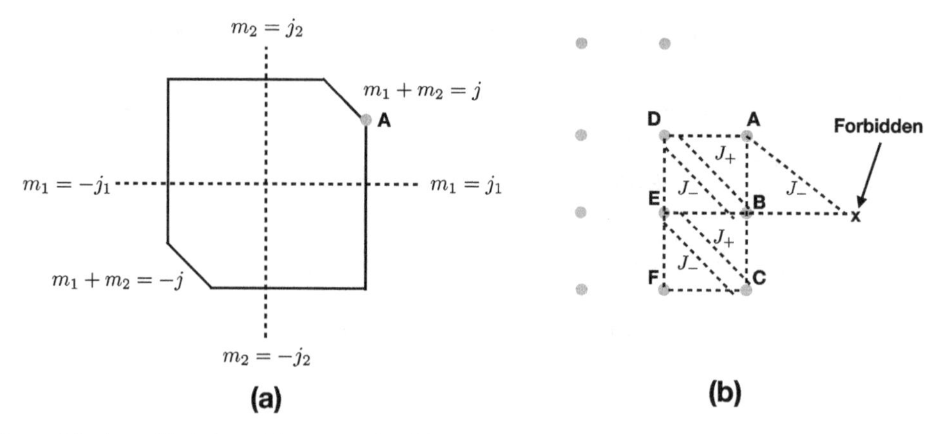

Figure 2.5. A graphical view of the manner in which the recursion relation (equation (2.71)) connects points in the m_1–m_2 plane, each point of which represents a Clebsch–Gordan coefficient. (Compare with figures 2.3(a)–(e).)

$(m_1 + 1, m_2)$ and $(m_1, m_2 + 1)$ to (m_1, m_2). To determine the Clebsch–Gordan coefficients defined by equation (2.71), the procedure is to draw the boundary, in the m_1–m_2 plane, defined by

$$-j_1 \leqslant m_1 \leqslant j_1,\ -j_2 \leqslant m_2 \leqslant j_2,\ -j \leqslant m_1 + m_2 \leqslant j \tag{2.76}$$

for *specified* j_1, j_2, and j. This is shown in figure 2.5. Then, starting at location A in figure 2.5, we determine which RHS points contribute to the LHS (cf. figure 2.4 and equation (2.71)). Evidently, because there is no point in the m_1–m_2 plane with $(m_1 = j_1 + 1, m_2)$, under the action of $\hat{J}_-$ the point B can only be reached from the point A. Thus, from the Clebsch–Gordan coefficient corresponding to the point A, we obtain the Clebsch–Gordan coefficient corresponding to the point B. Continuing, we then determine, using $\hat{J}_+$, the Clebsch–Gordan coefficient for point D from those at points A and B. This can be iterated, B and D giving E, B and E giving C, E and C giving F, and so on.

The literature contains many conventions for expressing ($SU(2)$) Clebsch–Gordan coefficients. Some of them are:

$\langle j_1 m_1 j_2 m_2 | jm \rangle$ *or* $(j_1 m_1 j_2 m_2 | jm)$ (note angular and curved brackets)

$\langle j_1 m_1 j_2 m_2 | j_1 j_2 jm \rangle$

$\langle j_1 j_2 m_1 m_2 | jm \rangle$

$\langle j_1 j_2 m_1 m_2 | j_1 j_2 jm \rangle$

$C(j_1 j_2 j;\ m_1 m_2 m)$

$C(j_1 j_2 j;\ m_1 m_2)$

$C_{j_1 j_2}(jm;\ m_1 m_2)$

$X(jm j_1 j_2 m_1)$

$C^{jm}_{j_1 m_1 j_2 m_2}$.

Clebsch–Gordan coefficients possess a high degree of permutational symmetry. Wigner introduced a form that facilitates the execution of these permutations by defining the *(Wigner) 3-j symbol:*

$$\begin{pmatrix} j_1 & j_2 & j_3 \\ m_1 & m_2 & m_3 \end{pmatrix} := (-1)^{j_1 - j_2 - m_3}(2j_3 + 1)^{-\frac{1}{2}}\langle j_1 m_1 j_2 m_2 | j_3, -m_3 \rangle. \tag{2.77}$$

Note: the $-m_3$ in the Clebsch–Gordan coefficient. The 3-j symbols possess the following symmetry properties:

(i)

$$m_1 + m_2 + m_3 = 0; \tag{2.78}$$

(ii)

$$\begin{pmatrix} j_1 & j_2 & j_3 \\ m_1 & m_2 & m_3 \end{pmatrix} = \begin{pmatrix} j_2 & j_3 & j_1 \\ m_2 & m_3 & m_1 \end{pmatrix} = \begin{pmatrix} j_3 & j_1 & j_2 \\ m_3 & m_1 & m_2 \end{pmatrix}; \tag{2.79}$$

(iii)

$$(-1)^{j_1 + j_2 + j_3}\begin{pmatrix} j_1 & j_2 & j_3 \\ m_1 & m_2 & m_3 \end{pmatrix} = \begin{pmatrix} j_2 & j_1 & j_3 \\ m_2 & m_1 & m_3 \end{pmatrix}; \tag{2.80}$$

(iv)

$$\begin{pmatrix} j_1 & j_2 & j_3 \\ m_1 & m_2 & m_3 \end{pmatrix} = (-1)^{j_1 + j_2 + j_3}\begin{pmatrix} j_1 & j_2 & j_3 \\ -m_1 & -m_2 & -m_3 \end{pmatrix}. \tag{2.81}$$

General expressions for some 3-j coefficients are given in table A.5. These are in a highly-compacted notation that requires the use of the permutational symmetry.

2.3 Spin–orbit coupling

Intrinsic spin and orbital angular momentum have the same units, namely $\hbar$, and appear in specific combinations in molecules, atoms, nuclei and hadrons. We describe states of specified spin and position by, e.g.

$$|\vec{r}, +\rangle := |\vec{r}\rangle \otimes |+\rangle, \tag{2.82}$$

where this is to be interpreted as describing a particle in a state of spin up along z located at the position $\vec{r}$. Spin operators only act on $\{|+\rangle, |-\rangle\}$; operators such as $\hat{r}$, $\hat{p}$, $\hat{L}$ only act on $\{|\vec{r}\rangle\}$. The space spanned by $\{|\vec{r}\rangle \otimes |+\rangle, |\vec{r}\rangle \otimes |-\rangle\}$ is called the *direct product* space. The spin operators commute with the space operators:

$$[\hat{S}_i, \hat{r}_j] = 0, [\hat{S}_i, \hat{p}_j] = 0, [\hat{S}_i, \hat{L}_j] = 0, \forall i, j. \tag{2.83}$$

The rotation operator for the coupled system has the form

$$\mathcal{D}(\hat{n}, \phi) = \exp\left\{-i\frac{\vec{J}\cdot\hat{n}}{\hbar}\phi\right\}, \tag{2.84}$$

where

$$\vec{J} = \vec{L} \otimes \hat{I}_{\text{spin}} + \hat{I}_{\text{space}} \otimes \vec{S} = \vec{L} + \vec{S}, \tag{2.85}$$

and $\hat{I}_{\text{spin}}$ is the identity operator acting only on the spin kets and $\hat{I}_{\text{space}}$ is the identity operator acting only on the space kets. Note, with respect to vector operators, the symbols being used indicate explicitly that they are operators; but usually the symbols do not indicate that they are operators. We ask the Reader to make the necessary inferences with respect to vector operators versus vectors. From $[\hat{S}_i, \hat{L}_j] = 0, \forall i, j,$

$$\mathcal{D}(\hat{n}, \phi) = \exp\left\{-i\frac{\vec{L}+\vec{S}}{\hbar}\cdot\hat{n}\phi\right\} = \exp\left\{-i\frac{\vec{L}\cdot\hat{n}}{\hbar}\phi\right\}\exp\left\{-i\frac{\vec{S}\cdot\hat{n}}{\hbar}\phi\right\}. \tag{2.86}$$

Space–spin wave functions can be written:

$$\langle\vec{r}; \pm|\alpha\rangle = \Psi_{\alpha\pm}(\vec{r}), \tag{2.87}$$

and are usually arranged in column matrix form

$$\begin{pmatrix} \langle\vec{r};+|\alpha\rangle \\ \langle\vec{r};-|\alpha\rangle \end{pmatrix} = \begin{pmatrix} \Psi_{\alpha+}(\vec{r}) \\ \Psi_{\alpha-}(\vec{r}) \end{pmatrix}. \tag{2.88}$$

Here, $|\Psi_{\alpha\pm}(\vec{r})|^2$ gives the probability for finding the particle at the position $\vec{r}$ with spin up or down. We can also write:

$$\langle\vec{r};+|\alpha\rangle = \Psi_{\alpha+}(\vec{r})\chi_+, \langle\vec{r};-|\alpha\rangle = \Psi_{\alpha-}(\vec{r})\chi_-. \tag{2.89}$$

Expressing $\Psi_{\alpha\pm}(\vec{r})$ as

$$\Psi_{\alpha\pm}(\vec{r}) = R_{\alpha\pm}(r)Y_{lm}(\theta, \phi)\chi_\pm, \tag{2.90}$$

we are confronted with the task of coupling spin to orbital angular momentum in

$$\langle\vec{r}; \pm|\alpha\rangle = R_{\alpha\pm}(r)Y_{lm}(\theta, \phi)\chi_\pm. \tag{2.91}$$

We know the following:

$$\hat{L}^2 Y_{lm}(\theta, \phi) = l(l+1)\hbar^2 Y_{lm}(\theta, \phi), \tag{2.92}$$

$$\hat{L}_z Y_{lm}(\theta, \phi) = m\hbar Y_{lm}(\theta, \phi), \tag{2.93}$$

$$\hat{S}^2\chi_\pm = \frac{3}{4}\hbar^2\chi_\pm, \tag{2.94}$$

$$\hat{S}_z\chi_\pm = \pm\frac{1}{2}\hbar\chi_\pm, \tag{2.95}$$

The $\left\{Y_{lm}(\theta, \phi)\chi_\pm\right\}$ are simultaneous eigenfunctions of the set of mutually commuting operators $\left\{\hat{L}^2, \hat{L}_z, \hat{S}^2, \hat{S}_z\right\}$. They are referred to as the uncoupled basis. We seek the coupled basis which are simultaneous eigenfunctions of the set of mutually commuting operators $\{\hat{J}^2, \hat{J}_z, \hat{L}^2, \hat{S}^2\}$. From the development of Clebsch–Gordan coefficients, the coupled basis can be directly constructed:

$$\begin{aligned}\mathcal{Y}_l^{j=l\pm\frac{1}{2},m}(\theta, \phi) = &\left\langle l, m-\frac{1}{2}, \frac{1}{2}\frac{1}{2}\middle| l\pm\frac{1}{2}, m\right\rangle Y_{l,m-\frac{1}{2}}(\theta, \phi)\chi_+ \\ &+ \left\langle l, m+\frac{1}{2}, \frac{1}{2}, -\frac{1}{2}\middle| l\pm\frac{1}{2}, m\right\rangle Y_{l,m+\frac{1}{2}}(\theta, \phi)\chi_-\end{aligned} \tag{2.96}$$

or

$$\mathcal{Y}_l^{j=l\pm\frac{1}{2},m}(\theta, \phi) := \begin{pmatrix}\left\langle l, m-\frac{1}{2}, \frac{1}{2}\frac{1}{2}\middle| l\pm\frac{1}{2}, m\right\rangle Y_{l,m-\frac{1}{2}}(\theta, \phi) \\ \left\langle l, m+\frac{1}{2}, \frac{1}{2}, -\frac{1}{2}\middle| l\pm\frac{1}{2}, m\right\rangle Y_{l,m+\frac{1}{2}}(\theta, \phi)\end{pmatrix}, \tag{2.97}$$

where *m is an odd-half integer*. The functions $\mathcal{Y}$ are called *spinor spherical harmonics* or *spin-angular functions*. The Clebsch–Gordan coefficients in equations (2.96) and (2.97) are:

$$\left\langle l, m-\frac{1}{2}, \frac{1}{2}\frac{1}{2}\middle| l\pm\frac{1}{2}, m\right\rangle = \pm\sqrt{\frac{l\pm m+\frac{1}{2}}{2l+1}}, \tag{2.98}$$

$$\left\langle l, m+\frac{1}{2}, \frac{1}{2}, -\frac{1}{2}\middle| l\pm\frac{1}{2}, m\right\rangle = \pm\sqrt{\frac{l\mp m+\frac{1}{2}}{2l+1}}. \tag{2.99}$$

The wave functions $\mathcal{Y}_l^{j=l\pm\frac{1}{2},m}(\theta, \phi)$ are, by construction, simultaneous eigenfunctions of $\hat{J}^2$, $\hat{J}_z$, $\hat{L}^2$, and $\hat{S}^2$. They are also eigenfunctions of the operator $\vec{L}\cdot\vec{S}$. This follows from $\vec{J} = \vec{L} + \vec{S}$ and:

$$\begin{aligned}\hat{J}^2 &= \vec{J}\cdot\vec{J} = (\vec{L}+\vec{S})\cdot(\vec{L}+\vec{S}) \\ &= \vec{L}\cdot\vec{L} + \vec{L}\cdot\vec{S} + \vec{S}\cdot\vec{L} + \vec{S}\cdot\vec{S} \\ &= \hat{L}^2 + \hat{S}^2 + 2\vec{L}\cdot\vec{S},\end{aligned} \tag{2.100}$$

whence

$$\vec{L} \cdot \vec{S} = \frac{1}{2}(\hat{J}^2 - \hat{L}^2 - \hat{S}^2). \tag{2.101}$$

The eigenvalues of $\vec{L} \cdot \vec{S}$ are:

$$\langle \vec{L} \cdot \vec{S} \rangle = \frac{\hbar^2}{2}\left\{ j(j+1) - l(l+1) - \frac{3}{4} \right\}; \tag{2.102}$$

and, for $j = l + \frac{1}{2}$

$$\langle \vec{L} \cdot \vec{S} \rangle = \frac{1}{2} l \hbar^2; \tag{2.103}$$

and, for $j = l - \frac{1}{2}$

$$\langle \vec{L} \cdot \vec{S} \rangle = -\frac{1}{2}(l+1)\hbar^2. \tag{2.104}$$

The operator $\vec{L} \cdot \vec{S}$ is called the *spin–orbit interaction* operator. It plays an important role in both atoms and nuclei. This is illustrated in figures 2.6(a) and (b).

2.4 Vector spherical harmonics

The vector spherical harmonics are very similar to the spinor spherical harmonics. They are defined by

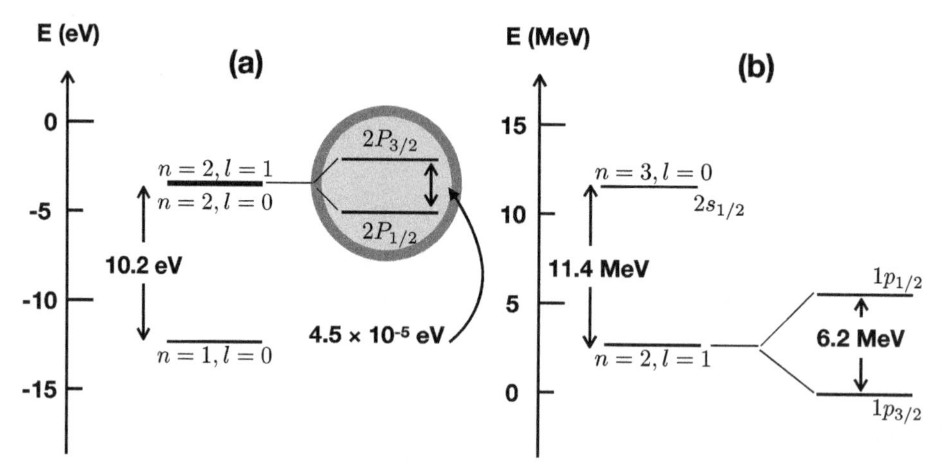

Figure 2.6. Spin–orbit interaction $A\vec{L} \cdot \vec{S}$ in: (a) the hydrogen atom ($A > 0$); (b) the nucleus ^{15}O ($A < 0$). The labelling quantum numbers are for single-particle motion in a central force field. The energy zero for ^{15}O is arbitrary. Note that the magnitude of the spin–orbit interaction in ^{15}O (and all other nuclei) is enormous: unlike in atoms, the origin of the spin–orbit interaction in nuclei is not electromagnetic but comes from the strong force. (Note: the conventional labelling for the hydrogen atom is nl_j and for nuclei is $n_r l_j$, cf. Volume 1, figures 12.6 and 12.7.)

$$Y_{LJM}(\theta, \phi) := \sum_{M,\nu} \langle Lm1\nu|JM\rangle Y_{LM}(\theta, \phi)\vec{\xi}_\nu, \ \nu = 0, \pm 1, \tag{2.105}$$

where

$$\vec{\xi}_0 = \vec{\xi}_z, \ \vec{\xi}_{\pm 1} = \mp\frac{(\vec{\xi}_x \pm i\vec{\xi}_y)}{\sqrt{2}}, \tag{2.106}$$

and $(\vec{\xi}_x, \vec{\xi}_y, \vec{\xi}_z)$ denote a triple of orthogonal Cartesian unit vectors. They are used in the multipole expansion of the electromagnetic field, where the $\vec{\xi}_\nu$ describe photon polarization (recall that the spin of the photon is $1\hbar$).

2.5 Clebsch–Gordan coefficients and rotation matrices

Clebsch–Gordan coefficients and rotation matrices are related. The connection can be stated as a theorem:

Theorem 2.2.

$$\mathcal{D}^{(j_1)}_{m_1m_1'}(R)\mathcal{D}^{(j_2)}_{m_2m_2'}(R) = \sum_j\sum_m\sum_{m'}\langle j_1m_1j_2m_2|jm\rangle\langle j_1m_1'j_2m_2'|jm'\rangle\mathcal{D}^{(j)}_{mm'}(R), \tag{2.107}$$

where the sum over j is from $|j_1 - j_2|$ to $j_1 + j_2$. (This is an explicit way of stating equation (2.45), i.e. the Clebsch–Gordan series.)

Proof. Consider

$$\begin{aligned}\langle j_1m_1j_2m_2|\mathcal{D}(R)|j_1m_1'j_2m_2'\rangle &= \langle j_1m_1|\mathcal{D}(R)|j_1m_1'\rangle\langle j_2m_2|\mathcal{D}(R)|j_2m_2'\rangle \\ &= \mathcal{D}^{(j_1)}_{m_1m_1'}(R)\mathcal{D}^{(j_2)}_{m_2m_2'}(R),\end{aligned} \tag{2.108}$$

i.e. the left-hand side of equation (2.107) is equal to $\langle j_1m_1j_2m_2|\mathcal{D}(R)|j_1m_1'j_2m_2'\rangle$. But this can also be written as

$$\begin{aligned}\langle j_1m_1j_2m_2|\mathcal{D}(R)|j_1m_1'j_2m_2'\rangle &= \sum_j\sum_m\sum_{j'}\sum_{m'}\langle j_1m_1j_2m_2|jm\rangle \\ &\quad\times\langle jm|\mathcal{D}(R)|j'm'\rangle\langle j'm'|j_1m_1'j_2m_2'\rangle \\ &= \sum_j\sum_m\sum_{j'}\sum_{m'}\langle j_1m_1j_2m_2|jm\rangle\mathcal{D}^{(j)}_{mm'}(R)\delta_{jj'} \\ &\quad\times\langle j_1m_1'j_2m_2'|j'm'\rangle,\end{aligned} \tag{2.109}$$

where the completeness relation has been used twice and the reality of the Clebsch–Gordan coefficients has been used. Evidently, this proves the theorem. □

A useful application of equation (2.107) is the simplification of an integral over the product of three spherical harmonics. This commonly occurs in problems involving electromagnetic transitions in atoms, molecules, and nuclei. From equation (1.229), i.e.

$$\mathcal{D}^{(l)}_{m0}(l)(\alpha, \beta, \gamma = 0) = \sqrt{\frac{4\pi}{2l+1}}\, Y^*_{lm}(\theta, \phi)|_{\theta=\beta, \phi=\alpha}, \tag{2.110}$$

and equation (2.107) for $j_1 = l_1$, $j_2 = l_2$, $j = l$, $m_1' = 0$, $m_2' = 0$, and hence $m = 0$, i.e.

$$\mathcal{D}^{(l_1)}_{m_10}(R)\mathcal{D}^{(l_2)}_{m_20}(R) = \sum_l\sum_m \langle l_1m_1l_2m_2|lm\rangle\langle l_10l_20|l0\rangle\mathcal{D}^{(l)}_{m0}(R), \tag{2.111}$$

substituting equation (2.110) in equation (2.111),

$$\begin{aligned} \therefore \sqrt{\frac{4\pi}{2l_1+1}}\, Y^*_{l_1m_1}(\theta, \phi)\sqrt{\frac{4\pi}{2l_2+1}}\, Y^*_{l_2m_2}(\theta, \phi) \\ = \sum_l\sum_m \langle l_1m_1l_2m_2|lm\rangle\langle l_10l_20|l0\rangle\sqrt{\frac{4\pi}{2l+1}}\, Y^*_{lm}(\theta, \phi). \end{aligned} \tag{2.112}$$

Rearranging and taking the complex conjugate, this gives

$$\begin{aligned} Y_{l_1m_1}(\theta, \phi)\, Y_{l_2m_2}(\theta, \phi) = \frac{\sqrt{(2l_1+1)(2l_2+1)}}{4\pi}\sum_l\sum_m \langle l_1m_1l_2m_2|lm\rangle \\ \times \langle l_10l_20|l0\rangle\sqrt{\frac{4\pi}{2l+1}}\, Y_{lm}(\theta, \phi). \end{aligned} \tag{2.113}$$

Then multiplying both sides of equation (2.113) by $Y^*_{\lambda\mu}(\theta, \phi)$ and integrating, the result is obtained:

$$\begin{aligned} \int Y^*_{\lambda\mu}(\theta, \phi)\, Y_{l_1m_1}(\theta, \phi)\, Y_{l_2m_2}(\theta, \phi)\mathrm{d}\Omega = \sqrt{\frac{(2l_1+1)(2l_2+2)}{4\pi(2\lambda+1)}} \\ \times \langle l_10l_20|\lambda0\rangle\langle l_1m_1l_2m_2|\lambda\mu\rangle, \end{aligned} \tag{2.114}$$

where the orthonormality of the spherical harmonics has been used.

Exercises

2.1. Show that $\hat{L}_z$ and $\hat{S}_z$ cannot be included in the mutually commuting set of operators $\{\hat{J}^2, \hat{J}_z, \hat{L}^2, \hat{S}^2\}$, $\vec{J} = \vec{L} + \vec{S}$, because they do not commute with $\hat{J}^2$.

2.2. Show that equation (2.45) holds using the brute-force method, i.e. compute the dimensions of the left-hand and right-hand sides. (Recall $\sum_{n=1}^k n = \frac{1}{2}k(k+1)$.)

2.3. Using the recursion relationship, equation (2.71), carry out the derivation of the general Clebsch–Gordan coefficients for spin–orbit coupling, equations (2.98) and (2.99).

2.4. Show that for $m_1 = m_2 = m_3 = 0$ and $j_1 + j_2 + j_3$ odd, all 3-j symbols are zero.

2.5. From table A.5, obtain the Clebsch–Gordan coefficients in tables A.1 and A.2.

2.6. What are the numerical values of:

(a) $\begin{pmatrix} \frac{3}{2} & \frac{3}{2} & 2 \\ \frac{1}{2} & \frac{1}{2} & -1 \end{pmatrix}$?

(b) $\begin{pmatrix} \frac{3}{2} & \frac{3}{2} & 0 \\ \frac{1}{2} & -\frac{1}{2} & 0 \end{pmatrix}$?

2.7. Express $|jmj_1j_2\rangle$ as a linear combination of the basis states $|j_1m_1j_2m_2\rangle$, for:

(a) $j_1 = 1, j_2 = 2, j = 1, m = 0$,

(b) $j_1 = 1, j_2 = 2, j = 2, m = 0$.

2.8. The operator for the quadrupole moment of a quantum mechanical body can be expressed as $r^2 Y_{20}(\theta, \phi)$ in the position representation. Compute the numerical value of the integral $\int_0^{\pi}\int_0^{2\pi} d\Omega Y^*_{LM}(\theta, \phi) Y_{20}(\theta, \phi) Y_{LM}(\theta, \phi)$, which appears in the expression from the expectation value of the quadrupole moment of the body in the state $|LM\rangle$.

2.6 The coupling of many spins and angular momenta and their recoupling

The coupling of many spins and angular momenta is quite straightforward. It is done sequentially, e.g. for $|j_1m_1\rangle$, $|j_2m_2\rangle$, $|j_3m_3\rangle$, first couple $|j_1m_1\rangle$ and $|j_2m_2\rangle$ to $|J_{12}M_{12}\rangle$ and then couple $|J_{12}M_{12}\rangle$ and $|j_3m_3\rangle$ to $|JM\rangle$:

$$|(j_1j_2)J_{12}, j_3; JM\rangle = \sum_{m_1,m_2} |j_1m_1j_2m_2j_3m_3\rangle \times \langle j_1m_1j_2m_2|J_{12}M_{12}\rangle\langle J_{12}M_{12}j_3m_3|JM\rangle. \tag{2.115}$$

However, the order of coupling can be '2 with 3 and then 1', viz.

$$|j_1(j_2j_3)J_{23}; JM\rangle = \sum_{m_2,M_{23}} |j_1m_1j_2m_2j_3m_3\rangle \times \langle j_2m_2j_3m_3|J_{23}M_{23}\rangle\langle J_{23}M_{23}j_1m_1|JM\rangle; \tag{2.116}$$

or it could be '1 with 3 and then 2'. These coupling schemes are different. For example, consider $j_1 = \frac{1}{2}$, $j_2 = \frac{3}{2}$, $j_3 = \frac{5}{2}$ (ignoring any special symmetrization, e.g. for fermions):

$$
\begin{aligned}
&\left(j_1 = \frac{1}{2}\right) \otimes \left(j_2 = \frac{3}{2}\right) \otimes \left(j_3 = \frac{5}{2}\right) \\
&= \{(J_{12} = 1) \oplus (J_{12} = 2)\} \otimes \left(j_3 = \frac{5}{2}\right) \\
&= (J_{12} = 1) \otimes \left(j_3 = \frac{5}{2}\right) \oplus (J_{12} = 2) \otimes \left(j_3 = \frac{5}{2}\right) \\
&= \left(J = \frac{3}{2}\right) \oplus \left(J = \frac{5}{2}\right) \oplus \left(J = \frac{7}{2}\right) \oplus \left(J = \frac{1}{2}\right) \\
&\quad \oplus \left(J = \frac{3}{2}\right) \oplus \left(J = \frac{5}{2}\right) \oplus \left(J = \frac{7}{2}\right) \oplus \left(J = \frac{9}{2}\right) \\
&= \left(J = \frac{1}{2}\right) \oplus \left(J = \frac{3}{2}\right)^2 \oplus \left(J = \frac{5}{2}\right)^2 \oplus \left(J = \frac{7}{2}\right)^2 \oplus \left(J = \frac{9}{2}\right),
\end{aligned}
\tag{2.117}
$$

or,

$$
\begin{aligned}
&\left(j_1 = \frac{1}{2}\right) \otimes \left(j_2 = \frac{3}{2}\right) \otimes \left(j_3 = \frac{5}{2}\right) \\
&= \left(j_1 = \frac{1}{2}\right) \otimes \{(J_{23} = 1) \oplus (J_{23} = 2) \oplus (J_{23} = 3) \oplus (J_{23} = 4)\}
\end{aligned}
\tag{2.118}
$$

$$
\begin{aligned}
\therefore &\left(j_1 = \frac{1}{2}\right) \otimes \left(j_2 = \frac{3}{2}\right) \otimes \left(j_3 = \frac{5}{2}\right) \\
&= \left(j_1 = \frac{1}{2}\right) \otimes (J_{23} = 1) \oplus \left(j_1 = \frac{1}{2}\right) \otimes (J_{23} = 2) \\
&\quad \oplus \left(j_1 = \frac{1}{2}\right) \otimes (J_{23} = 3) \oplus \left(j_1 = \frac{1}{2}\right) \otimes (J_{23} = 4) \\
&= \left(J = \frac{1}{2}\right) \oplus \left(J = \frac{3}{2}\right) \oplus \left(J = \frac{3}{2}\right) \oplus \left(J = \frac{5}{2}\right) \\
&\quad \oplus \left(J = \frac{5}{2}\right) \oplus \left(J = \frac{7}{2}\right) \oplus \left(J = \frac{7}{2}\right) \oplus \left(J = \frac{9}{2}\right).
\end{aligned}
\tag{2.119}
$$

The intermediate couplings above are clearly different. In actual physical situations, *interactions* may exist between j_1, j_2, and j_3, e.g. magnetic dipole–dipole interactions, and this can lead to a preference of coupling scheme. A good example is the hydrogen atom where spin–orbit coupling for the electron dominates and coupling the proton spin is secondary (weaker). By contrast, in positronium (a hydrogen-like bound state of an electron and a positron) spin–spin coupling dominates. This matters because the uncoupled or partially coupled bases are mixed by the interactions and the choice of intermediate coupling is usually the partially

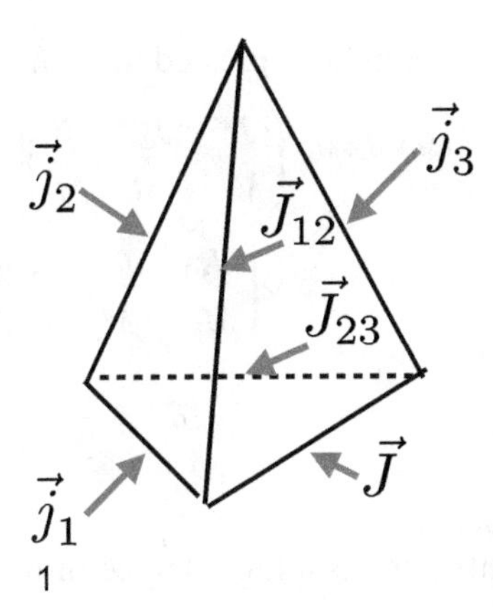

Figure 2.7. A depiction of $\vec{j}_1$, $\vec{j}_2$, and $\vec{j}_3$ to $\vec{J}$ showing the alternative intermediate couplings $\vec{J}_{12}$ and $\vec{J}_{23}$, cf. equation (2.120).

coupled basis which is least mixed *by the final coupling*. For example, the (electron) spin–orbit coupled basis in the hydrogen atom is only slightly mixed when it is coupled to the proton spin; if the proton and electron spins were to be coupled first, the subsequent coupling of the orbital angular momentum would mix states in the spin–spin basis more strongly.

Recoupling and its coefficients is defined with respect to the coupled bases, e.g. $|(j_1j_2)J_{12}, j_3; JM\rangle$, $|(j_2j_3)J_{23}, j_1; JM\rangle$, and $|(j_1j_3)J_{13}, j_2; JM\rangle$; or with respect to the coupling of four (or more) spin-angular momenta. Thus,

$$|(j_2j_3)J_{23}, j_1; JM\rangle = \Sigma_{J_{12}}|(j_1j_2)J_{12}, j_3; JM\rangle \times \langle(j_1j_2)J_{12}, j_3J|(j_2j_3)J_{23}, j_1J\rangle. \quad (2.120)$$

The coefficient $\langle(j_1j_2)J_{12}, j_3J|(j_2j_3)J_{23}, j_1J\rangle$ is called a *recoupling coefficient*. A useful representation of the vectors involved (not quantum mechanically correct) is provided by a tetrahedron (see figure 2.7).

2.6.1 6-*j* coefficients

The recoupling coefficient for three spin-angular momenta is presented in a number of various forms:

$$\langle(j_1j_2)J_{12}, j_3J|(j_2j_3)J_{23}, j_1J\rangle = \sqrt{(2J_{12}+1)(2J_{23}+1)}\, W(j_1j_2Jj_3; J_{12}J_{23}), \quad (2.121)$$

$$\langle(j_1j_2)J_{12}, j_3J|(j_2j_3)J_{23}, j_1J\rangle = (-1)^{j_1+j_2+j_3+J}\sqrt{(2J_{12}+1)(2J_{23}+1)}\begin{Bmatrix} j_1 & j_2 & J_{12} \\ j_3 & J & J_{23} \end{Bmatrix}. \quad (2.122)$$

The W are called *Racah W coefficients*; the {} are called *Wigner 6-j coefficients.*

These recoupling coefficients can be expressed in terms of coupling coefficients, e.g.

$$\sum_{M_1,M_2,M_3,m_1,m_2} (-1)^{J_1+J_2+J_3+M_1+M_2+M_3} \begin{pmatrix} J_1 & J_2 & j_3 \\ M_1 & -M_2 & M_3 \end{pmatrix} \begin{pmatrix} J_2 & J_3 & j_1 \\ M_2 & -M_3 & m_1 \end{pmatrix} \times \begin{pmatrix} J_3 & J_1 & j_2 \\ M_3 & -M_1 & m_2 \end{pmatrix} \begin{pmatrix} j_1 & j_2 & j'_3 \\ m_1 & m_2 & m'_3 \end{pmatrix} = \delta_{j_3 j'_3} \delta_{m_3 m'_3} \frac{1}{2j_3+1} \begin{Bmatrix} j_1 & j_2 & j_3 \\ J_1 & J_2 & J_3 \end{Bmatrix}, \tag{2.123}$$

where the '( )' are 3-j coefficients.

The recoupling coefficients possess a high degree of permutational symmetry, e.g.

$$\begin{Bmatrix} l_1 & l_2 & l_3 \\ \lambda_1 & \lambda_2 & \lambda_3 \end{Bmatrix} = \begin{Bmatrix} l_3 & l_1 & l_2 \\ \lambda_3 & \lambda_1 & \lambda_2 \end{Bmatrix}, \tag{2.124}$$

$$\begin{Bmatrix} l_1 & l_2 & l_3 \\ \lambda_1 & \lambda_2 & \lambda_3 \end{Bmatrix} = \begin{Bmatrix} \lambda_1 & \lambda_2 & l_3 \\ l_1 & l_2 & \lambda_3 \end{Bmatrix}. \tag{2.125}$$

The coupling of *four* angular momenta, j_1, j_2, j_3, j_4 can be carried out in a variety of ways, e.g.

$$|(j_1 j_2)J_{12}, (j_3 j_4)J_{34}; JM\rangle \quad \text{or} \quad |(j_1 j_3)J_{13}, (j_2 j_4)J_{24}; JM\rangle. \tag{2.126}$$

The recoupling coefficient between this pair is defined:

$$\langle (j_1 j_2)J_{12}, (j_3 j_4)J_{34}; JM | (j_1 j_3)J_{13}, (j_2 j_4)J_{24}; JM\rangle. \tag{2.127}$$

2.6.2 9-j coefficients

The recoupling coefficients for four spin-angular momenta are related to the so-called 9-j coefficients by:

$$\langle (j_1 j_2)J_{12}, (j_3 j_4)J_{34}; JM | (j_1 j_3)J_{13}, (j_2 j_4)J_{24}; JM\rangle = \sqrt{(2J_{12}+1)(2J_{34}+1)(2J_{13}+1)(2J_{24}+1)} \begin{Bmatrix} j_1 & j_2 & J_{12} \\ j_3 & j_4 & J_{34} \\ J_{13} & J_{24} & J \end{Bmatrix}, \tag{2.128}$$

where '{ }' is the 9-j coefficient.

An expression for the 9-j coefficient in terms of 3-j coefficients is:

$$\begin{pmatrix} J_{12} & J_{24} & J \\ M_{13} & M_{24} & M \end{pmatrix} \begin{Bmatrix} j_1 & j_2 & J_{12} \\ j_3 & j_4 & J_{34} \\ J_{13} & J_{24} & J \end{Bmatrix} = \sum_{m_1,m_2,m_3,m_4,M_{13},M_{24}} \begin{pmatrix} j_1 & j_2 & J_{12} \\ m_1 & m_2 & M_{12} \end{pmatrix} \begin{pmatrix} j_3 & j_4 & J_{34} \\ m_3 & m_4 & M_{34} \end{pmatrix} \times \begin{pmatrix} j_1 & j_3 & J_{13} \\ m_1 & m_3 & M_{13} \end{pmatrix} \begin{pmatrix} j_2 & j_4 & J_{24} \\ m_2 & m_4 & M_{24} \end{pmatrix} \begin{pmatrix} J_{12} & J_{34} & J \\ M_{12} & M_{34} & M \end{pmatrix}; \tag{2.129}$$

and in terms of the 6-j coefficients is:

$$\begin{Bmatrix} j_1 & j_2 & J_{12} \\ j_3 & j_4 & J_{34} \\ J_{13} & J_{24} & J \end{Bmatrix} = \sum_j (-1)^{2j}(2j+1) \begin{Bmatrix} j_1 & j_3 & J_{13} \\ J_{24} & J & j \end{Bmatrix} \times \begin{Bmatrix} j_2 & j_4 & J_{24} \\ j_3 & j & J_{34} \end{Bmatrix} \begin{Bmatrix} J_{12} & J_{34} & J \\ j & j_1 & j_2 \end{Bmatrix}. \quad (2.130)$$

The recoupling of four spin-angular momenta arises, for example, if it is desired to transform from 'J–J' to 'L–S' coupling in atoms.

Chapter 3

Vector and tensor operators

The vector and tensor character of operators is described. This is developed into identifiable commutator bracket relationships with the operator components of angular momentum. Irreducible tensor operators are introduced as outer products of bras and kets. The Wigner–Eckart theorem is proved using the outer-product structure of irreducible tensor operators. The projection theorem is proved.

Concepts: vector operators; tensor operators; spherical tensor operators; irreducible outer-product operators; Wigner–Eckart theorem; projection theorem.

All operators can be classified in terms of their tensor properties. The value of knowing these properties is that it tells us how the operator transforms under rotations. Knowing the transformation properties of an operator under rotations enables us to greatly simplify the computation of its matrix elements. The simplest possible tensorial classification of an operator is when it is a scalar or rank-zero tensor. Such an operator is invariant under rotations. The next simplest possibility is a vector or rank-one tensor operator.

3.1 Vector operators

Operators corresponding to $\vec{r}$, $\vec{p}$, $\vec{S}$, $\vec{L}$, $\vec{J}$ are vector operators. All of these possess the standard Cartesian form

$$\vec{A} = \vec{e}_x\hat{A}_x + \vec{e}_y\hat{A}_y + \vec{e}_z\hat{A}_z, \tag{3.1}$$

where $\vec{e}_x$, $\vec{e}_y$, $\vec{e}_z$ are unit Cartesian base vectors. We reiterate that vector quantities are given without explicit indication of whether or not they are operators, to avoid cumbersome notation. Thus, the reader must interpret when an operator has been introduced. The vector character of such an operator is manifest in the expectation values $\langle\alpha|\hat{A}_x|\alpha\rangle$, $\langle\alpha|\hat{A}_y|\alpha\rangle$, $\langle\alpha|\hat{A}_z|\alpha\rangle$, for some state $|\alpha\rangle$, which form a vector in $(3, \mathbb{R})$. Thus,

doi:10.1088/978-0-7503-2171-6ch3

$$\begin{pmatrix} {}_R\langle\alpha|\hat{A}_x|\alpha\rangle_R \\ {}_R\langle\alpha|\hat{A}_y|\alpha\rangle_R \\ {}_R\langle\alpha|\hat{A}_z|\alpha\rangle_R \end{pmatrix} = (R)\begin{pmatrix} \langle\alpha|\hat{A}_x|\alpha\rangle \\ \langle\alpha|\hat{A}_y|\alpha\rangle \\ \langle\alpha|\hat{A}_z|\alpha\rangle \end{pmatrix}. \tag{3.2}$$

For example, for $R = R_z(\phi)$

$$|\alpha\rangle_R = \exp\left\{\frac{-i\hat{J}_z\phi}{\hbar}\right\}|\alpha\rangle; \tag{3.3}$$

and

$$\begin{pmatrix} \langle\hat{A}_x\rangle_R \\ \langle\hat{A}_y\rangle_R \\ \langle\hat{A}_z\rangle_R \end{pmatrix} = \begin{pmatrix} \cos\phi & -\sin\phi & 0 \\ \sin\phi & \cos\phi & 0 \\ 0 & 0 & 1 \end{pmatrix}\begin{pmatrix} \langle\hat{A}_x\rangle \\ \langle\hat{A}_y\rangle \\ \langle\hat{A}_z\rangle \end{pmatrix}, \tag{3.4}$$

where $\langle\hat{A}_x\rangle := \langle\alpha|\hat{A}_x|\alpha\rangle$, $\langle\hat{A}_x\rangle_R :=_R \langle\alpha|\hat{A}_x|\alpha\rangle_R$, etc. Thus,

$$\langle\alpha|\exp\left\{\frac{i\hat{J}_z\phi}{\hbar}\right\}\hat{A}_x\exp\left\{\frac{-i\hat{J}_z\phi}{\hbar}\right\}|\alpha\rangle = \langle\hat{A}_x\rangle\cos\phi - \langle\hat{A}_y\rangle\sin\phi. \tag{3.5}$$

For small ϕ,

$$\langle\alpha|\left(\hat{I} + \frac{i\hat{J}_z\phi}{\hbar}\right)\hat{A}_x\left(\hat{I} - \frac{-i\hat{J}_z\phi}{\hbar}\right)|\alpha\rangle = \langle\hat{A}_x\rangle - \langle\hat{A}_y\rangle\phi, \tag{3.6}$$

$$\therefore \cancel{\langle\alpha|\hat{A}_x|\alpha\rangle} + \langle\alpha|\frac{i\hat{J}_z\phi}{\hbar}\hat{A}_x|\alpha\rangle - \langle\alpha|\hat{A}_x\frac{i\hat{J}_z\phi}{\hbar}|\alpha\rangle + O(\phi^2) = \cancel{\langle\hat{A}_x\rangle} - \langle\hat{A}_y\rangle\phi, \tag{3.7}$$

$$\therefore \frac{i\cancel{\phi}}{\hbar}\{\langle\alpha|\hat{J}_z\hat{A}_x|\alpha\rangle - \langle\alpha|\hat{A}_x\hat{J}_z|\alpha\rangle\} = -\langle\hat{A}_y\rangle\cancel{\phi}, \tag{3.8}$$

$$\therefore \langle\alpha|\left\{\frac{i}{\hbar}(\hat{J}_z\hat{A}_x - \hat{A}_x\hat{J}_z)\right\}|\alpha\rangle = \langle\alpha|(-\hat{A}_y)|\alpha\rangle, \tag{3.9}$$

and since $|\alpha\rangle$ is an arbitrary state,

$$\therefore [\hat{J}_z, \hat{A}_x] = i\hbar\hat{A}_y. \tag{3.10}$$

Proceeding similarly:

$$[\hat{J}_z, \hat{A}_y] = -i\hbar\hat{A}_x, \tag{3.11}$$

$$[\hat{J}_z, \hat{A}_z] = 0, \tag{3.12}$$

$$[\hat{J}_x, \hat{A}_x] = 0, \tag{3.13}$$

$$[\hat{J}_x, \hat{A}_y] = i\hbar\hat{A}_z, \tag{3.14}$$

$$[\hat{J}_x, \hat{A}_z] = -i\hbar\hat{A}_y, \tag{3.15}$$

$$[\hat{J}_y, \hat{A}_x] = -i\hbar\hat{A}_z, \tag{3.16}$$

$$[\hat{J}_y, \hat{A}_y] = 0, \tag{3.17}$$

$$[\hat{J}_y, \hat{A}_z] = i\hbar\hat{A}_x. \tag{3.18}$$

Note the cyclic permutational structure of the indices, i.e.

$$[\hat{J}_i, \hat{A}_j] = i\hbar\varepsilon_{ijk}\hat{A}_k. \tag{3.19}$$

Equations (3.10)–(3.18) provide a definition of a vector operator. It is an alternative definition to equation (3.2). It can be noted that the case $\vec{A} = \vec{J}$, i.e.

$$[\hat{J}_x, \hat{J}_y] = i\hbar\hat{J}_z, \quad [\hat{J}_y, \hat{J}_z] = i\hbar\hat{J}_x, \quad [\hat{J}_z, \hat{J}_x] = i\hbar\hat{J}_y, \tag{3.20}$$

is just a special case of equations (3.10)–(3.18). We can also write immediately,

$$[\hat{J}_x, \hat{y}] = i\hbar\hat{z}, \text{ etc.}, \tag{3.21}$$

$$[\hat{J}_x, \hat{p}_y] = i\hbar\hat{p}_z, \text{ etc.} \tag{3.22}$$

3.2 Tensor operators

Cartesian tensor operators of rank r can be introduced in a manner similar to Cartesian vector operators:

$$\hat{\mathbf{T}}^{(r)} = \sum_{ijk\cdots} \hat{T}_{ijk\ldots}\vec{e}_i\vec{e}_j\vec{e}_k\ldots, \; i, j, k \in x, y, z, \tag{3.23}$$

where the $\hat{T}_{ijk\ldots}$ are operators and the $\vec{e}$ are unit Cartesian vectors. For example,

$$\hat{\mathbf{T}}^{(2)} = \sum_{ij} \hat{T}_{ij}\vec{e}_i\vec{e}_j, \; i, j \in x, y, z, \tag{3.24}$$

$$\therefore \hat{\mathbf{T}}^{(2)} = \hat{T}_{xx}\vec{e}_x\vec{e}_x + \hat{T}_{xy}\vec{e}_x\vec{e}_y + \cdots + \hat{T}_{zz}\vec{e}_z\vec{e}_z, \tag{3.25}$$

which was introduced in the form

$$T_{ij} = U_iV_j \tag{3.26}$$

in equation (1.79). Here, the T_{ij} are operators.

The transformation of the T_{ij} under rotations, equation (1.80), was shown to be reducible, equations (1.81)–(1.90). Thus, a scalar T (equation (1.81)), a vector $\vec{A}$ (equations (1.82)–(1.84)), and a five-component quantity S_{ij} (equations (1.85)–(1.90)) were obtained. These linear combinations of the T_{ij} are simply related to the spherical harmonics. For T, let $U_1 = V_1 = x$, $U_2 = V_2 = y$, $U_3 = V_3 = z$, then

$$T = r^2 \sim Y_{00}. \tag{3.27}$$

For $\vec{A}$, let $A_1 = A\sin\theta\cos\phi$, $A_2 = \sin\theta\sin\phi$, $A_3 = A\cos\theta$, then

$$A_1 \pm iA_2 = A\sin\theta e^{\pm i\phi} \sim Y_{1,\pm 1}(\theta, \phi), \tag{3.28}$$

$$A_3 = A\cos\theta \sim Y_{10}(\theta, \phi). \tag{3.29}$$

For the S_{ij}, again let $U_1 = V_1 = x$, $U_2 = V_2 = y$, $U_3 = V_3 = z$, then

$$S_{12} = xy = r^2\sin^2\theta\cos\phi\sin\phi, \tag{3.30}$$

$$S_{23} = yz = r^2\sin\theta\cos\theta\sin\phi, \tag{3.31}$$

$$S_{31} = zx = r^2\sin\theta\cos\theta\cos\phi, \tag{3.32}$$

$$S_{11} = x^2 - \frac{r^2}{3} = r^2\sin^2\theta\cos^2\phi - \frac{r^2}{3}, \tag{3.33}$$

$$S_{22} = y^2 - \frac{r^2}{3} = r^2\sin^2\theta\sin^2\phi - \frac{r^2}{3}; \tag{3.34}$$

whence,

$$S_{11} - S_{22} \pm 2iS_{12} = r^2\sin^2\theta e^{\pm 2i\phi} \sim r^2 Y_{2,\pm 2}(\theta, \phi), \tag{3.35}$$

$$S_{31} \pm iS_{23} = r^2\sin\theta\cos\theta e^{\pm i\phi} \sim r^2 Y_{2,\pm 1}(\theta, \phi), \tag{3.36}$$

$$S_{11} + S_{22} = \frac{-r^2}{3}(3\cos^2\theta - 1) \sim r^2 Y_{2,0}(\theta, \phi). \tag{3.37}$$

The spherical harmonics were shown to transform as

$$Y_{lm}(\theta_R, \phi_R) = \sum_{m'} \mathcal{D}^{(l)*}_{mm'}(R)\, Y_{lm'}(\theta, \phi) \tag{3.38}$$

(cf. equation (1.256)). Thus, we define a spherical tensor operator as one that transforms under rotations as

$$\left(\hat{T}^{(k)}_q\right)_R := \sum_{q'=-k}^{+k} \mathcal{D}^{(k)*}_{qq'}(R)\hat{T}^{(k)}_{q'}, \tag{3.39}$$

i.e.

$$\mathcal{D}^{\dagger}(R)\hat{T}_q^{(k)}\mathcal{D}(R) = \sum_{q'=-k}^{+k} \mathcal{D}_{qq'}^{(k)*}(R)\hat{T}_{q'}^{(k)}, \tag{3.40}$$

or

$$\mathcal{D}(R)\hat{T}_q^{(k)}\mathcal{D}^{\dagger}(R) = \sum_{q'=-k}^{+k} \mathcal{D}_{qq'}^{(k)}(R)\hat{T}_{q'}^{(k)}. \tag{3.41}$$

Equations (3.39)–(3.41) provide definitions of a *spherical tensor operator of rank k*.

By considering the infinitesimal form of equation (3.41), we can obtain, as we did for vector operators, an algebraic definition of spherical tensor operators, viz.

$$\left(\hat{I} - \frac{i\vec{J}\cdot\vec{n}}{\hbar}\phi\right)\hat{T}_q^{(k)}\left(\hat{I} + \frac{i\vec{J}\cdot\vec{n}}{\hbar}\phi\right) = \sum_{q'}\hat{T}_{q'}^{(k)}\langle kq'|\left(\hat{I} - \frac{i\vec{J}\cdot\vec{n}}{\hbar}\phi\right)|kq\rangle, \tag{3.42}$$

$$\therefore \left[\vec{J}\cdot\vec{n},\, \hat{T}_q^{(k)}\right] = \sum_{q'}\hat{T}_{q'}^{(k)}\langle kq'|\vec{J}\cdot\vec{n}|kq\rangle. \tag{3.43}$$

Then, from equation (3.43) for $\vec{n} = \vec{e}_z$

$$\begin{aligned}\left[\hat{J}_z,\, \hat{T}_q^{(k)}\right] &= \sum_{q'}\hat{T}_{q'}^{(k)}\langle kq'|\hat{J}_z|kq\rangle \\ &= \sum_{q'}\hat{T}_{q'}^{(k)} q\hbar\langle kq'|kq\rangle,\end{aligned} \tag{3.44}$$

$$\therefore \left[\hat{J}_z,\, \hat{T}_q^{(k)}\right] = \hbar q\hat{T}_q^{(k)}. \tag{3.45}$$

Similarly, from equation (3.43), for $\vec{n} = \frac{1}{\sqrt{2}}(\vec{e}_x \pm i\vec{e}_y)$,

$$\left[\hat{J}_\pm,\, \hat{T}_q^{(k)}\right] = \hbar\sqrt{(k \mp q)(k \pm q + 1)}\,\hat{T}_{q\pm 1}^{(k)}. \tag{3.46}$$

An alternative view of spherical tensor operators can be obtained by considering the general decomposition of an operator over a complete set of states $\{|p\rangle, p = 1, \ldots, n\}$:

$$\hat{A} = \sum_{p=1}^{n}\sum_{q=1}^{n}|p\rangle\,\langle p|\hat{A}|q\rangle\,\langle q|, \tag{3.47}$$

$$\therefore \hat{A} = \sum_{p,q=1}^{n} A_{pq}|p\rangle\langle q|. \tag{3.48}$$

Thus, any operator $\hat{A}$ can be decomposed into a linear combination of 'elementary' or 'unit' operators $e_{pq} = |p\rangle\langle q|$ made of outer products of bras and kets. This can be done in any basis which is complete. Of particular interest is such a decomposition over a basis which has well-defined symmetry properties with respect to a group, e.g. $\{|jm\rangle, j = 0, \frac{1}{2}, 1, \cdots; m = -j, -j+1, \ldots, +j\}$ and $SU(2)$.

If the elementary operators are combined with amplitudes equal to the vector coupling coefficients of the group, the operators so formed transform as irreducible representations of the group. Thus, for $SU(2)$, from the transformation properties of $|jm\rangle$ and $\langle jm|$, equations (1.244) and (1.248), we can construct the *irreducible spherical tensor operators*:

$$\hat{T}(j_1j_2)_q^{(k)} = \sum_{m_1m_2} \langle j_1m_1j_2m_2|kq\rangle |j_1m_1\rangle\langle j_2, -m_2|(-1)^{j_2-m_2}, \tag{3.49}$$

where $j_1 + j_2 \geqslant k \geqslant |j_1 - j_2|$, $m_1 + m_2 = q$, and the phase factor $(-1)^{j_2-m_2}$ is obtained and discussed in section 1.13. This construction is seen to fulfil the definition given in equation (3.41)

$$\begin{aligned} &\mathcal{D}(R)\hat{T}(j_1j_2)_q^{(k)}\mathcal{D}^\dagger(R) \\ &= \sum_{m_1,m_2} \langle j_1m_1j_2m_2|kq\rangle \mathcal{D}(R)|j_1m_1\rangle\langle j_2, -m_2|\mathcal{D}^\dagger(R)(-1)^{j_2-m_2} \\ &= \sum_{m_1,m_2} \langle j_1m_1j_2m_2|kq\rangle \sum_{m'_1} |j_1m'_1\rangle \mathcal{D}^{(j_1)}_{m'_1m_1}(R) \sum_{m'_2} \langle j_2, -m'_2|\mathcal{D}^{(j_2)}_{m'_2m_2}(R)(-1)^{j_2-m'_2}, \end{aligned} \tag{3.50}$$

where equations (1.244) and (1.248) have been used; hence using equation (2.107) (theorem 2.2)

$$\begin{aligned} \mathcal{D}(R)\hat{T}(j_1j_2)_q^{(k)}\mathcal{D}^\dagger(R) = &\sum_{m_1,m_2,m'_1,m'_2,m,m',j} \langle j_1m_1j_2m_2|kq\rangle\langle j_1m'_1j_2m'_2|jm'\rangle \\ &\times \langle j_1m_1j_2m_2|jm\rangle \mathcal{D}^{(j)}_{m'm}|j_1m'_1\rangle\langle j_2, -m'_2|(-1)^{j_2-m'_2}, \end{aligned} \tag{3.51}$$

and from the orthonormality of the Clebsch–Gordan coefficients, equation (2.66) and (2.69),

$$\begin{aligned} &\therefore \mathcal{D}(R)\hat{T}(j_1j_2)_q^{(k)}\mathcal{D}^\dagger(R) \\ &= \sum_{m'_1,m'_2,m} \langle j_1m'_1j_2m'_2|km'\rangle \mathcal{D}^{(k)}_{m'q}(R)|j_1m'_1\rangle\langle j_2, -m'_2|(-1)^{j_2-m'_2}, \end{aligned} \tag{3.52}$$

whence

$$\mathcal{D}(R)\hat{T}(j_1j_2)_q^{(k)}\mathcal{D}^\dagger(R) = \sum_{m'} \mathcal{D}^{(k)}_{m'q}(R)\hat{T}(j_1j_2)^{(k)}_{m'}, \tag{3.53}$$

where equation (3.49) has been used on the right-hand side. Equation (3.41) then follows by identifying $m'(=-k, -k+1, \ldots, +k)$ with q'.

To test the construction manifest in equation (3.49), consider the possibilities for $j_1 = \frac{1}{2}, j_2 = \frac{1}{2}$:

(a)

$$\hat{T}\left(\frac{1}{2}\frac{1}{2}\right)_0^{(0)} = \sum_{m_1, m_2 = -m_1} \left\langle \frac{1}{2}m_1\frac{1}{2}m_2 \,\middle|\, 00 \right\rangle \left| \frac{1}{2}m_1 \right\rangle \left\langle \frac{1}{2}, -m_2 \right| (-1)^{\frac{1}{2}-m_2}, \tag{3.54}$$

$$\begin{aligned} \therefore \hat{T}\left(\frac{1}{2}\frac{1}{2}\right)_0^{(0)} &= \left\langle \frac{1}{2}\frac{1}{2}\frac{1}{2}, -\frac{1}{2} \,\middle|\, 00 \right\rangle \left| \frac{1}{2}\frac{1}{2} \right\rangle \left\langle \frac{1}{2}\frac{1}{2} \right| (-1)^{\frac{1}{2}+\frac{1}{2}} \\ &+ \left\langle \frac{1}{2}, -\frac{1}{2}, \frac{1}{2}\frac{1}{2} \,\middle|\, 00 \right\rangle \left| \frac{1}{2}, -\frac{1}{2} \right\rangle \left\langle \frac{1}{2}, -\frac{1}{2} \right| (-1)^{\frac{1}{2}-\frac{1}{2}}, \end{aligned} \tag{3.55}$$

and from table A.1

$$\left\langle \frac{1}{2}, -\frac{1}{2}, \frac{1}{2}, \frac{1}{2} \,\middle|\, 00 \right\rangle = -\frac{1}{\sqrt{2}}, \left\langle \frac{1}{2}\frac{1}{2}\frac{1}{2}, -\frac{1}{2} \,\middle|\, 00 \right\rangle = \frac{1}{\sqrt{2}}, \tag{3.56}$$

$$\therefore \hat{T}\left(\frac{1}{2}\frac{1}{2}\right)_0^{(0)} = \frac{-1}{\sqrt{2}}\left\{ \left| \frac{1}{2}\frac{1}{2} \right\rangle \left\langle \frac{1}{2}\frac{1}{2} \right| + \left| \frac{1}{2}, -\frac{1}{2} \right\rangle \left\langle \frac{1}{2}, -\frac{1}{2} \right| \right\}, \tag{3.57}$$

$$\therefore \hat{T}\left(\frac{1}{2}\frac{1}{2}\right)_0^{(0)} = \frac{-1}{\sqrt{2}}\hat{I}, \tag{3.58}$$

i.e. a scalar (where the resolution of the identity over the basis $\{|\frac{1}{2}\frac{1}{2}\rangle, |\frac{1}{2}, -\frac{1}{2}\rangle\}$ has been used).

(b)

$$\begin{aligned} \hat{T}\left(\frac{1}{2}\frac{1}{2}\right)_1^{(1)} &= \sum_{m_1 m_2} \left\langle \frac{1}{2}m_1\frac{1}{2}m_2 \,\middle|\, 11 \right\rangle \left| \frac{1}{2}m_1 \right\rangle \left\langle \frac{1}{2}, -m_2 \right| (-1)^{\frac{1}{2}-m_2} \\ &= \left\langle \frac{1}{2}\frac{1}{2}\frac{1}{2}\frac{1}{2} \,\middle|\, 11 \right\rangle \left| \frac{1}{2}\frac{1}{2} \right\rangle \left\langle \frac{1}{2}, -\frac{1}{2} \right| (-1)^{\frac{1}{2}-\frac{1}{2}} \\ &= \left| \frac{1}{2}\frac{1}{2} \right\rangle \left\langle \frac{1}{2}, -\frac{1}{2} \right|. \end{aligned} \tag{3.59}$$

$$\therefore \hat{T}\left(\frac{1}{2}\frac{1}{2}\right)_1^{(1)} = |+\rangle\langle-|, \tag{3.60}$$

where $|+\rangle := |\frac{1}{2}\frac{1}{2}\rangle$ and $|-\rangle := |\frac{1}{2}, -\frac{1}{2}\rangle$.

(c)

$$\hat{T}\left(\frac{1}{2}\frac{1}{2}\right)^{(1)}_0 = \sum_{m_1 m_2}\left\langle \frac{1}{2}m_1\frac{1}{2}m_2 \,\middle|\, 10\right\rangle \left|\frac{1}{2}m_1\right\rangle\left\langle\frac{1}{2}, -m_2\right| (-1)^{\frac{1}{2}-m_2}$$

$$= \left\langle \frac{1}{2}\frac{1}{2}\frac{1}{2}, -\frac{1}{2} \,\middle|\, 10\right\rangle \left|\frac{1}{2}\frac{1}{2}\right\rangle\left\langle\frac{1}{2}\frac{1}{2}\right| (-1)^{\frac{1}{2}+\frac{1}{2}} \tag{3.61}$$

$$+ \left\langle \frac{1}{2}, -\frac{1}{2}, \frac{1}{2}\frac{1}{2} \,|10\right\rangle \left|\frac{1}{2}, -\frac{1}{2}\right\rangle\left\langle\frac{1}{2}, -\frac{1}{2}\right| (-1)^{\frac{1}{2}-\frac{1}{2}}.$$

and from table A.1

$$\left\langle \frac{1}{2}\frac{1}{2}\frac{1}{2}, -\frac{1}{2} \,\middle|\, 10\right\rangle = \frac{1}{\sqrt{2}}, \left\langle \frac{1}{2}, -\frac{1}{2}, \frac{1}{2}\frac{1}{2} \,\middle|\, 10\right\rangle = \frac{1}{\sqrt{2}}, \tag{3.62}$$

$$\therefore \hat{T}\left(\frac{1}{2}\frac{1}{2}\right)^{(1)}_0 = \frac{1}{\sqrt{2}}\left\{-\left|\frac{1}{2}\frac{1}{2}\right\rangle\left\langle\frac{1}{2}\frac{1}{2}\right| + \left|\frac{1}{2}, -\frac{1}{2}\right\rangle\left\langle\frac{1}{2}, -\frac{1}{2}\right|\right\}, \tag{3.63}$$

$$\therefore \hat{T}\left(\frac{1}{2}\frac{1}{2}\right)^{(1)}_0 = \frac{1}{\sqrt{2}}\{|-\rangle\langle -| - |+\rangle\langle +|\}. \tag{3.64}$$

(d)

$$\hat{T}\left(\frac{1}{2}\frac{1}{2}\right)^{(1)}_{-1} = \sum_{m_1 m_2}\left\langle \frac{1}{2}m_1\frac{1}{2}m_2 \,\middle|\, 1, -1\right\rangle \left|\frac{1}{2}m_1\right\rangle\left\langle\frac{1}{2}, -m_2\right| (-1)^{\frac{1}{2}-m_2}$$

$$= \left\langle \frac{1}{2}, -\frac{1}{2}, \frac{1}{2}, -\frac{1}{2} \,\middle|\, 1, -1\right\rangle \left|\frac{1}{2}, -\frac{1}{2}\right\rangle\left\langle\frac{1}{2}\frac{1}{2}\right| (-1)^{\frac{1}{2}-\frac{1}{2}} \tag{3.65}$$

$$= -\left|\frac{1}{2}, -\frac{1}{2}\right\rangle\left\langle\frac{1}{2}\frac{1}{2}\right|.$$

$$\therefore \hat{T}\left(\frac{1}{2}\frac{1}{2}\right)^{(1)}_{-1} = -|-\rangle\langle +|. \tag{3.66}$$

The results of (b), (c) and (d),

$$\hat{T}^{(1)}_1 = |+\rangle\langle -|, \; \hat{T}^{(1)}_0 = \frac{1}{\sqrt{2}}(|-\rangle\langle -| - |+\rangle\langle +|), \; \hat{T}^{(1)}_{-1} = -|-\rangle\langle +|, \tag{3.67}$$

have the matrix representation

$$\hat{T}^{(1)}_1 \leftrightarrow \begin{pmatrix} 0 & 1 \\ 0 & 0 \end{pmatrix}, \; \hat{T}^{(1)}_0 \leftrightarrow \frac{1}{\sqrt{2}}\begin{pmatrix} -1 & 0 \\ 0 & 1 \end{pmatrix}, \; \hat{T}^{(1)}_{-1} \leftrightarrow \begin{pmatrix} 0 & 0 \\ -1 & 0 \end{pmatrix}. \tag{3.68}$$

in the $\{|+\rangle, |-\rangle\}$ basis, whence for

$$\hat{T}_x := -\frac{\left(\hat{T}_{-1}^{(1)} - \hat{T}_1^{(1)}\right)}{\sqrt{2}} \leftrightarrow \frac{1}{\sqrt{2}}\begin{pmatrix} 0 & 1 \\ 1 & 0 \end{pmatrix} = \frac{1}{\sqrt{2}}\sigma_x, \tag{3.69}$$

$$\hat{T}_y := -i\frac{\left(\hat{T}_{-1}^{(1)} + \hat{T}_1^{(1)}\right)}{\sqrt{2}} \leftrightarrow \frac{1}{\sqrt{2}}\begin{pmatrix} 0 & -i \\ i & 0 \end{pmatrix} = \frac{1}{\sqrt{2}}\sigma_y, \tag{3.70}$$

$$\hat{T}_z := -\hat{T}_0^{(1)} \leftrightarrow \frac{1}{\sqrt{2}}\begin{pmatrix} 1 & 0 \\ 0 & -1 \end{pmatrix} = \frac{1}{\sqrt{2}}\sigma_z, \tag{3.71}$$

the rank-1 tensor operator (i.e. vector operator) character is revealed.

Spherical tensor operators can be formed by 'coupling' together other spherical tensor operators. This is proved by theorem 3.1.

Theorem 3.1. *For $\hat{X}_{q_1}^{(k_1)}$, $\hat{Z}_{q_2}^{(k_2)}$ which are irreducible SU(2) tensors of rank k_1 and k_2, respectively*:

$$\hat{T}_q^{(k)} = \sum_{q_1}\sum_{q_2}\langle k_1q_1k_2q_2|kq\rangle \hat{X}_{q_1}^{(k_1)}\hat{Z}_{q_2}^{(k_2)} \tag{3.72}$$

is an irreducible SU(2) tensor of rank k.

Proof.

$$\begin{aligned} \mathcal{D}(R)\hat{T}_q^{(k)}\mathcal{D}^\dagger(R) &= \sum_{q_1,q_2}\langle k_1q_1k_2q_2|kq\rangle \mathcal{D}(R)\hat{X}_{q_1}^{(k_1)}\mathcal{D}^\dagger(R)\mathcal{D}(R)\hat{Z}_{q_2}^{(k_2)}\mathcal{D}^\dagger(R) \\ &= \sum_{q_1,q_2}\langle k_1q_1k_2q_2|kq\rangle \sum_{q'_1}\hat{X}_{q'_1}^{(k_1)}\mathcal{D}_{q'_1q_1}^{(k_1)}(R)\sum_{q'_2}\hat{Z}_{q'_2}^{(k_2)}\mathcal{D}_{q'_2q_2}^{(k_2)}(R), \end{aligned} \tag{3.73}$$

where equation (3.41) has been used. Then using equation (2.107) (theorem 2.2):

$$\begin{aligned} &\sum_{q_1q_2q'_1q'_2}\langle k_1q_1k_2q_2|kq\rangle \hat{X}_{q'_1}^{(k_1)}\hat{Z}_{q'_2}^{(k_2)}\mathcal{D}_{q'_1q_1}^{(k_1)}(R)\mathcal{D}_{q'_2q_2}^{(k_2)}(R) \\ &= \sum_{q_1q_2q'_1q'_2}\langle k_1q_1k_2q_2|kq\rangle \hat{X}_{q'_1}^{(k_1)}\hat{Z}_{q'_2}^{(k_2)}\sum_{k''q'q''}\langle k_1q'_1k_2q'_2|k''q'\rangle\langle k_1q_1k_2q_2|k''q''\rangle \mathcal{D}_{q'q''}^{(k'')}(R), \end{aligned} \tag{3.74}$$

$$\begin{aligned} \therefore &\sum_{q_1q_2q'_1q'_2}\langle k_1q_1k_2q_2|kq\rangle \hat{X}_{q'_1}^{(k_1)}\hat{Z}_{q'_2}^{(k_2)}\mathcal{D}_{q'_1q_1}^{(k_1)}(R)\mathcal{D}_{q'_2q_2}^{(k_2)}(R) \\ &= \sum_{k''q'_1q'_2q'q''}\delta_{kk''}\delta_{qq''}\langle k_1q'_1k_2q'_2|k''q'\rangle \mathcal{D}_{q'q''}^{(k'')}(R)\hat{X}_{q'_1}^{(k_1)}\hat{Z}_{q'_2}^{(k_2)}, \end{aligned} \tag{3.75}$$

where the orthonormality of the Clebsch–Gordan coefficients, equation (2.69), has been used, whence

$$= \sum_{q'} \left(\sum_{q'_1 q'_2} \langle k_1 q'_1 k_2 q'_2 | k q' \rangle \hat{X}^{(k_1)}_{q'_1} \hat{Z}^{(k_2)}_{q'_2} \right) \mathcal{D}^{(k)}_{q'q}(R) \tag{3.76}$$

$$= \sum_{q'} \hat{T}^{(k)}_{q'} \mathcal{D}^{(k)}_{q'q}(R). \tag{3.77}$$

Thus, $T^{(k)}_q$, defined by equation (3.72), obeys equation (3.53). □

The statement of theorem 3.1 should be compared with the definition of the Clebsch–Gordan coefficients, viz.

$$|jm\rangle = \sum_{m_1} \sum_{m_2} |j_1 m_1 j_2 m_2\rangle \langle j_1 m_1 j_2 m_2 | jm\rangle, \tag{3.78}$$

cf.

$$\hat{T}^{(j)}_m = \sum_{m_1} \sum_{m_2} \hat{X}^{(j_1)}_{m_1} \hat{Z}^{(j_2)}_{m_2} \langle j_1 m_1 j_2 m_2 | jm\rangle. \tag{3.79}$$

This similarity exists because both spherical tensor operators and angular momentum eigenkets are irreducible representations of SO(3).

As an example of the construction of higher-rank spherical tensor operators out of lower rank spherical tensor operators, consider the operators:

$$\hat{U}_0 := \hat{U}_z, \ \hat{U}_{\pm 1} := \hat{U}_x \pm i\hat{U}_y, \tag{3.80}$$

$$\hat{V}_0 := \hat{V}_z, \ \hat{V}_{\pm 1} := \hat{V}_x \pm i\hat{V}_y, \tag{3.81}$$

where $\vec{U} = (\hat{U}_x, \hat{U}_y, \hat{U}_z)$, $\vec{V} = (\hat{V}_x, \hat{V}_y, \hat{V}_z)$. Equations (3.80) and (3.81) define two spherical tensors of rank 1. Then we can use theorem 3.1 to form:

$$\begin{aligned} \hat{T}^{(2)}_{\pm 2} &= \langle 1, \pm 1, 1, \pm 1 | 2, \pm 2\rangle \hat{U}_{\pm 1} \hat{V}_{\pm 1} = \hat{U}_{\pm 1} \hat{V}_{\pm 1}, \\ \hat{T}^{(2)}_{\pm 1} &= \langle 1, \pm 1, 1, 0 | 2, \pm 1\rangle \hat{U}_{\pm 1} \hat{V}_0 + \langle 101, \pm 1 | 2, \pm 1\rangle \hat{U}_0 \hat{V}_{\pm 1} \end{aligned} \tag{3.82}$$

$$= \frac{1}{\sqrt{2}} (\hat{U}_{\pm 1} \hat{V}_0 + \hat{U}_0 \hat{V}_{\pm 1}), \tag{3.83}$$

and

$$\begin{aligned} \hat{T}^{(2)}_0 &= \langle 111, -1 | 2, 0\rangle \hat{U}_{+1} \hat{V}_{-1} + \langle 1010 | 20\rangle \hat{U}_0 \hat{V}_0 + \langle 1, -111 | 20\rangle \hat{U}_{-1} \hat{V}_{+1} \\ &= \frac{1}{\sqrt{6}} (\hat{U}_{+1} \hat{V}_{-1} + 2\hat{U}_0 \hat{V}_0 + \hat{U}_{-1} \hat{V}_{+1}). \end{aligned} \tag{3.84}$$

Spherical tensor operators possess a structure (with respect to rotations) that greatly simplifies the computation of their matrix elements in many cases. This simplification is embodied in the *Wigner–Eckart theorem*. This can be proved in a variety of ways: the method used here is not unique.

3.3 Matrix elements of spherical tensor operators and the Wigner–Eckart theorem

Theorem 3.2. *(The Wigner–Eckart Theorem.)* If an operator, $\hat{T}_q^{(k)}$ can be expressed in the form $\hat{T}^{(k)}\sum_{j_1j_2}\hat{T}(j_1j_2)_q^{(k)}$, where $\hat{T}^{(k)}$ is rotationally invariant and $\hat{T}(j_1j_2)_q^{(k)}$ is given by equation (3.49), then

$$\langle\alpha'; j'm'|\hat{T}_q^{(k)}|\alpha; jm\rangle = C \times \langle jmkq|j'm'\rangle, \tag{3.85}$$

where α', α are quantum numbers other than the $SU(2)$ quantum numbers (j, m) that are needed to label the states, and C is a rotationally invariant quantity that depends on α', α, j', j, and k.

Proof.

$$\begin{aligned}\hat{T}_q^{(k)} &= \hat{T}^{(k)} \sum_{j_1, j_2, m_1, m_2} \langle j_1 m_1 j_2 m_2|kq\rangle|j_1 m_1\rangle\langle j_2, -m_2|(-1)^{j_2-m_2} \\ &= \sum_{j_1, j_2, m_1, m_2} \langle j_1 m_1 j_2 m_2|kq\rangle|j_1 m_1\rangle\hat{T}^{(k)}\langle j_2, -m_2|(-1)^{j_2-m_2}.\end{aligned} \tag{3.86}$$

Further, the state $|\alpha; jm\rangle$ can be written $|\alpha j\rangle \otimes |jm\rangle$ where $|\alpha j\rangle$ is rotationally invariant. (An example of $T_q^{(k)}$ would be $R_{nl}(r)Y_{lm}(\theta, \phi)$, in a position representation.) Then,

$$\begin{aligned}\langle\alpha'; j'm'|\hat{T}_q^{(k)}|\alpha; jm\rangle = &\sum_{j_1, j_2, m_1, m_2} \langle j_1 m_1 j_2 m_2|kq\rangle\langle\alpha' j'| \otimes \langle j'm'|j_1 m_1\rangle \\ &\times \hat{T}^{(k)}\langle j_2, -m_2|jm\rangle \otimes |\alpha j\rangle(-1)^{j_2-m_2},\end{aligned} \tag{3.87}$$

$$\therefore \langle\alpha'; j'm'|\hat{T}_q^{(k)}|\alpha; jm\rangle = \langle j'm'j, -m|kq\rangle\langle\alpha' j'|\hat{T}^{(k)}|\alpha j\rangle(-1)^{j+m}. \tag{3.88}$$

To permute the Clebsch–Gordan coefficient, using equation (2.77):

$$\langle jm'j, -m|kq\rangle = \sqrt{2k+1}(-1)^{-j'+j-q}\begin{pmatrix} j' & j & k \\ m' & -m & -q \end{pmatrix}, \tag{3.89}$$

and

$$\begin{pmatrix} j' & j & k \\ m' & -m & -q \end{pmatrix} = (-1)^{j'+j+k}\begin{pmatrix} j' & j & k \\ -m' & m & q \end{pmatrix} = (-1)^{j'+j+k}\begin{pmatrix} j & k & j' \\ m & q & -m' \end{pmatrix}. \tag{3.90}$$

$$\therefore \langle j'm'j, -m|kq\rangle = \frac{\sqrt{2k+1}}{\sqrt{2j'+1}}(-1)^{-j'+j-q+j'+j+k+j-k+m'}\langle jmkq|j'm'\rangle, \tag{3.91}$$

$$\therefore \langle \alpha'; j'm'|\hat{T}_q^{(k)}|\alpha; jm\rangle = \langle jmkq|j'm'\rangle(-1)^{2j}\sqrt{\frac{2k+1}{2j'+1}}\langle \alpha'j'|\hat{T}^{(k)}|\alpha j\rangle, \tag{3.92}$$

where $m' = m + q$ and $(-1)^{2(j-m)} = +1$ have been used to simplify the phase. □

The Wigner–Eckart theorem can provide an enormous saving in computational labour in many cases. Manifestly, from the Clebsch–Gordan coefficient $\langle jmkq|j'm'\rangle$ it follows that:

(a)

$$m' = m + q; \tag{3.93}$$

(b)

$$j + k \geqslant j' \geqslant |j - k|. \tag{3.94}$$

These are called *selection rules*. Thus, for a tensor operator of rank zero, $k = 0$, $q = 0$, and from (i) and (ii) $j' = j$, $m' = m$. For a vector operator, it can be expressed as a linear combination of rank-1 spherical tensor operators and $k = 1$, $q = 0, \pm 1$. Then, from (i) and (ii) $\Delta j \equiv j' - j = 0, \pm 1$, $\Delta m = m' - m = 0, \pm 1$, and for $j = 0$, $\Delta j = 1$. These are the selection rules for electric dipole radiation.

A more subtle consequence of the Wigner–Eckart theorem is that all matrix elements of a spherical tensor operator (k) between states belonging to two $SU(2)$ irreps (j and j') are proportional to each other: they differ in the numerical values of the Clebsch–Gordan coefficients; the constant of proportionality depends only on j, j', and k and is independent of m, m', and q. This constant of proportionality is called a *reduced matrix element*. Often, one is only interested in ratios of transition matrix elements between two $SU(2)$ irreps for a particular rank of spherical tensor operator and these ratios are simple ratios of Clebsch–Gordan coefficients, i.e. the reduced matrix elements cancel and, thus, do not need to be computed.

The constant called the reduced matrix element is usually written $\langle \alpha'j'||\hat{T}^{(k)}||\alpha j\rangle$. An often used convention is to 'absorb' the factor $(-1)^{2j}\sqrt{2k+1}$ into $\langle \alpha'j'||\hat{T}^{(k)}||\alpha j\rangle$, i.e.

$$(-1)^{2j}\sqrt{\frac{2k+1}{2j'+1}}\langle \alpha'j'|\hat{T}^{(k)}|\alpha j\rangle := \frac{\langle \alpha'j'||\hat{T}^{(k)}||\alpha j\rangle}{\sqrt{2j'+1}}. \tag{3.95}$$

The factor $\sqrt{2j'+1}$ on the right-hand side of equation (3.95) is retained explicitly: this ensures a symmetrical relationship between $\langle \alpha'; j'm'|\hat{T}_q^{(k)}|\alpha; jm\rangle$ and $\langle \alpha; jm|\hat{T}_q^{(k)}|\alpha'; j'm'\rangle^*$, where

$$\left(\hat{T}_q^{(k)}\right)^\dagger = (-1)^q \hat{T}_q^{(k)} \tag{3.96}$$

and

$$\langle\alpha' j'||\hat{T}^{(k)}||\alpha j\rangle^* = (-1)^{j-j'}\langle\alpha j||\hat{T}^{(k)}||\alpha' j'\rangle. \tag{3.97}$$

If the numerical value of $\langle\alpha'; j'm'|\hat{T}_q^{(k)}|\alpha; jm\rangle$ is needed, then $\langle\alpha' j'||\hat{T}^{(k)}||\alpha j\rangle$ must still be computed. This can be done by standard techniques of integration or operator algebra.

A useful theorem that follows from the Wigner–Eckart theorem is the *projection theorem*. The proof is given as theorem 3.3.

Theorem 3.3. *(The projection theorem.)* For a vector operator $\hat{V}_q$, $q = 0, \pm 1$:

$$\langle\alpha'; jm'|\hat{V}_q|\alpha; jm\rangle = \frac{\langle\alpha'; jm|\vec{J}\cdot\vec{V}|\alpha; jm\rangle}{\hbar^2 j(j+1)}\langle jm'|\hat{J}_q|jm\rangle \tag{3.98}$$

(note: no j').

Proof. Expressing $\vec{J}$ in spherical tensor form:

$$\hat{J}_{\pm 1} = \mp\frac{1}{\sqrt{2}}(\hat{J}_x \pm i\hat{J}_y) = \mp\frac{1}{\sqrt{2}}\hat{J}_\pm, \tag{3.99}$$

$$\hat{J}_0 = \hat{J}_z, \tag{3.100}$$

then

$$\langle\alpha'; jm|\vec{J}\cdot\vec{V}|\alpha; jm\rangle = \langle\alpha'; jm|\hat{J}_0\hat{V}_0 - \hat{J}_{+1}\hat{V}_{-1} - \hat{J}_{-1}\hat{V}_{+1}|\alpha; jm\rangle, \tag{3.101}$$

$$\begin{aligned}\therefore \langle\alpha'; jm|\vec{J}\cdot\vec{V}|\alpha; jm\rangle &= m\hbar\langle\alpha'; jm|\hat{V}_0|\alpha; jm\rangle \\ &+ \frac{\hbar}{\sqrt{2}}\sqrt{(j+m)(j-m+1)}\,\langle\alpha'; jm-1|\hat{V}_{-1}|\alpha; jm\rangle \\ &- \frac{\hbar}{\sqrt{2}}\sqrt{(j-m)(j+m+1)}\,\langle\alpha'; jm+1|\hat{V}_{+1}|\alpha; jm\rangle.\end{aligned} \tag{3.102}$$

Then, using the Wigner–Eckart theorem:

$$\langle\alpha'; jm|\hat{V}_0|\alpha; jm\rangle = \langle jm10|jm\rangle\frac{\langle\alpha'; j||V||\alpha; j\rangle}{\sqrt{2j+1}}, \tag{3.103}$$

$$\langle\alpha'; j, m-1|\hat{V}_{-1}|\alpha; jm\rangle = \langle jm1, -1|j, m-1\rangle\frac{\langle\alpha'; j||V||\alpha; j\rangle}{\sqrt{2j+1}}, \tag{3.104}$$

and

$$\langle\alpha'; j, m+1|\hat{V}_{+1}|\alpha; jm\rangle = \langle jm11|jm+1\rangle\frac{\langle\alpha'; j||V||\alpha; j\rangle}{\sqrt{2j+1}}. \tag{3.105}$$

Thus,

$$\langle\alpha'; jm|\vec{J}\cdot\vec{V}|\alpha; jm\rangle = C_{jm}\langle\alpha'; j||V||\alpha; j\rangle, \tag{3.106}$$

where

$$\begin{aligned} C_{jm} = \frac{\hbar}{\sqrt{2j+1}}\Big\{ & m\langle jm10|jm\rangle + \frac{1}{\sqrt{2}}\sqrt{(j+m)(j-m+1)}\langle jm1,-1|j,m-1\rangle \\ & - \frac{1}{\sqrt{2}}\sqrt{(j-m)(j+m+1)}\langle jm11|j,m+1\rangle\Big\}, \end{aligned} \tag{3.107}$$

i.e. C_{jm} is independent of α, α', and $\vec{V}$, and the reduced matrix elements of $\hat{V}_0$, $\hat{V}_{\pm 1}$ are equal. In fact, we can argue that C_{jm} must be independent of m because $\vec{J}\cdot\vec{V}$ is a scalar operator. Hence,

$$\langle\alpha'; jm|\vec{J}\cdot\vec{V}|\alpha; jm\rangle = C_j\langle\alpha'; j||\vec{V}||\alpha; j\rangle. \tag{3.108}$$

Then, for $\vec{V} = \vec{J}$, $\alpha' = \alpha$

$$\langle\alpha; jm|J^2|\alpha; jm\rangle = C_j\langle\alpha; j||\vec{J}||\alpha; j\rangle. \tag{3.109}$$

Now, applying the Wigner–Eckart theorem to $\hat{V}_q$ and $\hat{J}_q$:

$$\frac{\langle\alpha'; jm'|\hat{V}_q|\alpha; jm\rangle}{\langle\alpha; jm'|\hat{J}_q|\alpha; jm\rangle} = \frac{\langle\alpha'; j||\vec{V}||\alpha; j\rangle}{\langle\alpha; j||\vec{J}||\alpha; j\rangle}. \tag{3.110}$$

But, from equations (3.108) and (3.109),

$$\frac{\langle\alpha'; j||\vec{V}||\alpha; j\rangle}{\langle\alpha; j||\vec{J}||\alpha; j\rangle} = \frac{\langle\alpha'; jm|\vec{J}\cdot\vec{V}|\alpha; jm\rangle}{\langle\alpha; jm|\hat{\vec{J}}^2|\alpha; jm\rangle} \tag{3.111}$$

and

$$\langle\alpha; jm|\hat{\vec{J}}^2|\alpha; jm\rangle = j(j+1)\hbar^2, \tag{3.112}$$

$$\therefore \langle\alpha'; jm'|\hat{V}_q|\alpha; jm\rangle = \frac{\langle\alpha'; jm|\vec{J}\cdot\vec{V}|\alpha; jm\rangle}{\hbar^2 j(j+1)}\langle jm'|\hat{J}_q|jm\rangle. \tag{3.113}$$

□

Exercises

3.1. By considering matrix elements $\langle j'm'|\hat{X}|jm\rangle$, where $\hat{X} := [\hat{J}_z, \hat{T}_q^{(k)}] - \hbar q \hat{T}_q^{(k)}$, show that $m' = m + q$.

3.2. The magnetic moment operator for an atom has the form

$$\vec{\mu} = \frac{-e}{2m_e c}(g_l \vec{L} + g_S \vec{S}), \tag{3.114}$$

where $\vec{L}$ and $\vec{S}$ are the total orbital angular momentum and total intrinsic spin of the atom, $-e$ and m_e are the charge and mass of the electron, c is the velocity of light and g_L and g_S are parameters (g factors) with the approximate values $g_L = 1$, $g_S = 2$. Show that

$$\langle \alpha; JM'|\vec{\mu}|\alpha; JM\rangle = \frac{-e}{2m_e c} g_{eff}\langle JM'|\vec{J}|JM\rangle, \tag{3.115}$$

where

$$g_{eff} = \left\{1 + \frac{J(J+1) - L(L+1) + S(S+1)}{2J(J+1)}\right\}, \tag{3.116}$$

which is called the *Landé g factor*. (Hint: use the projection theorem.)

IOP Publishing

Quantum Mechanics for Nuclear Structure, Volume 2
An intermediate level view
Kris Heyde and John L Wood

Chapter 4

Identical particles

An introduction to the representation of quantum mechanical many-body states is given. The fundamental notion of symmetrization and antisymmetrization of many-particle states is defined. The concept of a Slater determinant is sketched. The occupation number representation or second quantization is developed in detail for both states and operators. The use of anticommutator brackets is demonstrated. Representations of many-boson and many-fermion systems are illustrated. Particularly, the handling of 'condensed' systems—especially many-fermion systems that exhibit superconductivity or superfluidity—is introduced. This is done using 'quasi-spin' and its $SU(2)$ structure. The fundamental nature of quantum correlations is illustrated; the emergence of an energy gap and the concept of Pauli blocking is clarified. Boson-like behaviour of many-fermion systems is explained. Bardeen–Cooper–Schrieffer (BCS) theory of superconductivity is derived. The Lipkin model, an exactly solvable many-fermion system, is described and solved.

Concepts: symmetrization and antisymmetrization; Slater determinants; occupation number representation or second quantization; bosons and fermions; field operators; many-particle Hamiltonians; Feynman graphs; normal ordering; pairing; coherent correlations; Cooper pairs; quasispin algebra; correlation blocking; dynamical symmetry; BCS theory; Lipkin model.

The most extraordinary thing about identical particles is their indistinguishability. We know of no way to fix the identity of, e.g. an electron in a multi-electron system so that we can keep track of it. This leads to the idea of *permutation symmetry* in systems of two or more identical particles: the interchange of two identical particles cannot result in any physically observable change. We have in mind here that the particles occupy well-defined quantum states with specified space and spin degrees of freedom. Thus, for the two-particle state $\psi(1, 2)$ and the permutation operator $\hat{P}_{12}$ we can write

$$\hat{P}_{12}\psi(1, 2) = c\psi(2, 1) \tag{4.1}$$

doi:10.1088/978-0-7503-2171-6ch4

and

$$\hat{P}_{12}\psi(2, 1) = c\psi(1, 2), \tag{4.2}$$

whence

$$\hat{P}_{12}^2\psi(1, 2) = c^2\psi(1, 2); \tag{4.3}$$

and so we must have

$$c^2 = 1, \qquad c = \pm 1. \tag{4.4}$$

Further, for any operator $\hat{A}(1, 2)$ acting on $\psi(1, 2)$, we must have

$$\hat{P}_{12}\hat{A}(1, 2) = \hat{A}(2, 1) = \hat{A}(1, 2). \tag{4.5}$$

Two types of wave function for states of a system with many identical particles thus arise,

$$P_{12}\psi(1, 2) = +\psi(1, 2), \tag{4.6}$$

called *symmetric*, and

$$P_{12}\psi(1, 2) = -\psi(1, 2), \tag{4.7}$$

called *antisymmetric*. Both types are found in nature. Particles of the first type are called *bosons* and particles of the second type are called *fermions*.

4.1 Slater determinants

To ensure indistinguishability and symmetry or antisymmetry, for a system of two identical particles we write

$$\psi(1, 2) = \frac{1}{\sqrt{2}}\{\psi_1(1)\psi_2(2) + \psi_2(1)\psi_1(2)\} \tag{4.8}$$

for bosons, and

$$\psi(1, 2) = \frac{1}{\sqrt{2}}\{\psi_1(1)\psi_2(2) - \psi_2(1)\psi_1(2)\} \tag{4.9}$$

for fermions. A convenient short hand notation for equations (4.8) and (4.9) is

$$\psi(1, 2) = \frac{1}{\sqrt{2}}\begin{vmatrix} \psi_1(1) & \psi_2(1) \\ \psi_1(2) & \psi_2(2) \end{vmatrix}_{\pm}, \tag{4.10}$$

where the $\pm$ define the signs in the determinantal expansion ('–' for the conventional definition).

For three or more identical particles the correct prescription is similar to two identical particles, e.g.

$$P_{12}\psi(1, 2, 3) = \pm\psi(2, 1, 3); \tag{4.11}$$

thus (cf. equations (4.8) and (4.9)), e.g.

$$\begin{aligned}\psi(1, 2, 3) = \psi_1(1)\psi_2(2)\psi_3(3) - \psi_1(2)\psi_2(1)\psi_3(3)\\ + \psi_1(3)\psi_2(1)\psi_3(2) + \psi_1(3)\psi_2(2)\psi_3(1)\end{aligned} \tag{4.12}$$

is an inadmissible three-particle wave function because

$$\begin{aligned}P_{12}\psi(1, 2, 3) = \psi_1(2)\psi_2(1)\psi_3(3) - \psi_1(1)\psi_2(2)\psi_3(3)\\ + \psi_1(3)\psi_2(2)\psi_3(1) + \psi_1(3)\psi_2(1)\psi_3(2),\end{aligned} \tag{4.13}$$

$$\therefore P_{12}\psi(1, 2, 3) \neq c\psi(1, 2, 3). \tag{4.14}$$

Such states are not known in nature. The admissible wave functions for, e.g. three identical particles are

$$\psi(1, 2, 3) = \frac{1}{\sqrt{3!}}\begin{vmatrix}\psi_1(1) & \psi_2(1) & \psi_3(1)\\ \psi_1(2) & \psi_2(2) & \psi_3(2)\\ \psi_1(3) & \psi_2(3) & \psi_3(3)\end{vmatrix}_{\pm} : \tag{4.15}$$

these are called *Slater determinants*, and for n identical particles they provide a standard $n \times n$ determinantal code for writing down identical many-particle wave functions. Equation (4.15) can be expanded (for the minus sign or fermionic case):

$$\begin{aligned}\psi(1, 2, 3) \;=\; & \frac{1}{\sqrt{6}}\{\psi_1(1)\psi_2(2)\psi_3(3) - \psi_1(1)\psi_3(2)\psi_2(3) - \psi_2(1)\psi_1(2)\psi_3(3)\\ & + \psi_2(1)\psi_3(2)\psi_1(3) + \psi_3(1)\psi_1(2)\psi_2(3) - \psi_3(1)\psi_2(2)\psi_1(3)\}.\end{aligned} \tag{4.16}$$

Evidently, even for three identical particles the Slater determinant notation is very cumbersome. The problem with this notation is that we are labelling particles that actually cannot have labels attached to them (they are indistinguishable). Thus, to compensate for this we have to introduce symmetrization (for bosons) or anti-symmetrization (for fermions). A much more elegant language for handling these systems is the *occupation number representation*, also called the *Fock representation*, or *second quantization*. It is based on the use of creation and annihilation operators; and it has its origin in the idea of the creation and annihilation of quanta in the harmonic oscillator (cf. Volume 1, chapter 5). This does not imply that the number of particles is changing, only that the occupancies of states are changing. In place of equation (4.8) we write

$$\psi(1, 2) \leftrightarrow a_1^\dagger a_2^\dagger|0\rangle \text{ or } a_1^\dagger a_2^\dagger|00\rangle, \tag{4.17}$$

where $|0\rangle$ is the zero particle vacuum state. The symmetrization requirement for equation (4.8) is fulfilled by the condition

$$a_1^\dagger a_2^\dagger|0\rangle = a_2^\dagger a_1^\dagger|0\rangle \tag{4.18}$$

and, hence

$$[a_1^\dagger, a_2^\dagger] = 0. \tag{4.19}$$

For equation (4.9) the antisymmetrization requirement is fulfilled by the condition

$$a_1^\dagger a_2^\dagger = -a_2^\dagger a_1^\dagger, \tag{4.20}$$

i.e.

$$a_1^\dagger a_2^\dagger + a_2^\dagger a_1^\dagger := \{a_1^\dagger, a_2^\dagger\} = 0, \tag{4.21}$$

where $\{a_1^\dagger, a_2^\dagger\}$ is an *anticommutator bracket*.

4.2 The occupation number representation for bosons

For a system of identical bosons, we define the state of the system where there are n_1 bosons in state 1, n_2 bosons in state 2, etc., by

$$|n_1 n_2 \ldots n_i \ldots\rangle := \frac{(a_1^\dagger)^{n_1}}{\sqrt{n_1!}} \frac{(a_2^\dagger)^{n_2}}{\sqrt{n_2!}} \cdots \frac{(a_i^\dagger)^{n_i}}{\sqrt{n_i!}} \cdots |0\rangle, \tag{4.22}$$

i.e.

$$|n_1 n_2 \ldots n_i \ldots\rangle = \prod_i \frac{(a_i^\dagger)^{n_i}}{\sqrt{n_i!}} |0\rangle, \tag{4.23}$$

where

$$\left[a_i^\dagger, a_j^\dagger\right] = 0, \ \forall\, i, j. \tag{4.24}$$

Further,

$$[a_i, a_j] = 0, \ \forall\, i, j, \tag{4.25}$$

and

$$\left[a_i, a_j^\dagger\right] = \delta_{ij}, \ \forall\, i, j. \tag{4.26}$$

It immediately follows from the algebra of the one-dimensional harmonic oscillator that:

$$a_i^\dagger |n_1 n_2 \ldots n_i \ldots\rangle = \sqrt{n_i + 1}\, |n_1 n_2 \ldots (n_i + 1) \ldots\rangle, \tag{4.27}$$

$$a_i |n_1 n_2 \ldots n_i \ldots\rangle = \sqrt{n_i}\, |n_1 n_2 \ldots (n_i - 1) \ldots\rangle, \tag{4.28}$$

$$a_i^\dagger a_i |n_1 n_2 \ldots n_i \ldots\rangle = n_i |n_1 n_2 \ldots n_i \ldots\rangle, \tag{4.29}$$

$$\langle n_1 n_2 \ldots n_i \ldots | n_1 n_2 \ldots n_i \ldots\rangle = 1. \tag{4.30}$$

Note that this language keeps track of the states occupied, their occupancies, and their symmetrization with respect to particle interchange, i.e. only those properties of the many-particle system that are definable: there is no (artificial) labelling of particles.

We can also define the particle number operator, $\hat{N}_i$,

$$\hat{N}_i := a_i^\dagger a_i, \tag{4.31}$$

and the total particle number operator

$$\hat{N} := \sum_i \hat{N}_i. \tag{4.32}$$

Evidently,

$$\hat{N}|n_1 n_2 \dots n_i \dots\rangle = \left(\sum_i n_i\right)|n_1 n_2 \dots n_i \dots\rangle, \tag{4.33}$$

i.e. $|n_1 n_2 \dots n_i \dots\rangle$ is an eigenstate of $\hat{N}$. However, for a state

$$|\alpha\rangle = \sum_{n_1, n_2, \dots, n_i, \dots} c_{n_1 n_2 \dots n_i \dots}|n_1 n_2 \dots n_i \dots\rangle, \tag{4.34}$$

this will not (in general) be an eigenstate of $\hat{N}$. (There will be circumstances when our state of knowledge will be imprecise with respect to the occupancy number.)

4.3 The occupation number representation for fermions

For a system of identical fermions, we define the state of the system in a manner similar to that for bosons, but with $n_i = 0$ or 1, $\forall i$, viz.

$$|n_1 n_2 \dots n_i \dots\rangle := \prod_i (a_i^\dagger)^{n_i}|0\rangle, \; n_i = 0 \text{ or } 1, \quad \forall i, \tag{4.35}$$

where

$$\left\{a_i^\dagger, a_j^\dagger\right\} := 0, \quad \forall i. \tag{4.36}$$

Further,

$$\{a_i, a_j\} := 0, \quad \forall i, j, \tag{4.37}$$

and

$$\left\{a_i, a_j^\dagger\right\} := \delta_{ij}, \quad \forall i, j. \tag{4.38}$$

Equations (4.36)–(4.38) have remarkable consequences:

(a) For equation (4.36) with $i = j$,

$$a_j^\dagger a_j^\dagger + a_j^\dagger a_j^\dagger = 0, \tag{4.39}$$

$$\therefore \left(a_j^\dagger\right)^2 = 0. \tag{4.40}$$

Thus, we cannot create two identical fermions in the same state. (Non-identical fermions, both in the state j, would be described by, e.g. $a_j^\dagger$, $b_j^\dagger$, etc.) Similarly,

$$(a_j)^2 = 0. \tag{4.41}$$

(b) Consider

$$\hat{N}_j = a_j^\dagger a_j, \tag{4.42}$$

then

$$\hat{N}_j^2 = a_j^\dagger a_j a_j^\dagger a_j; \tag{4.43}$$

and from equation (4.38) with $i = j$,

$$\begin{aligned} \therefore \hat{N}_j^2 &= a_j^\dagger(1 - a_j^\dagger a_j)a_j \\ &= a_j^\dagger a_j - a_j^\dagger a_j^\dagger a_j a_j \to 0 \\ &= \hat{N}_j. \end{aligned} \tag{4.44}$$

Thus, the eigenvalues of $\hat{N}_j$ must be 0 or 1. Equation (4.44) expresses the *idempotency* of $\hat{N}_j$.

The two prototype operations on the states of a many-fermion system are:

$$a_i|n_1 n_2 \ldots n_i \ldots\rangle = a_i\{(a_1^\dagger)^{n_1}(a_2^\dagger)^{n_2}\ldots(a_i^\dagger)^{n_i}\ldots\}|0\rangle \tag{4.45}$$

and

$$a_i^\dagger|n_1 n_2 \ldots n_i \ldots\rangle = a_i^\dagger\{(a_1^\dagger)^{n_1}(a_2^\dagger)^{n_2}\ldots(a_i^\dagger)^{n_i}\ldots\}|0\rangle. \tag{4.46}$$

From equation (4.45), using equation (4.36),

$$\begin{aligned} a_i|n_1 n_2 \ldots n_i \ldots\rangle &= (-1)^{n_1}(a_1^\dagger)^{n_1}(a_i)(a_2^\dagger)^{n_2}\ldots(a_i^\dagger)^{n_i}\ldots|0\rangle \\ &= (-1)^{n_1+n_2}(a_1^\dagger)^{n_1}(a_2^\dagger)^{n_2}(a_i)\ldots(a_i^\dagger)^{n_i}\ldots|0\rangle \\ &= (-1)^{\sum_{j=1}^{i-1} n_j}(a_1^\dagger)^{n_1}(a_2^\dagger)^{n_2}\ldots(a_i)(a_i^\dagger)^{n_i}\ldots|0\rangle, \end{aligned} \tag{4.47}$$

and

$$a_i(a_i^\dagger)^{n_i} = a_i,\ n_i = 0; \qquad a_i(a_i^\dagger)^{n_i} = 1 - a_i^\dagger a_i,\quad n_i = 1; \tag{4.48}$$

$$\therefore a_i|n_1 n_2 \ldots n_i \ldots\rangle = \theta_i n_i|n_1 n_2 \ldots n_i - 1 \ldots\rangle, \tag{4.49}$$

where

$$\theta_i := (-1)^{\sum_{j=1}^{i-1} n_j}. \tag{4.50}$$

Similarly,

$$a_i^\dagger |n_1 n_2 \ldots n_i \ldots\rangle = \theta_i (1 - n_i) |n_1 n_2 \ldots 1 - n_i \ldots\rangle. \tag{4.51}$$

4.4 Hamiltonians and other operators in the occupation number representation

The total Hamiltonian for a many-particle system, in the absence of interactions between the particles, is simply the sum of the single-particle Hamiltonians:

$$\hat{\mathcal{H}} = \sum_k E_k \hat{N}_k = \sum_k E_k a_k^\dagger a_k. \tag{4.52}$$

This is not very interesting. The occupation number representation doesn't do anything for us. We would have to solve for the E_k by standard (one-body) methods. The occupation number representation only becomes useful when we have interactions between the particles in a many-body system.

We must find out how to handle interaction operators in the occupation number representation. We are familiar with operators in single-particle quantum mechanics and so we start from there. Single-particle quantum mechanics is just a special case of many-particle quantum mechanics: just one particle is present.

We define the *wave function operator* or *field operator*:

$$\Psi_{op}(\vec{r}) := \sum_k a_k u_k(\vec{r}), \tag{4.53}$$

where the a_k are annihilation operators. The $a_k^\dagger$ and a_k operate only in the many-particle ket spaces, they do not operate on the $u_k(\vec{r})$. The many-particle ket space is usually termed *Fock space*. The $\{u_k(\vec{r})\}$ are a complete set of position eigenfunctions. The operator $\Psi_{op}(\vec{r})$ can be interpreted as annihilating a particle at the point $\vec{r}$ in physical space. The conjugate of $\Psi_{op}(\vec{r})$ is $\Psi_{op}^\dagger(\vec{r})$. It creates a particle at the point $\vec{r}$. The term 'second quantization' becomes evident in the separate identities of the $u_k(\vec{r})$, which are defined by the process of the quantization, and the a_k, $a_k^\dagger$ which manifestly describe quanta (particles or field quanta).

Then, we must explore the properties of $\Psi_{op}(\vec{r})$. With one-particle systems in mind, consider

$$\begin{aligned} \Psi_{op}(\vec{r})|00\ldots0\underset{\substack{\uparrow\\ n\text{th}}}{1}0\ldots\rangle &= \sum_k a_k u_k(\vec{r})|00\ldots0\underset{\substack{\uparrow\\ n\text{th}}}{1}0\ldots\rangle, \\ &= u_n(\vec{r})|0\rangle. \end{aligned} \tag{4.54}$$

Then for a general one-particle state $|\alpha\rangle$,

$$\begin{aligned}\Psi_{op}(\vec{r})|\alpha\rangle &= \Psi_{op}(\vec{r})\sum_n c_n|00\ldots0\underset{\substack{\uparrow \\ n\text{th}}}{1}0\ldots\rangle \\ &= \left\{\sum_n c_n u_n(\vec{r})\right\}|0\rangle \\ &= \psi_\alpha(\vec{r})|0\rangle,\end{aligned} \tag{4.55}$$

where

$$\psi_\alpha(\vec{r}) = \sum_n c_n u_n(\vec{r}); \tag{4.56}$$

and

$$\langle 0|\Psi_{op}(\vec{r})|\alpha\rangle = \psi_\alpha(\vec{r}). \tag{4.57}$$

It can be shown that the field operator is independent of any basis of reference (recall, it is the operator that annihilates a particle at the point $\vec{r}$). It can also be shown that

$$\int \Psi^\dagger_{op}(\vec{r})\Psi_{op}(\vec{r})\mathrm{d}\vec{r} = \hat{N}_{\text{total}}. \tag{4.58}$$

We are now in a position to construct operators for many-particle systems from single-particle operators and the field operator. Consider the single-particle operator $\hat{A}$. We define

$$\hat{\mathcal{A}} := \int \Psi^\dagger_{op}(\vec{r})\hat{A}\Psi_{op}(\vec{r})\mathrm{d}\vec{r}, \tag{4.59}$$

where $\hat{\mathcal{A}}$ replaces $\hat{A}$ as the single-particle operator (describing the dynamical quantity A) when going to a many-particle system. To see the validity of this, we expand $\Psi_{op}(\vec{r})$ in terms of eigenfunctions of $\hat{A}$:

$$\therefore\ \Psi_{op}(\vec{r}) = \sum_n a_n u_n(\vec{r}), \tag{4.60}$$

where

$$\hat{A}u_n(\vec{r}) = \alpha_n u_n(\vec{r}). \tag{4.61}$$

$$\begin{aligned}\therefore \int \Psi^\dagger_{op}(\vec{r})\hat{A}\Psi_{op}(\vec{r})\mathrm{d}\vec{r} &= \sum_{i,j}\int a_i^\dagger u_i^*(\vec{r})\hat{A}a_j u_j(\vec{r})\mathrm{d}\vec{r} \\ &= \sum_{i,j} a_i^\dagger a_j \int u_i^*(\vec{r})\hat{A}u_j(\vec{r})\mathrm{d}\vec{r} \\ &= \sum_{i,j} a_i^\dagger a_j \alpha_j \delta_{ij} \\ &= \sum_i \alpha_i a_i^\dagger a_i,\end{aligned} \tag{4.62}$$

$$\hat{\mathcal{A}} = \sum_i \alpha_i a_i^\dagger a_i. \tag{4.63}$$

Equation (4.52) is formally identical to equation (4.63).

More generally, for any single-particle operator $\hat{V}$ in the basis $\{u_n(\vec{r})\}$:

$$\begin{aligned}\int \Psi_{op}^\dagger(\vec{r})\hat{V}\Psi_{op}(\vec{r})\mathrm{d}\vec{r} &= \sum_{i,j}\int a_i^\dagger u_i^*(\vec{r})\hat{V}a_j u_j(\vec{r})\mathrm{d}\vec{r}\\ &= \sum_{i,j} a_i^\dagger a_j \int u_i^*(\vec{r})\hat{V}u_j(\vec{r})\mathrm{d}\vec{r}\\ &= \sum_{i,j} a_i^\dagger a_j V_{ij},\end{aligned} \tag{4.64}$$

where

$$V_{ij} = \int u_i^*(\vec{r})\hat{V}u_j(\vec{r})\mathrm{d}\vec{r}. \tag{4.65}$$

Equation (4.64) is a one-body interaction potential in the occupation number representation, and its components can be represented graphically as shown in figure 4.1.

In a similar manner, using two-particle states, we can define a *two-particle field operator*: e.g. for

$$\psi_{ij}(1, 2) = \frac{1}{\sqrt{2}}\{u_i(1)u_j(2) - u_i(2)u_j(1)\}, \tag{4.66}$$

we define

$$\Psi_{op}(1, 2) := \frac{1}{\sqrt{2}}\sum_{i,j}\{a_i u_i(1)a_j u_j(2) - a_i u_i(2)a_j u_j(1)\}. \tag{4.67}$$

$$\therefore \Psi_{op}(1, 2) = \sum_{i,j} a_i a_j \psi_{ij}(1, 2). \tag{4.68}$$

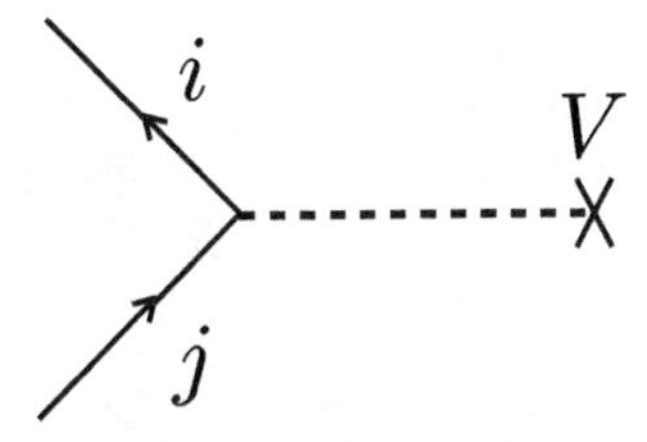

Figure 4.1. Graphical representation for the scattering of a particle from state j to state i by a one-body potential V. This is called a *Feynman–Goldstone diagram* or *Feynman graph*. The initial state is at the bottom of the graph and the final state is at the top.

(It is implicit that the u here are functions of $\vec{r}$.) Consider then

$$\begin{aligned}\Psi_{op}(1, 2)|00\ldots \underset{\substack{\uparrow \\ l\text{th}}}{1} \ldots \underset{\substack{\uparrow \\ m\text{th}}}{1} \ldots\rangle &= \sum_{i,j} a_i a_j \psi(1, 2)|00\ldots \underset{\substack{\uparrow \\ l\text{th}}}{1} \ldots \underset{\substack{\uparrow \\ m\text{th}}}{1} \ldots\rangle \\ &= \psi_{lm}(1, 2)|0\rangle\end{aligned} \tag{4.69}$$

and for a general two-particle state $|\beta\rangle$,

$$\begin{aligned}\Psi_{op}(1, 2)|\beta\rangle &= \Psi_{op}(1, 2)\sum_{lm} c_{lm}|00\ldots \underset{\substack{\uparrow \\ l\text{th}}}{1} \ldots \underset{\substack{\uparrow \\ m\text{th}}}{1} \ldots\rangle \\ &= \left\{\sum_{lm} c_{lm}\psi_{lm}(1, 2)\right\}|0\rangle \\ &= \psi_\beta(1, 2)|0\rangle,\end{aligned} \tag{4.70}$$

where

$$\psi_\beta(1, 2) = \sum c_{lm}\psi_{lm}(1, 2), \tag{4.71}$$

and the $\psi_{lm}(1, 2)$ are defined by equation (4.66) (with $i = l, j = m$);

$$\therefore \langle 0|\Psi_{op}(1, 2)|\beta\rangle = \psi_\beta(1, 2). \tag{4.72}$$

We can construct two-body interaction potentials in a manner analogous to our one-body interaction potentials. For any two-particle operator $\hat{V}$ in the basis $\psi_{ij}(1, 2)$:

$$\begin{aligned}\int \Psi^\dagger_{op}(1, 2)\hat{V}\Psi_{op}(1, 2)\mathrm{d}\vec{r}_1\, \mathrm{d}\vec{r}_2 &= \int \sum_{ijkl} a_j^\dagger a_i^\dagger \psi^*_{ij}(1, 2)\hat{V} a_k a_l \psi_{kl}(1, 2)\mathrm{d}\vec{r}_1\, \mathrm{d}\vec{r}_2 \\ &= \sum_{\substack{ijkl \\ i<j,k<l}} a_j^\dagger a_i^\dagger V_{ijkl} a_k a_l,\end{aligned} \tag{4.73}$$

where the condition $i < j$, $k < l$ prevents double counting. Equation (4.73) is a two-body interaction potential in the occupation number representation and its components can be represented graphically as shown in figure 4.2.

Figure 4.2 can be viewed as a scattering of the two particles from states k, l to states i, j. It can also be viewed as the two particles being annihilated in states k and l and recreated in states i and j. It can further be viewed as a mixing of two-particle configurations. The amplitude or matrix element for the process is V_{ijkl}.

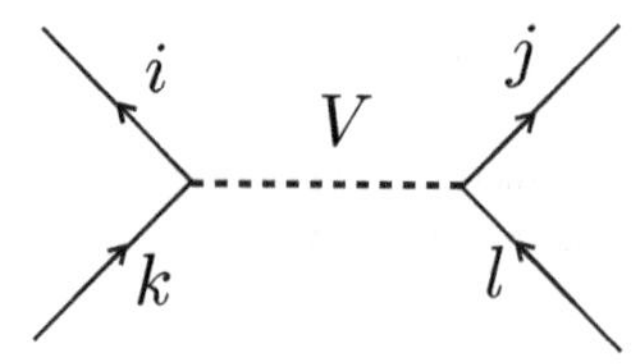

Figure 4.2. Graphical representation for a two-body scattering by the two-body interaction potential V.

A three-particle interaction would be written

$$\sum_{\substack{ijklmn \\ i<j<k,\, l<m<n}} V_{ijklmn} a_k^\dagger a_j^\dagger a_i^\dagger a_l a_m a_n. \tag{4.74}$$

A commonly occurring Hamiltonian form is

$$\mathcal{H} = \sum_k E_k a_k^\dagger a_k + \sum_{i<j,k<l} V_{ijkl} a_j^\dagger a_i^\dagger a_k a_l, \tag{4.75}$$

where the first sum is over single-particle energies and the second sum is over two-body interactions. Note that the number of particles is conserved (there are equal numbers of annihilation and creation operators).

The procedure for carrying out computations is as follows:

(a) Compute the E_k and V_{ijkl} using standard 'one-body' techniques.

(b) Write each many-body matrix element so that the many-particle kets are expressed as creation operators acting on the vacuum (and the bras are the corresponding adjoints). For example,

$$\langle n_1' n_2' | V_{12} a_1^\dagger a_2 | n_1 n_2 \rangle = V_{12} \langle 0 | \frac{(a_2)^{n_2'}}{\sqrt{n_2'!}} \frac{(a_1)^{n_1'}}{\sqrt{n_1'!}} a_1^\dagger a_2 \frac{(a_1^\dagger)^{n_1}}{\sqrt{n_1!}} \frac{(a_2^\dagger)^{n_2}}{\sqrt{n_2!}} | 0 \rangle. \tag{4.76}$$

(c) Put the creation and annihilation operators in *normal order*, i.e. all the creation operators to the left and all the annihilation operators to the right. For example, for equation (4.76) with $n_1' = 2$, $n_2' = 0$, $n_1 = 1$, $n_2 = 1$, for bosons,

$$\begin{aligned} V_{12} \langle 0 | \frac{(a_1)^2}{\sqrt{2!}} a_1^\dagger a_2 a_1^\dagger a_2^\dagger | 0 \rangle &= \frac{V_{12}}{\sqrt{2}} \langle 0 | a_1 a_1 a_1^\dagger a_1^\dagger a_2 a_2^\dagger | 0 \rangle \\ &= \frac{V_{12}}{\sqrt{2}} \langle 0 | a_1 (1 + a_1^\dagger a_1) a_1^\dagger (1 + \cancelto{0}{a_2^\dagger a_2}) | 0 \rangle \\ &= \frac{V_{12}}{\sqrt{2}} \left\{ \langle 0 | a_1 a_1^\dagger | 0 \rangle + \langle 0 | a_1 a_1^\dagger a_1 a_1^\dagger | 0 \rangle \right\}, \end{aligned}$$

$$\begin{aligned} \therefore \frac{V_{12}}{\sqrt{2}} \langle 0 | (a_1)^2 a_1^\dagger a_2 a_1^\dagger a_2^\dagger | 0 \rangle &= \frac{V_{12}}{\sqrt{2}} \left\{ \langle 0 | (1 + \cancelto{0}{a_1^\dagger a_1}) | 0 \rangle + \langle 0 | (1 + a_1^\dagger a_1)(1 + \cancelto{0}{a_1^\dagger a_1}) | 0 \rangle \right\} \\ &= \frac{V_{12}}{\sqrt{2}} \left\{ \langle 0 | 0 \rangle + \langle 0 | 0 \rangle + \langle 0 | \cancelto{0}{a_1^\dagger a_1} | 0 \rangle \right\} \\ &= \sqrt{2} V_{12}. \end{aligned} \tag{4.77}$$

4.4.1 Exercises

4-1. From equation (4.76), put the operators in the normal order and hence compute the matrix elements for:

	n_1'	n_2'	n_1	n_2
(*i*)	3	0	2	1
(*ii*)	4	0	3	1
(*iii*)	1	1	0	2
(*iv*)	1	2	0	3
(*v*)	1	3	0	4
(*vi*)	2	1	1	2.

4-2. Show that

$$\int \Psi_{op}^{\dagger}(\vec{r})\Psi_{op}(\vec{r})\mathrm{d}\vec{r} = \hat{N}_{\text{total}} \tag{4.78}$$

(cf. equation (4.58)).

4.5 Condensed states (superconductors and superfluids)

Many-particle quantum systems are capable of exhibiting an extraordinary array of behaviour—witness chemistry and condensed matter physics. The present section considers some idealised systems with fluid properties, i.e. the particles are free to move throughout a specified region of space, but forces are acting between them. This is distinct from solids where at least some of the particles are localised in small subregions of the space occupied by the system, e.g. as in a crystalline lattice.

4.5.1 Two fermions in a degenerate set of levels with a pairing force

Consider a system of two fermions that are free to move in a set of $2N$ states which are all degenerate in energy and which are characterised by a quantum number k, where $k = \pm 1, \pm 2, \ldots, \pm N$. (The quantum number k might label states of well-defined linear momentum or states of well-defined z-component of angular momentum.) Further, there is a pairing force acting between the fermions such that the pairing matrix elements, V_{ijrs}, are

$$\begin{aligned} V_{ijrs} = V; \quad & r = k, s = -k, i = k', j = -k', \\ & k, k' = 1, 2, \ldots, N, \\ V_{ijrs} = 0, \quad & \text{otherwise.} \end{aligned} \tag{4.79}$$

Then consider the case with $N = 2$. In the occupation number representation the basis states are

$$a_i^{\dagger} a_j^{\dagger}|0\rangle, \; i, j = +2, +1, -1, -2, \; i \neq j,$$

i.e. $a^\dagger_{+2}a^\dagger_{+1}|0\rangle$, $a^\dagger_{+2}a^\dagger_{-1}|0\rangle$, $a^\dagger_{+2}a^\dagger_{-2}|0\rangle$, $a^\dagger_{+1}a^\dagger_{-1}|0\rangle$, $a^\dagger_{+1}a^\dagger_{-2}|0\rangle$, and $a^\dagger_{-1}a^\dagger_{-2}|0\rangle$. The Hamiltonian is

$$\hat{H} = \varepsilon \sum_{k=-2}^{k=2} a^\dagger_k a_k + V \sum_{k',k=1}^{2} a^\dagger_{k'} a^\dagger_{-k'} a_{-k} a_k, \tag{4.80}$$

where ε is the energy of the degenerate states. The matrix elements of $\hat{H}$ contain one-body contributions of the form $\langle 0|a_j a_i(\varepsilon \sum_{k=-2}^{2} a^\dagger_k a_k) a^\dagger_i a^\dagger_j|0\rangle$ and two-body contributions of the form $\langle 0|a_j a_i(V \sum_{k',\,k=1}^{2} a^\dagger_{k'} a^\dagger_{-k'} a_{-k} a_k) a^\dagger_i a^\dagger_j|0\rangle$. These yield, via the standard process of normal ordering, the Hamiltonian

$$\hat{H} \leftrightarrow \begin{pmatrix} 2\varepsilon & 0 & 0 & 0 & 0 & 0 \\ 0 & 2\varepsilon & 0 & 0 & 0 & 0 \\ 0 & 0 & 2\varepsilon + V & V & 0 & 0 \\ 0 & 0 & V & 2\varepsilon + V & 0 & 0 \\ 0 & 0 & 0 & 0 & 2\varepsilon & 0 \\ 0 & 0 & 0 & 0 & 0 & 2\varepsilon \end{pmatrix}, \tag{4.81}$$

where the ordering of the base states is $(i, j) = (+2, +1)$, $(+2, -1)$, $(+2, -2)$, $(+1, -1)$, $(+1, -2)$, $(-1, -2)$.

We focus on the $(+2, -2)$, $(+1, -1)$ subspace for which the Hamiltonian matrix is

$$\hat{H} \leftrightarrow \begin{pmatrix} 2\varepsilon + V & V \\ V & 2\varepsilon + V \end{pmatrix} := \begin{pmatrix} \varepsilon' & V \\ V & \varepsilon' \end{pmatrix}. \tag{4.82}$$

For an attractive interaction, $V < 0$,

$$\varepsilon' = 2\varepsilon - V, \tag{4.83}$$

and if we were to neglect the off-diagonal matrix elements the energy spectrum for $\hat{H}$ would be as depicted in the middle of figure 4.3. The results of diagonalizing the Hamiltonian, equation (4.82), is (cf. Volume 1, chapter 6, section 6.3)

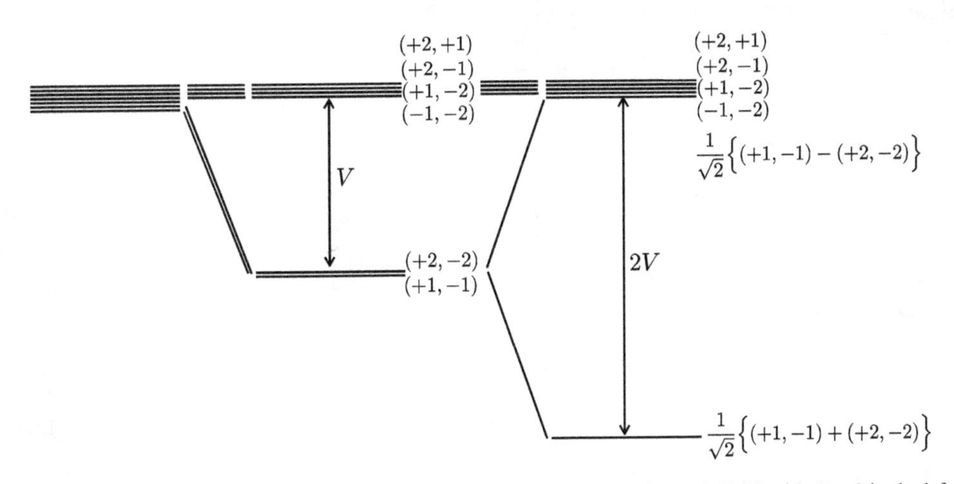

Figure 4.3. Depiction of the spectrum of the Hamiltonians, equations (4.81) and (4.82) with $V = 0$ in the left-hand panel, with non-zero diagonal V in the middle panel, and the Hamiltonian equation (4.82) in the right-hand panel (after diagonalization).

$$\lambda = \varepsilon' \pm V, \text{ i. e. } \lambda = 2\varepsilon, 2\varepsilon - 2V, \tag{4.84}$$

which can be depicted as shown on the right-hand side of figure 4.3.

The eigenstates resulting from the diagonalization of equation (4.82) are:

$$|\lambda = 2\varepsilon - 2V\rangle = \frac{1}{\sqrt{2}}\{a_{+1}^{\dagger}a_{-1}^{\dagger}|0\rangle + a_{+2}^{\dagger}a_{-2}^{\dagger}|0\rangle\}, \tag{4.85}$$

$$|\lambda = 2\varepsilon\rangle = \frac{1}{\sqrt{2}}\{a_{+1}^{\dagger}a_{-1}^{\dagger}|0\rangle - a_{+2}^{\dagger}a_{-2}^{\dagger}|0\rangle\}. \tag{4.86}$$

We call the symmetric combination, equation (4.85), a *coherent* superposition and the antisymmetric combination, equation (4.86), an *incoherent* superposition. For $V < 0$, the coherent superposition has the lowest energy. This is a significant result. We can regard V as a small attractive two-body potential superimposed on the one-body Hamiltonian. In figure 4.3, the two configurations that can benefit from this particular attractive force are shown with their gain in binding energy. This is readily understood in classical physical terms In figure 4.3, one particular linear combination of the six basis states alone has increased binding energy. This additional gain in binding energy is called *correlation energy*. It has no classical analog. This is a characteristic feature of a *quantum fluid* with an attractive *pairing force*. The ground state of the system is called a *condensate*. It is characterised by being separated from the other states by an *energy gap* even when the unperturbed states are all degenerate.

4.5.2 Many fermions in a degenerate set of levels with a pairing force: the quasispin formalism

Systems with pairing forces can be described in an elegant and economical way using the so-called *quasispin formalism*. For a set of states $k = \pm 1, \pm 2, \pm 3, \ldots, \pm N$, we define:

$$S_{k_+} := a_k^{\dagger}a_{-k}^{\dagger}, \tag{4.87}$$

$$S_{k_-} := a_{-k}a_k, \tag{4.88}$$

$$S_{k_0} := \frac{1}{2}(a_k^{\dagger}a_k + a_{-k}^{\dagger}a_{-k} - 1). \tag{4.89}$$

These fermion operators obey the following commutator (not anticommutator) bracket relations:

$$[S_{k_0}, S_{k_+}] = S_{k_+}, \tag{4.90}$$

$$[S_{k_0}, S_{k_-}] = -S_{k_-}, \tag{4.91}$$

$$[S_{k_+}, S_{k_-}] = 2S_{k_0}, \tag{4.92}$$

i.e. S_{k_+}, S_{k_-} and S_{k_0} define an $su(2)$ algebra.

For example, consider

$$\begin{aligned}
[S_{k_0}, S_{k_+}] &= \left[\frac{1}{2}(a_k^\dagger a_k + a_{-k}^\dagger a_{-k} - 1), a_k^\dagger a_{-k}^\dagger\right] \\
&= \frac{1}{2}\Big\{a_k^\dagger a_k a_k^\dagger a_{-k}^\dagger + a_{-k}^\dagger a_{-k} a_k^\dagger a_{-k}^\dagger - a_k^\dagger a_{-k}^\dagger \\
&\quad - a_k^\dagger a_{-k}^\dagger a_k^\dagger a_k - a_k^\dagger a_{-k}^\dagger a_{-k}^\dagger a_{-k} + a_k^\dagger a_{-k}^\dagger\Big\} \\
&= \frac{1}{2}\Big\{a_k^\dagger a_{-k}^\dagger (1 - a_k^\dagger a_k) + a_k^\dagger a_{-k}^\dagger (1 - a_{-k}^\dagger a_{-k}) \\
&\quad + a_{-k}^\dagger \underbrace{a_k^\dagger a_k^\dagger}_{0} a_k - a_k^\dagger \underbrace{a_{-k}^\dagger a_{-k}^\dagger}_{0} a_{-k}\Big\} \\
&= \frac{1}{2}\Big\{a_k^\dagger a_{-k}^\dagger + a_{-k}^\dagger \underbrace{a_k^\dagger a_k^\dagger}_{0} a_k + a_k^\dagger a_{-k}^\dagger - a_k^\dagger \underbrace{a_{-k}^\dagger a_{-k}^\dagger}_{0} a_{-k}\Big\} \\
&= a_k^\dagger a_{-k}^\dagger,
\end{aligned} \tag{4.93}$$

$$\therefore [S_{k_0}, S_{k_+}] = S_{k_+}. \tag{4.94}$$

Further, we define:

$$S_+ := \sum_{k=1}^{N} S_{k_+}, \tag{4.95}$$

$$S_- := \sum_{k=1}^{N} S_{k_-}, \tag{4.96}$$

$$S_0 := \sum_{k=1}^{N} S_{k_0}. \tag{4.97}$$

These fermion operators obey the commutator bracket relations:

$$[S_0, S_+] = S_+, \tag{4.98}$$

$$[S_0, S_-] = -S_-, \tag{4.99}$$

$$[S_+, S_-] = 2S_0, \tag{4.100}$$

i.e. again an $su(2)$ algebra—this is the quasispin algebra. For example, consider

$$\begin{aligned}
[S_+, S_-] &= \left[\sum_k S_{k_+}, \sum_l S_{l_-}\right] \\
&= \sum_{k,l} [S_{k_+}, S_{l_-}].
\end{aligned} \tag{4.101}$$

For $k \neq l$

$$[a_k^\dagger a_{-k}^\dagger, a_{-l}a_l] = a_k^\dagger a_{-k}^\dagger a_{-l}a_l - a_l a_{-l} a_k^\dagger a_{-k}^\dagger \\ = a_k^\dagger a_{-k}^\dagger a_{-l}a_l - a_k^\dagger a_{-k}^\dagger a_{-l}a_l \tag{4.102}$$

$$= 0. \tag{4.103}$$

For $k = l$

$$[S_{k_+}, S_{k_-}] = 2S_{k_0},$$

$$\therefore [S_+, S_-] = \sum_k 2S_{k_0} \\ = 2S_0. \tag{4.104}$$

For the Hamiltonian, equation (4.80), we set the energy zero at $\varepsilon = 0$ (this is arbitrary, and simplifies the details). Further, we take an attractive pairing force, i.e. $V < 0$. Then,

$$\hat{H} = -V \sum_{k,l=1}^{N} a_k^\dagger a_{-k}^\dagger a_{-l} a_l. \tag{4.105}$$

This Hamiltonian can be written using the quasispin language as

$$\hat{H} = -VS_+S_-. \tag{4.106}$$

The algebraic structure of the Hamiltonian is immediately revealed by recognising the form of equation (4.106) and the commutator brackets as being identical to the problem of spin and angular momentum (cf. Volume 1, chapter 11) in quantum mechanics, i.e. compare:

$$[S_0, S_+] = S_+ \ \text{cf.} \ [L_z, L_+] = \hbar L_+, \tag{4.107}$$

$$[S_0, S_-] = -S_- \ \text{cf.} \ [L_z, L_-] = -\hbar L_-, \tag{4.108}$$

$$[S_+, S_-] = 2S_0 \ \text{cf.} \ [L_+, L_-] = 2\hbar L_z. \tag{4.109}$$

$$S^2 = S_0^2 - S_0 + S_+S_- \ \text{cf.} \ L^2 = L_z^2 - L_z\hbar + L_+L_-, \tag{4.110}$$

$$[S^2, S_+] = 0 \ \text{cf.} \ [L^2, L_+] = 0, \tag{4.111}$$

$$[S^2, S_-] = 0 \ \text{cf.} \ [L^2, L_-] = 0, \tag{4.112}$$

$$[S^2, S_-] = 0 \ \text{cf.} \ [L^2, L_-] = 0, \tag{4.113}$$

$$S^2|sm_s\rangle = s(s+1)|sm_s\rangle \ \text{cf.} \ L^2|lm\rangle = l(l+1)\hbar^2|lm\rangle, \\ s = 0, 1, 2, \dots, \qquad l = 0, 1, 2, \dots, \\ m_s = -s, -s+1, \dots, +s, \qquad m = -l, -l+1, \dots, +l, \tag{4.114}$$

$$S_0|sm_s\rangle = m_s|sm_s\rangle \text{ cf. } L_z|lm\rangle = m\hbar|lm\rangle. \tag{4.115}$$

Then, from equation (4.110)

$$\hat{H} = -V\{S^2 - S_0^2 + S_0\}, \tag{4.116}$$

and

$$\hat{H}|sm_s\rangle = -V\{s(s+1) - m_s^2 + m_s\}|sm_s\rangle, \tag{4.117}$$

i.e. we have a closed-form expression for the eigenvalues of $\hat{H}$: it only remains to interpret the quasispin quantum numbers s and m_S.

To interpret s and m_S, first note that

$$S_0 = \frac{1}{2}\sum_{k=1}^{N}(\hat{N}_k + \hat{N}_{-k} - 1), \tag{4.118}$$

i.e. it is related to the number of particles in the system: let this be n, whence

$$n = \sum_{k=1}^{N}(n_k + n_{-k}). \tag{4.119}$$

Evidently, from equations (4.119), (4.118) and (4.115), together with the recognition that in equation (4.118), $\sum_{k=1}^{N} 1 = N$, we have

$$m_s = \frac{1}{2}(n - N). \tag{4.120}$$

Then, for $n = 0$,

$$m_s = -\frac{1}{2}N \tag{4.121}$$

and for $n = 2N$ (the maximum possible value of n in a many-fermion system with states $k = \pm 1, \pm 2, \cdots, \pm N$)

$$m_s = \frac{1}{2}N. \tag{4.122}$$

It follows that

$$s = \frac{1}{2}N, \tag{4.123}$$

i.e. the quasispin quantum number s describes the number of pair states in the system and the quasispin quantum number m_S is related to the number of fermion pairs in the system. We can rewrite the eigenvalues of $\hat{H}$, equation (4.117), as

$$E(n, N) = -V\left\{\frac{N}{2}\left(\frac{N}{2} + 1\right) - \left(\frac{n - N}{2}\right)^2 + \left(\frac{n - N}{2}\right)\right\}, \tag{4.124}$$

whence

$$E(n, N) = -\frac{V}{4}n(2N - n + 2). \tag{4.125}$$

For the two-particle problem solved earlier, N was taken to be 2, thus

$$E(2, 2) = -\frac{V}{4}2(4 - 2 + 2) = -2V. \tag{4.126}$$

We can depict the energy eigenvalues, equation (4.125), graphically as shown in figure 4.4. A number of observations can be made with respect to figure 4.4:

(a) Although the binding energy $(-VN)$ for one pair $(N = 2)$ and N pairs $(n = 2N)$ is the same, the contributions are very different. For one pair, an amount $-V$ comes from 'diagonal' pairing and $-V(N - 1)$ comes from off-diagonal pairing (pairing correlations). For N pairs, all the binding energy comes from diagonal pairing: correlations are completely *blocked* (all states are occupied).

(b) *Pauli blocking* is revealed in the saturation of the binding energy at $n = N + 1$: As more and more particles are added, they 'get in the way of each other' with respect to correlations. In other words, the possibilities for scattering (correlations) decreases with increasing particle number because an increasing number of states are occupied which blocks the scattering of fermions as a result of the Pauli exclusion principle.

The eigenvalues that we have obtained are for the ground-state binding energy as a function of particle number n. We next consider excited states in a system with specified particle number n. Excited states are formed by 'breaking' pairs, i.e. by forming states such as $(+k, +l)$, $k \neq l$: for such configurations, the pair does not

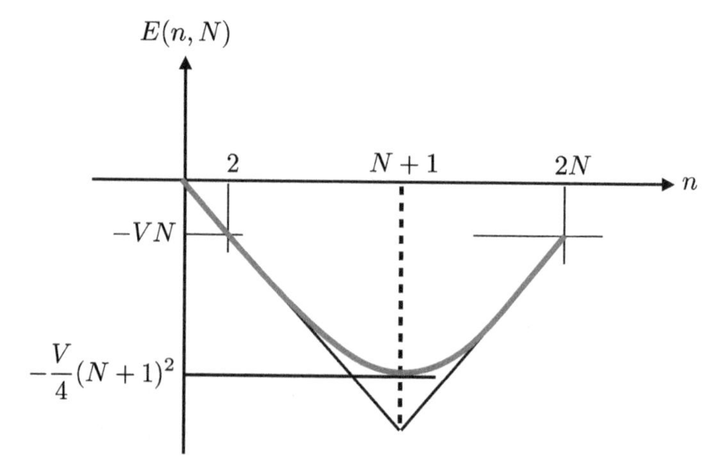

Figure 4.4. A graphical view of the energy eigenvalues of the Hamiltonian (cf. equations (4.106), (4.117), and (4.125)) for n particles occupying a set of $2N$ degenerate states and interacting through a pairing force (equation (4.79)).

contribute to the binding energy. However, this subtracts more than their diagonal pairing energy from the total binding energy because of blocking.

Consider a one broken-pair excitation. Each fermion is in a single-particle state, e.g. $+p$ and $+q$ for the pair. Thus, the pair states $(+p, -p)$ and $(+q, -q)$ are blocked in the correlations of other pairs. This effectively reduces N by 2. We call the number of unpaired fermions the *seniority* and denote the seniority of the system by the quantum number v. Hence for $\frac{v}{2}$ broken pairs in a system of n fermions occupying states $(+k, -k)$, $k = 1, 2, \ldots, N$, N is effectively reduced by v. Thus, equation (4.123) is modified to read

$$s = \frac{1}{2}(N - v), \tag{4.127}$$

and $E(n, N)$ (equation (4.124)) is replaced by

$$\begin{aligned} E(n, v, N) = -V\Bigg\{&\left(\frac{N - v}{2}\right)\left(\frac{N - v}{2} + 1\right) \\ &- \left(\frac{n - v - (N - v)}{2}\right)^2 + \left(\frac{n - v - (N - v)}{2}\right)\Bigg\}, \end{aligned} \tag{4.128}$$

whence

$$E(n, v, N) = -\frac{V}{4}(n - v)(2N - n - v + 2). \tag{4.129}$$

Equation (4.129) readily leads to the excitation spectrum of the system:

(a) For zero-broken pairs ($v = 0$),

$$E(n, 0, N) = -\frac{V}{4}n(2N - n + 2), \tag{4.130}$$

cf. equation (4.125).

(b) For one broken pair ($v = 2$),

$$E(n, 2, N) = -\frac{V}{4}(n - 2)(2N - n). \tag{4.131}$$

Thus,

$$\begin{aligned} E(n, 2, N) - E(n, 0, N) &= -\frac{V}{4}\{\cancel{2nN} - \cancel{n^2} - 4N + \cancel{2n} - \cancel{2nN} \\ &\quad + \cancel{n^2} - \cancel{2n}\} \\ &= VN, \end{aligned} \tag{4.132}$$

i.e. the excitation energy of the one broken pair states relative to the zero-broken pair state is *independent of the particle number n.*

(c) For two broken pairs ($v = 4$)

$$E(n, 4, N) = -\frac{V}{4}(n - 4)(2N - n + 2) \tag{4.133}$$

and

$$E(n, 4, N) - E(n, 0, N) = 2V(N - 1), \tag{4.134}$$

which, again, is independent of n.

We can depict the energy eigenvalues, equation (4.129), graphically as shown in figure 4.5. The most remarkable result is that the excitation spectrum is independent of n. This outcome is not at all evident in the formulation of the problem. It has its origin in the quasispin quantum number $s = \frac{1}{2}(N - v)$ and $m_s = \frac{1}{2}(n - N)$: a change of seniority (an excitation) is a change in s, i.e. it is a change of $SU(2)$ (quasispin) irrep. A change of particle number is a change of m_s, which is confined to a single quasispin irrep. The excitation spectrum for a given N is shown in figure 4.6.

Observations regarding figure 4.6:

(a) The binding energy is not shown: zero binding energy corresponds to $(n, v) = (0, 0), (2, 2), (4, 4), \cdots$.
(b) The excitations are independent of n.
(c) Successive excitations for a given n are compressed, i.e. VN, $V(N - 2)$, $V(N - 4), \cdots$. This reflects Pauli blocking of successive broken pairs.
(d) Each value of v labels a quasispin irrep: recall $s = \frac{1}{2}(N - v)$, $m_s = \frac{1}{2}(n - N)$.
(e) The system has an $SU(2)$ *dynamical symmetry*: the binding energies vary with n and v; but v is a good quantum number.

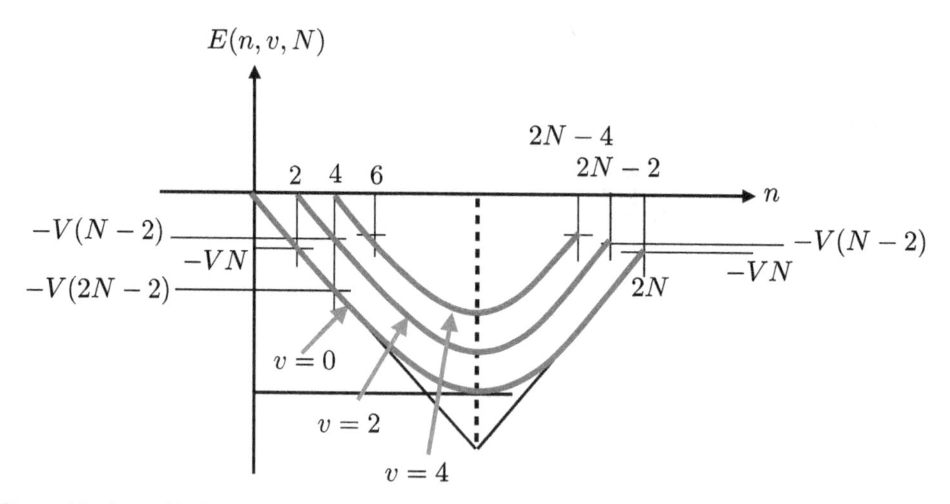

Figure 4.5. A graphical view of the energy eigenvalues of a system of n particles with seniority v (v is the number of unpaired particles) occupying a set of $2N$ degenerate states and interacting through a pairing force (equation (4.79)). The excitation spectrum for a given n is determined by a vertical 'cut' through the red lines, the lowest red line corresponding to the ground state.

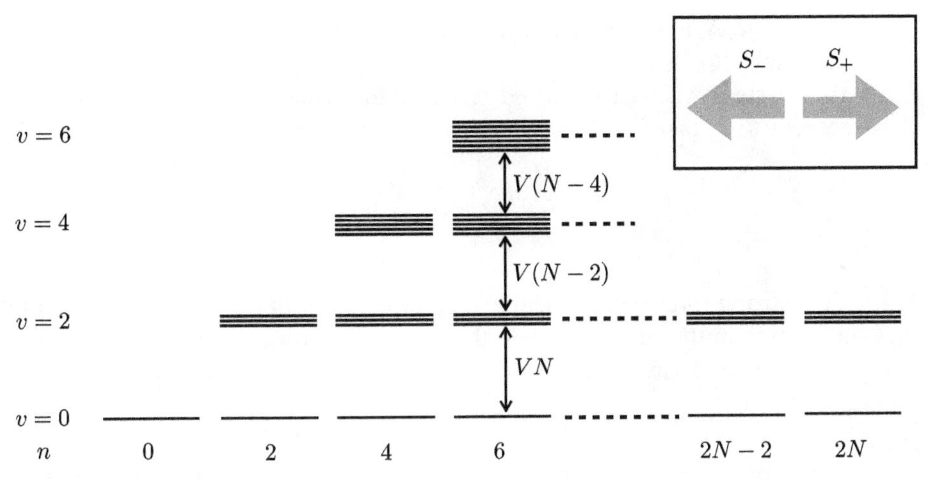

Figure 4.6. The excitation spectra of a system, with the Hamiltonian $H = -VS_+S_-$, containing n particles with seniority v occupying $2N$ degenerate states. Details are discussed in the text.

We are now in a position to summarise the physics of the system: a system of n fermions occupying N degenerate pair states $(+k, -k)$, $k = 1, 2, \ldots, N$ and interacting through a pairing force of the type given by equation (4.79) will exhibit the following features:

(a) An *energy gap* in the excitation spectrum will be present. The magnitude of the gap is proportional to N, and is independent of n.

(b) The ground state of the system is a highly *correlated state*. To see this, consider the addition of a pair of particles to the system, as described by the operator S_+:

$$S_+ = \sum_{k=1}^{N} a_k^\dagger a_{-k}^\dagger. \tag{4.135}$$

This will add (create) a pair to the system with equal probability amplitudes for being in all pair configurations.

(c) Although the system is a many-fermion system, the correlated pairs do not behave like fermions! Consider $(S_+)^2|0\rangle$: if the pair created by S_+ is fermionic, then $(S_+)^2|0\rangle$ should be zero. But

$$(S_+)^2|0\rangle = \left(\sum_{k=1}^{N} a_k^\dagger a_{-k}^\dagger\right)\left(\sum_{l=1}^{N} a_l^\dagger a_{-l}^\dagger\right)|0\rangle. \tag{4.136}$$

The product of the two sums gives N^2 terms, each of which contains four creation operators. Of these, N contain products of the type $(a_k^\dagger)^2(a_{-k}^\dagger)^2$ which are zero. Thus, only a fraction $\frac{1}{N}$ of these terms are lost due to the fermionic character of $a_k^\dagger$ and $a_{-k}^\dagger$. For very large N this is negligible. *Note:*

very large N is a very large number of states not a very large number of particles.

(d) In a system with an unpaired fermion the situation is very different with respect to approximate bosonic behaviour. Consider the operator

$$\mathcal{S}_{+i} \equiv a_i^\dagger \sum_{\substack{k=1,\\ k\neq i}}^{N} a_k^\dagger a_{-k}^\dagger. \tag{4.137}$$

The pairing force will not scatter the unpaired fermion from the state i. Only a fermion in the state $-i$ can do this and, by definition, there is no fermion in the state $-i$. Evidently,

$$(\mathcal{S}_{+i})^2|0\rangle = 0. \tag{4.138}$$

Thus, an aggregate of fermions with correlated pairs (for the pairing force, equation (4.79)) behaves like a fermion if there is an odd number of fermions.

This system is the prototype of systems that exhibit superconductivity and superfluidity. Such systems have an energy gap which prevents dissipation of the current flow (no resistance) or the mass flow (no viscosity) at low temperature. That is, there is no energy (thermal) available to excite the system out of its ground state which would lead to dissipation of the supercurrent or super-flow. The 'flow' is manifest in the highly-correlated nature of the ground state: it extends throughout the bulk of the material. Addition of pairs at 'one end' and removal of pairs at the 'other end' can be made, effecting current (or mass) flow.

4.5.3 BCS theory

The foregoing section contains all the physics of superconductivity or superfluidity in a many-fermion system with an attractive pairing force. The energy gap prevents dissipation of 'super-flow', i.e. it results in zero resistance or zero viscosity. The highly correlated pairs (Cooper pairs) give the extraordinary coherence properties of these systems.

However, the Hamiltonian solved in the previous section is simplistic. All the states are degenerate in energy and all the interactions are equal in strength.

Consider a relaxation of the degeneracy of the ε_k and of the identicality of the $V_{kk'}$ in

$$H = \sum_{k=1}^{N}(\varepsilon_k - \lambda)(a_k^\dagger a_k + a_{-k}^\dagger a_{-k}) - \sum_{k,k'=1}^{N} V_{kk'} a_k^\dagger a_{-k}^\dagger a_{-k'} a_{k'}; \tag{4.139}$$

where λ has been introduced as a reference energy, with respect to which the ε_k are measured.

The two-body term can be greatly simplified by considering the operator identity

$$AB \equiv (A - \langle A\rangle)(B - \langle B\rangle) + A\langle B\rangle + B\langle A\rangle - \langle A\rangle\langle B\rangle, \tag{4.140}$$

where $\langle A\rangle$ and $\langle B\rangle$ are average values (numbers) with respect to some state: in the present case—the ground state of the system. If the quantum fluctuations of A about $\langle A\rangle$ or B about $\langle B\rangle$ are small, $(A - \langle A\rangle)(B - \langle B\rangle) \approx 0$, and

$$AB = A\langle B\rangle + B\langle A\rangle + \text{constant}. \tag{4.141}$$

Thus,

$$\sum_{k,k'=1}^{N} V_{kk'}(a_k^\dagger a_{-k}^\dagger a_{-k}' a_k') \approx \sum_{k=1}^{N} \Delta_k(a_k^\dagger a_{-k}^\dagger + a_{-k}a_k), \tag{4.142}$$

where

$$\Delta_k := \sum_{k'=1}^{N} V_{kk'}\langle a_{-k'}a_{k'}\rangle = \sum_{k'=1}^{N} V_{kk'}\langle a_{k'}^\dagger a_{-k'}^\dagger\rangle \tag{4.143}$$

($\langle\psi|AB|\psi\rangle = \langle\chi|\psi\rangle = \langle\psi|\chi\rangle^* = \langle\psi|B^\dagger A^\dagger|\psi\rangle^* = \langle\psi|B^\dagger A^\dagger|\psi\rangle$ if $\langle\ \rangle$ is real).

The Hamiltonian can be written

$$H = \sum_{k=1}^{N} H_k + \text{constant}, \tag{4.144}$$

where

$$H_k = (\varepsilon_k - \lambda)(a_k^\dagger a_k + a_{-k}^\dagger a_{-k}) - \Delta_k(a_k^\dagger a_{-k}^\dagger + a_{-k}a_k). \tag{4.145}$$

Then, using the quasispin operators

$$S_{k_+} = a_k^\dagger a_{-k}^\dagger, \quad S_{k_-} = a_{-k}a_k, \quad S_{k_3} = \frac{1}{2}(a_k^\dagger a_k + a_{-k}^\dagger a_{-k} - 1), \tag{4.146}$$

and defining

$$S_{k_\pm} := S_{k_1} \pm iS_{k_2}, \tag{4.147}$$

we obtain

$$S_{k_1} = \frac{1}{2}(S_{k_+} + S_{k_-}) = \frac{1}{2}(a_k^\dagger a_{-k}^\dagger + a_{-k}a_k), \tag{4.148}$$

$$S_{k_2} = -\frac{i}{2}(S_{k_+} - S_{k_-}) = -\frac{i}{2}(a_k^\dagger a_{-k}^\dagger + a_{-k}a_k), \tag{4.149}$$

and

$$[S_{k_\alpha}, S_{k_\beta}] = i\varepsilon_{\alpha\beta\gamma}\delta_{k\gamma}, \quad \alpha, \beta, \gamma = 1, 2, 3; \tag{4.150}$$

H_k can be written as

$$H_k = (\varepsilon_k - \lambda)(2S_{k_3} + 1) - 2\Delta_k S_{k_1}. \tag{4.151}$$

This Hamiltonian can be diagonalized by a rotation, cf. $H = \alpha L_z - \beta L_x$ and

$$H' = \mathcal{D}(y, \theta) H \mathcal{D}^\dagger(y, \theta) = \gamma L_z. \tag{4.152}$$

Thus, introducing

$$U_k = \exp\{i\theta_k S_{k_2}\}, \tag{4.153}$$

then (defining $\tilde{\varepsilon}_k := \varepsilon_k - \lambda$)

$$U_k H_k U_k^\dagger = 2\tilde{\varepsilon}_k U_k S_{k_3} U_k^\dagger - 2\Delta_k U_k S_{k_1} U_k^\dagger + \tilde{\varepsilon}_k. \tag{4.154}$$

Using the Baker–Campbell–Hausdorff lemma

$$\begin{aligned} e^{i\theta_k S_{k_2}} S_{k_3} e^{-i\theta_k S_{k_2}} &= S_{k_3} + i\theta_k \overbrace{[S_{k_2}, S_{k_3}]}^{iS_{k_1}} - \frac{\theta_k^2}{2!}\left[S_{k_2}, \overbrace{[S_{k_2}, S_{k_3}]}^{iS_{k_1}}\right] + \cdots \\ &= S_{k_3}\left\{1 - \frac{\theta_k^2}{2!} + \cdots\right\} + S_{k_1}\left\{-\theta_k + \frac{\theta_k^3}{3!} + \cdots\right\} \\ &= \cos\theta_k S_{k_3} - \sin\theta_k S_{k_1} \end{aligned} \tag{4.155}$$

and

$$e^{i\theta_k S_{k_2}} S_{k_1} e^{-i\theta_k S_{k_2}} = \cos\theta_k S_{k_1} + \sin\theta_k S_{k_3}. \tag{4.156}$$

Hence,

$$U_k H_k U_k^\dagger = 2\tilde{\varepsilon}_k(\cos\theta_k S_{k_3} + \sin\theta_k S_{k_1}) - 2\Delta_k(\cos\theta_k S_{k_1} + \sin\theta_k S_{k_3}) + \tilde{\varepsilon}_k \tag{4.157}$$

and for

$$\tilde{\varepsilon}_k \sin\theta_k = -\Delta_k \cos\theta_k, \text{ i.e. } \tan\theta_k = -\frac{\Delta_k}{\tilde{\varepsilon}_k}, \tag{4.158}$$

and

$$\cos\theta_k = \frac{\tilde{\varepsilon}_k}{\sqrt{\tilde{\varepsilon}_k^2 + \Delta_k^2}}, \ \sin\theta_k = \frac{-\Delta_k}{\sqrt{\tilde{\varepsilon}_k^2 + \Delta_k^2}}, \tag{4.159}$$

$$\therefore\ U_k H_k U_k^\dagger = 2\frac{(\tilde{\varepsilon}_k^2 + \Delta_k^2)}{\sqrt{\tilde{\varepsilon}_k^2 + \Delta_k^2}} S_{k_3} + \tilde{\varepsilon}_k = 2E_k S_{k_3} + \tilde{\varepsilon}_k, \ E_k = \sqrt{\tilde{\varepsilon}_k^2 + \Delta_k^2}. \tag{4.160}$$

Further,

$$\begin{aligned} U_k = e^{i\theta_k S_{k_2}} &= \cos\frac{\theta_k}{2} + i(2S_{k_2})\sin\frac{\theta_k}{2} \\ &= \cos\frac{\theta_k}{2} + \sin\frac{\theta_k}{2}(a_k^\dagger a_{-k}^\dagger - a_{-k}a_k). \end{aligned} \tag{4.161}$$

The new vacuum state for the system is given by

$$|\tilde{0}\rangle = \prod_{k=1}^{N} U_k|0\rangle = \prod_{k=1}^{N}\left\{\cos\frac{\theta_k}{2} + \sin\frac{\theta_k}{2}(a_k^\dagger a_{-k}^\dagger - a_{-k}a_k)\right\}|0\rangle, \tag{4.162}$$

but $a_{-k}a_k|0\rangle = 0$, whence

$$|\tilde{0}\rangle = \prod_{k=1}^{N}(U_k + V_k a_k^\dagger a_{-k}^\dagger)|0\rangle, \tag{4.163}$$

where

$$U_k := \cos\frac{\theta_k}{2}, \tag{4.164}$$

$$V_k := \sin\frac{\theta_k}{2}, \tag{4.165}$$

$$U_k^2 + V_k^2 = 1. \tag{4.166}$$

This is the celebrated BCS approximation for the ground state of a pair-correlated system.

The procedure by which this solution was obtained is illustrated in figure 4.7. From equations (4.153) and (4.162) it can be seen that the BCS wave function is a (quasispin) $SU(2)$ coherent state, cf. section 1.8.

4.6 The Lipkin model

The Lipkin model, as the title of the original paper [1] suggests, was developed to provide tests of many-body quantum mechanical approximation methods. It was not intended as a realistic description of any many-body system. Nevertheless, it has received wide attention from theorists working in many areas of physics, and it has great pedagogical value. It is the latter feature and the group theoretical aspects that are emphasized here.

The Hamiltonian for the Lipkin model is

$$\begin{aligned} H = &\frac{1}{2}\varepsilon \sum_{p=1,\sigma=\pm 1}^{n} \sigma a_{p\sigma}^\dagger a_{p\sigma} + \frac{1}{2}V\sum_{pp'\sigma} a_{p\sigma}^\dagger a_{p'\sigma}^\dagger a_{p'-\sigma}a_{p-\sigma} \\ &+ \frac{1}{2}W\sum_{pp'\sigma} a_{p\sigma}^\dagger a_{p'-\sigma}^\dagger a_{p'\sigma}a_{p-\sigma}, \end{aligned} \tag{4.167}$$

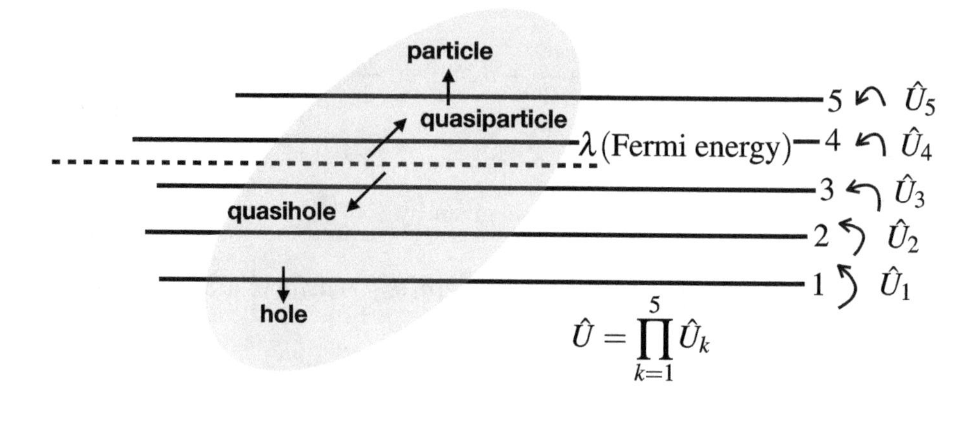

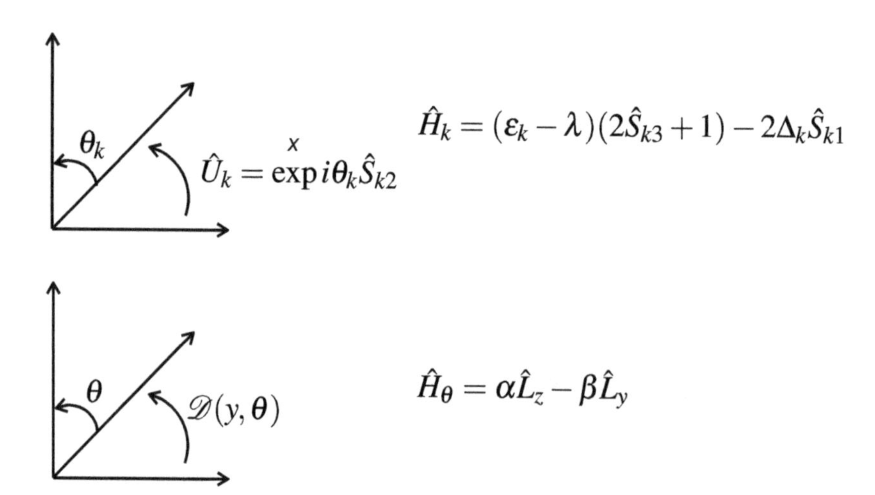

Figure 4.7. A schematic view of a many-fermion system, possessing pairing correlations, represented by a quasispin algebra. 'Holes' are quasispin down, 'particles' are quasispin up. The analog of rotating spins of magnetic moments in magnetic domains is sketched.

where (as indicated) $p = 1, 2, \ldots, n$, $p' = 1, 2, \ldots, n$, $\sigma = \pm 1$, ε, V, and W are (real) constants with dimensions of energy. The first term is a sum of single-particle energies; the second and third terms are sums over two-body interaction energies. The model can be depicted as shown in figure 4.8. The model is for a many-fermion system containing n fermions (the number of states is $2n$).

The Hamiltonian can be reformulated as

$$H = \varepsilon K_0 + \frac{1}{2}V(K_+^2 + K_-^2) + \frac{1}{2}W(K_+K_- + K_-K_+), \tag{4.168}$$

where the operators K_z, $K_\pm$ are of the 'quasi-spin type' and are defined by

$$K_+ := \sum_{p=1}^{n} a_{p+1}^\dagger a_{p-1}, \tag{4.169}$$

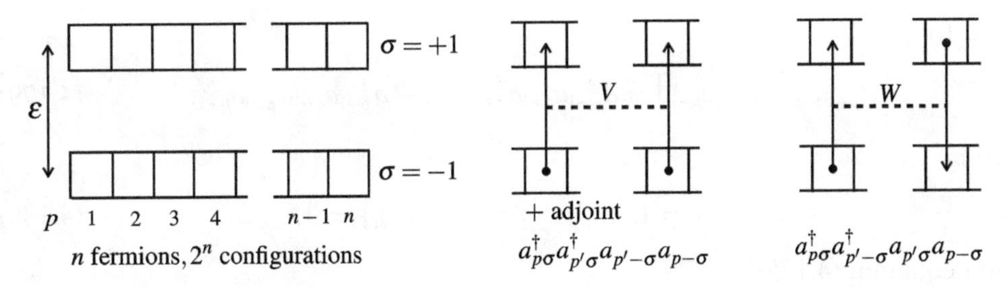

Figure 4.8. A diagrammatic representation of the Lipkin model. On the left are the available single-fermion states. The lower block have energy $-\frac{1}{2}\varepsilon$, the upper block have energy $+\frac{1}{2}\varepsilon$. There are n fermions and $2n$ single-particle states which results in 2^n n-fermion configurations. The two-body interactions are shown on the right in a manner suggestive of Feynman diagrams.

$$K_- := \sum_{p=1}^{n} a^\dagger_{p-1}a_{p+1}, \tag{4.170}$$

$$K_0 := \frac{1}{2}\sum_{p\sigma}^{n} \sigma a^\dagger_{p\sigma}a_{p\sigma}. \tag{4.171}$$

Note: throughout the '+1' and '−1' are indices separate from 'p' and label $\sigma = \pm 1$ states. The operators $K_\pm$, K_0 obey

$$[K_0, K_\pm] = \pm K_\pm,\ [K_+, K_-] = 2K_0, \tag{4.172}$$

i.e. they define an su(2) algebra.

To show that $[K_0, K_+] = K_+$:

$$[K_0, K_+] = \left[\left(\frac{1}{2}\sum_p a^\dagger_{p+1}a_{p+1} - a^\dagger_{p-1}a_{p-1}\right), \sum_q a^\dagger_{q+1}a_{q-1}\right]; \tag{4.173}$$

for $p \neq q$,

$$\left[a^\dagger_{p+1}a_{p+1}, a^\dagger_{q+1}a_{q-1}\right] = a^\dagger_{p+1}a_{p+1}a^\dagger_{q+1}a_{q-1} - a^\dagger_{q+1}a_{q-1}a^\dagger_{p+1}a_{p+1}, \tag{4.174}$$

using

$$\{a_i, a_j\} = \left\{a_i, a_j^\dagger\right\} = \left\{a_i^\dagger, a_j\right\} = \left\{a_i^\dagger, a_j^\dagger\right\} = 0, \text{ for } i \neq j, \tag{4.175}$$

$$\therefore \left[a^\dagger_{p+1}a_{p+1}, a^\dagger_{q+1}a_{q-1}\right] = \cancel{a^\dagger_{q+1}a_{q-1}a^\dagger_{p+1}a_{p+1}} - \cancel{a^\dagger_{q+1}a_{q-1}a^\dagger_{p+1}a_{p+1}}, \tag{4.176}$$

$$\therefore \left[a^\dagger_{p+1}a_{p+1}, a^\dagger_{q+1}a_{q-1}\right] = 0, \tag{4.177}$$

and similarly

$$\left[a^\dagger_{p-1}a_{p-1}, a^\dagger_{q+1}a_{q-1}\right] = 0; \tag{4.178}$$

and for $p = q$,

$$\left[a^{\dagger}_{p+1}a_{p+1},\, a^{\dagger}_{p+1}a_{p-1}\right] = a^{\dagger}_{p+1}a_{p+1}a^{\dagger}_{p+1}a_{p-1} - a^{\dagger}_{p+1}a_{p-1}a^{\dagger}_{p+1}a_{p+1}, \tag{4.179}$$

using

$$\{a_i,\, a_i^{\dagger}\} = 1,\ \{a_i^{\dagger},\, a_i^{\dagger}\} = 0,\ \{a_i,\, a_i\} = 0, \tag{4.180}$$

and equation (4.175),

$$\therefore \left[a^{\dagger}_{p+1}a_{p+1}, a^{\dagger}_{p+1}a_{p-1}\right] = a^{\dagger}_{p+1}(1 - a^{\dagger}_{p+1}a_{p+1})a_{p-1} + \cancelto{0}{a^{\dagger}_{p+1}a^{\dagger}_{p+1}}\, a_{p-1}a_{p+1}, \tag{4.181}$$

$$\therefore \left[a^{\dagger}_{p+1}a_{p+1}, a^{\dagger}_{p+1}a_{p-1}\right] = a^{\dagger}_{p+1}a_{p-1} - \cancelto{0}{a^{\dagger}_{p+1}a^{\dagger}_{p+1}}\, a_{p+1}a_{p-1}, \tag{4.182}$$

$$\therefore \left[a^{\dagger}_{p+1}a_{p+1},\, a^{\dagger}_{p+1}a_{p-1}\right] = a^{\dagger}_{p+1}a_{p-1}, \tag{4.183}$$

and

$$\left[a^{\dagger}_{p-1}a_{p-1},\, a^{\dagger}_{p+1}a_{p-1}\right] = -a^{\dagger}_{p+1}a_{p-1}. \tag{4.184}$$

$$\therefore \left[\frac{1}{2}(a^{\dagger}_{p+1}a_{p+1} - a^{\dagger}_{p-1}a_{p-1}),\, a^{\dagger}_{p+1}a_{p-1}\right] = a^{\dagger}_{p+1}a_{p-1}, \tag{4.185}$$

$$\therefore \sum_{p}\left[\frac{1}{2}(a^{\dagger}_{p+1}a_{p+1} - a^{\dagger}_{p-1}a_{p-1}),\, a^{\dagger}_{p+1}a_{p-1}\right] = \sum_{p} a^{\dagger}_{p+1}a_{p-1} = K_+, \tag{4.186}$$

whence

$$[K_0,\, K_+] = K_+. \tag{4.187}$$

Similarly,

$$[K_0,\, K_-] = -K_-. \tag{4.188}$$

To show that $[K_+, K_-] = 2K_0$:

$$[K_+,\, K_-] = \left[\sum_{p} a^{\dagger}_{p+1}a_{p-1},\, \sum_{q} a^{\dagger}_{q-1}a_{q+1}\right]; \tag{4.189}$$

for $p \neq q$,

$$\left[a^{\dagger}_{p+1}a_{p-1},\, a^{\dagger}_{q-1}a_{q+1}\right] = a^{\dagger}_{p+1}a_{p-1}a^{\dagger}_{q-1}a_{q+1} - a^{\dagger}_{q-1}a_{q+1}a^{\dagger}_{p+1}a_{p-1}, \tag{4.190}$$

$$\therefore \left[a^{\dagger}_{p+1}a_{p-1},\, a^{\dagger}_{q-1}a_{q+1}\right] = \cancel{a^{\dagger}_{q-1}a_{q+1}a^{\dagger}_{p+1}a_{p-1}} - \cancel{a^{\dagger}_{q-1}a_{q+1}a^{\dagger}_{p+1}a_{p-1}}, \tag{4.191}$$

$$\therefore \left[a^{\dagger}_{p+1}a_{p-1},\, a^{\dagger}_{q-1}a_{q+1}\right] = 0; \tag{4.192}$$

and for $p = q$,

$$\left[a^{\dagger}_{p+1}a_{p-1},\, a^{\dagger}_{p-1}a_{p+1}\right] = a^{\dagger}_{p+1}a_{p-1}a^{\dagger}_{p-1}a_{p+1} - a^{\dagger}_{p-1}a_{p+1}a^{\dagger}_{p+1}a_{p-1}, \tag{4.193}$$

$$\therefore \left[a^{\dagger}_{p+1}a_{p-1},\, a^{\dagger}_{p-1}a_{p+1}\right] = a^{\dagger}_{p+1}a_{p+1}(1 - a^{\dagger}_{p-1}a_{p-1}) - a^{\dagger}_{p-1}a_{p-1}(1 - a^{\dagger}_{p+1}a_{p+1}), \tag{4.194}$$

$$\begin{aligned}\therefore \left[a^{\dagger}_{p+1}a_{p-1},\, a^{\dagger}_{p-1}a_{p+1}\right] = a^{\dagger}_{p+1}a_{p+1} &- \cancel{a^{\dagger}_{p+1}a_{p+1}a^{\dagger}_{p-1}a_{p-1}} - a^{\dagger}_{p-1}a_{p-1} \\ &+ \cancel{a^{\dagger}_{p-1}a_{p-1}a^{\dagger}_{p+1}a_{p+1}},\end{aligned} \tag{4.195}$$

$$\therefore \left[a^{\dagger}_{p+1}a_{p-1},\, a^{\dagger}_{p-1}a_{p+1}\right] = a^{\dagger}_{p+1}a_{p+1} - a^{\dagger}_{p-1}a_{p-1}, \tag{4.196}$$

$$\therefore [K_+, K_-] = 2K_0. \tag{4.197}$$

The $su(2)$ basis states $|km_k\rangle$:

$$K^2|km_k\rangle = k(k+1)|km_k\rangle, \tag{4.198}$$

$$K_0|km_k\rangle = m_k|km_k\rangle, \tag{4.199}$$

$$K_\pm|km_k\rangle = \sqrt{(k \mp m_k)(k \pm m_k + 1)}\,|km_k \pm 1\rangle, \tag{4.200}$$

can be interpreted using

$$K_0 = \sum_p \frac{1}{2}\left(a^{\dagger}_{p+1}a_{p+1} - a^{\dagger}_{p-1}a_{p-1}\right), \tag{4.201}$$

whence

$$m_k^{\max} = \frac{1}{2}n, \quad m_k^{\min} = -\frac{1}{2}n, \tag{4.202}$$

i.e. $m_k^{\max}$ corresponds to all n fermions in $\sigma = +1$ states, $m_l^{\min}$ corresponds to all n fermions in $\sigma = -1$ states, and m_k gives the number of 'particle–hole' excitations. Evidently,

$$k = \frac{1}{2}n. \tag{4.203}$$

The number of configurations arising for a given value of n (i.e. n equals the number of fermions and half the number of single-particle states) is quite subtle. These are given in table 4.1 for values of n up to $n = 8$. Recall that for a given value of k,

$$m_k = k, k-1, \ldots, -k, \tag{4.204}$$

Table 4.1. The number of configurations in the Lipkin model as a function of n.

n	No. of configurations	Irrep labels k	Dimension of ground-state irrep
1	2	$\frac{1}{2}$	2
2	4	1, 0	3
3	8	$\frac{3}{2}, \frac{1}{2}, \frac{1}{2}$	4
4	16	2,1,1,1,0,0	5
5	32	$\frac{5}{2}, (\frac{3}{2})^4, (\frac{1}{2})^5$	6
6	64	$3, (2)^5, (1)^9, (0)^5$	7
7	128	$\frac{7}{2}, (\frac{5}{2})^6, \cdots$	8
8	256	$4, (3)^7, \cdots$	9

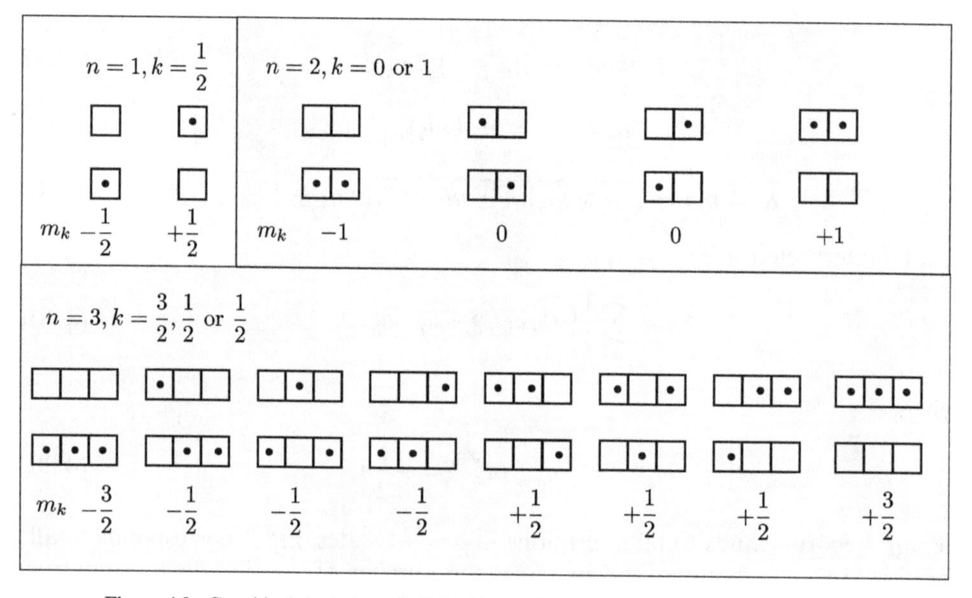

Figure 4.9. Graphical depiction of all Lipkin model configurations for n = 1, 2 and 3.

i.e. $2k + 1$ possible configurations. These configurations are depicted, for n = 1, 2, and 3, in figure 4.9. For n = 1 we could represent the $SU(2)$ basis states:

$$|km_k\rangle = |1, +1\rangle = \left|\begin{matrix}\boxed{\bullet}\boxed{\bullet}\\ \boxed{}\boxed{}\end{matrix}\right\rangle, \tag{4.205}$$

$$|10\rangle = \frac{1}{\sqrt{2}}\left|\begin{matrix}\boxed{\bullet}\boxed{}\\ \boxed{}\boxed{\bullet}\end{matrix}\right\rangle + \frac{1}{\sqrt{2}}\left|\begin{matrix}\boxed{}\boxed{\bullet}\\ \boxed{\bullet}\boxed{}\end{matrix}\right\rangle, \tag{4.206}$$

$$|00\rangle = \frac{1}{\sqrt{2}}\left|\begin{array}{|c|c|}\hline \bullet & \\ \hline & \bullet \\ \hline\end{array}\right\rangle - \frac{1}{\sqrt{2}}\left|\begin{array}{|c|c|}\hline & \bullet \\ \hline \bullet & \\ \hline\end{array}\right\rangle, \tag{4.207}$$

$$|1,-1\rangle = \left|\begin{array}{|c|c|}\hline & \\ \hline \bullet & \bullet \\ \hline\end{array}\right\rangle. \tag{4.208}$$

For $n = 2$, the operators K_+, K_-, and K_0 are:

$$K_+ = a_{1+1}^{\dagger}a_{1-1} + a_{2+1}^{\dagger}a_{2-1}, \tag{4.209}$$

$$K_- = a_{1-1}^{\dagger}a_{1+1} + a_{2-1}^{\dagger}a_{2+1}, \tag{4.210}$$

$$K_0 = \frac{1}{2}(a_{1+1}^{\dagger}a_{1+1} - a_{1-1}^{\dagger}a_{1-1} + a_{2+1}^{\dagger}a_{2+1} - a_{2-1}^{\dagger}a_{2-1}). \tag{4.211}$$

Evidently,

$$K_+|10\rangle = \sqrt{2}\,|1, +1\rangle \tag{4.212}$$

and

$$K_+|00\rangle = 0. \tag{4.213}$$

Solutions to the Lipkin model Hamiltonian (equation (4.167)) can be considered for a number of special cases:

(a) $V = 0$.

$$H_I = \varepsilon K_0 + \frac{1}{2}W(K_+K_- + K_-K_+), \tag{4.214}$$

and recalling equation (5.41),

$$\therefore\ H_I = \varepsilon K_0 + W(K^2 - K_0^2), \tag{4.215}$$

$$\therefore\ H_I|km_k\rangle = \varepsilon m_k + W[k(k+1) - m_k^2], \tag{4.216}$$

i.e. H_I shifts and splits but does not mix the $SU(2)$ basis states.

(b) $W = 0$.

$$H_{II} = \varepsilon K_0 + \frac{1}{2}V(K_+^2 + K_-^2). \tag{4.217}$$

The terms containing K_+^2 and K_-^2 mix states within an irrep, but they do not mix states from different irreps. This leads to exact solutions for $n = 1, 2, 3$ 4, 6, and 8. To proceed, the matrix elements of K_0, K_+^2 and K_-^2 in the $|km_k\rangle$ basis are needed:

$$\langle km_k|K_0|km_k\rangle = m_k, \tag{4.218}$$

$$\begin{aligned}\langle km_k + 2 \mid K_+^2 \mid km_k\rangle = &\sqrt{(k - m_k)(k + m_k + 1)}\\ &\times\sqrt{(k - m_k - 1)(k + m_k + 2)},\end{aligned} \tag{4.219}$$

$$\langle km_k - 2 \mid K_-^2 \mid km_k \rangle = \sqrt{(k + m_k)(k - m_k + 1)} \times \sqrt{(k + m_k - 1)(k - m_k + 2)}\,. \tag{4.220}$$

It then remains to construct the Hamiltonian matrix for a given value of n (recall $k = \frac{1}{2}n$) and to solve its secular equation. This task is greatly simplified by the $SU(2)$ structure of the problem. First, for a given value of n, the possible configurations can be grouped into $SU(2)$ irreps (given in table 4.1). Thus, for the 128 configurations corresponding to $n = 8$, for example, the largest $SU(2)$ irrep is only of dimension 9, i.e. it has $k = 4$. But there is a further simplification: K_+^2 and K_-^2 only mix configurations differing in m_k by ±2, that is to say the matrices to be diagonalized can be further reduced such that for a given matrix the m_k values are either all 'even' or all 'odd'. Thus for the energy eigenvalues λ:

(a) $n = 1$, $k = \frac{1}{2}$,

$$\lambda = \pm\frac{1}{2}\varepsilon. \tag{4.221}$$

(b) $n = 2$, $k = 1$ or 0,

$k = 0$,

$$\lambda = 0, \tag{4.222}$$

$k = 1$, $m_k^{\text{'even'}} = 0$,

$$\lambda = 0, \tag{4.223}$$

$k = 1$, $m_k^{\text{'odd'}} = \pm 1$,

$$\begin{pmatrix} \langle 11|H|11\rangle & \langle 11|H|1,-1\rangle \\ \langle 1,-1|H|11\rangle & \langle 1,-1|H|1,-1\rangle \end{pmatrix} = \begin{pmatrix} \varepsilon & V \\ V & -\varepsilon \end{pmatrix}, \tag{4.224}$$

$$\therefore \begin{vmatrix} \varepsilon - \lambda & V \\ V & -\varepsilon - \lambda \end{vmatrix} = 0, \tag{4.225}$$

$$\therefore \lambda = \pm\sqrt{\varepsilon^2 + V^2}\,. \tag{4.226}$$

(c) $n = 3$, $k = \frac{3}{2}, \frac{1}{2}$ or $\frac{1}{2}$,

$k = \frac{1}{2}$ (twice),

$$\lambda = \pm\frac{1}{2}\varepsilon \text{ (twice)}. \tag{4.227}$$

$k = \frac{3}{2}$, $m_k^{\text{'even'}} = +\frac{3}{2}, -\frac{1}{2}$,

$$\begin{pmatrix} \left\langle \frac{3}{2}\frac{3}{2} \middle| H \middle| \frac{3}{2}\frac{3}{2} \right\rangle & \left\langle \frac{3}{2}\frac{3}{2} \middle| H \middle| \frac{3}{2}, -\frac{1}{2} \right\rangle \\ \left\langle \frac{3}{2}, -\frac{1}{2} \middle| H \middle| \frac{3}{2}\frac{3}{2} \right\rangle & \left\langle \frac{3}{2}, -\frac{1}{2} \middle| H \middle| \frac{3}{2}, -\frac{1}{2} \right\rangle \end{pmatrix} = \begin{pmatrix} \frac{3}{2}\varepsilon & \sqrt{3}\,V \\ \sqrt{3}\,V & -\frac{1}{2}\varepsilon \end{pmatrix}, \tag{4.228}$$

$$\therefore \begin{vmatrix} \frac{3}{2}\varepsilon - \lambda & \sqrt{3}\,V \\ \sqrt{3}\,V & -\frac{1}{2}\varepsilon - \lambda \end{vmatrix} = 0, \tag{4.229}$$

$$\therefore \lambda = \frac{1}{2}\varepsilon \pm \sqrt{\varepsilon^2 + 3V^2}. \tag{4.230}$$

$k = \frac{3}{2}$, $m_k^{\text{'odd'}} = +\frac{1}{2}, -\frac{3}{2}$,

$$\begin{pmatrix} \left\langle \frac{3}{2}\frac{1}{2} \middle| H \middle| \frac{3}{2}\frac{1}{2} \right\rangle & \left\langle \frac{3}{2}\frac{1}{2} \middle| H \middle| \frac{3}{2}, -\frac{3}{2} \right\rangle \\ \left\langle \frac{3}{2}, -\frac{3}{2} \middle| H \middle| \frac{3}{2}\frac{1}{2} \right\rangle & \left\langle \frac{3}{2}, -\frac{3}{2} \middle| H \middle| \frac{3}{2} - \frac{3}{2} \right\rangle \end{pmatrix} = \begin{pmatrix} \frac{1}{2}\varepsilon & \sqrt{3}\,V \\ \sqrt{3}\,V & -\frac{3}{2}\varepsilon \end{pmatrix}, \tag{4.231}$$

and this is identical to $k = \frac{3}{2}$, $m_k^{\text{'even'}}$ with $\varepsilon \to -\varepsilon$,

$$\therefore \lambda = -\frac{1}{2}\varepsilon \pm \sqrt{\varepsilon^2 + 3V^2}. \tag{4.232}$$

(d) $n = 4$, $k = 2, 1, 1, 1, 0, 0$,
$k = 0$ (twice),

$$\lambda = 0 \text{ twice}. \tag{4.233}$$

$k = 1$ (three times),

$$\lambda = 0, \pm\sqrt{\varepsilon^2 + V^2} \text{(adopted from equations (4.223) and (4.266))}. \tag{4.234}$$

$k = 2$, $m_k^{\text{'even'}} = +2, 0, -2$,

$$\begin{pmatrix} \langle 22|H|22\rangle & \langle 22|H|20\rangle & \langle 22|H|2,-2\rangle \\ \langle 20|H|22\rangle & \langle 20|H|20\rangle & \langle 20|H|2,-2\rangle \\ \langle 2,-2|H|22\rangle & \langle 2,-2|H|20\rangle & \langle 2,-2|H|2,-2\rangle \end{pmatrix} = \begin{pmatrix} 2\varepsilon & \sqrt{6}\,V & 0 \\ \sqrt{6}\,V & 0 & \sqrt{6}\,V \\ 0 & \sqrt{6}\,V & -2\varepsilon \end{pmatrix}, \tag{4.235}$$

$$\therefore \begin{vmatrix} 2\varepsilon - \lambda & \sqrt{6}\,V & 0 \\ \sqrt{6}\,V & -\lambda & \sqrt{6}\,V \\ 0 & \sqrt{6}\,V & -2\varepsilon - \lambda \end{vmatrix} = 0, \tag{4.236}$$

$$\therefore \lambda = 0, \pm 2\sqrt{\varepsilon^2 + 3V^2}. \tag{4.237}$$

$k = 2$, $m_k^{\text{'odd'}} = +1, -1$,

$$\begin{pmatrix} \langle 21|H|21\rangle & \langle 21|H|2, -1\rangle \\ \langle 2, -1|H|21\rangle & \langle 2, -1|H|2, -1\rangle \end{pmatrix} = \begin{pmatrix} \varepsilon & 3V \\ 3V & -\varepsilon \end{pmatrix}, \tag{4.238}$$

$$\therefore \begin{vmatrix} \varepsilon - \lambda & 3V \\ 3V & -\varepsilon - \lambda \end{vmatrix} = 0, \tag{4.239}$$

$$\therefore \lambda = \pm\sqrt{\varepsilon^2 + 9V^2}. \tag{4.240}$$

Note that the lowest (ground-state) energies occur, in each case, for the maximum value of k. Thus, for $n \geqslant 5$ only the maximum value of k is considered (the lower values of k have already been solved for lower n).

(e) $n = 5$, $k^{\max} = \frac{5}{2}$,

$k = \frac{5}{2}$, $m_k^{\text{'even'}} = +\frac{5}{2}, +\frac{1}{2}, -\frac{3}{2}$,

$$\begin{pmatrix} \frac{5}{2}\varepsilon & \sqrt{10}\,V & 0 \\ \sqrt{10}\,V & \frac{1}{2}\varepsilon & \sqrt{18}\,V \\ 0 & \sqrt{18}\,V & -\frac{3}{2}\varepsilon \end{pmatrix} \leftrightarrow H \to \text{must be solved numerically.} \tag{4.241}$$

$k = \frac{5}{2}$, $m_k^{\text{'odd'}} = +\frac{3}{2}, -\frac{1}{2}, -\frac{5}{2}$,

$$\begin{pmatrix} \frac{3}{2}\varepsilon & \sqrt{18}\,V & 0 \\ \sqrt{18}\,V & -\frac{1}{2}\varepsilon & \sqrt{10}\,V \\ 0 & \sqrt{10}\,V & -\frac{5}{2}\varepsilon \end{pmatrix} \leftrightarrow H \to \text{must be solved numerically.} \tag{4.242}$$

(f) $n = 6$, $k^{\max} = 3$, $k = 3$, $m_k^{\text{'even'}} = +3, +1, -1, -3$,

$$\begin{pmatrix} 3\varepsilon & \sqrt{15}\,V & 0 & 0 \\ \sqrt{15}\,V & \varepsilon & 6V & 0 \\ 0 & 6V & -\varepsilon & \sqrt{15}\,V \\ 0 & 0 & \sqrt{15}\,V & -3\varepsilon \end{pmatrix} \leftrightarrow H, \tag{4.243}$$

$$\lambda = \pm\sqrt{5\varepsilon^2 + 33V^2 \pm \sqrt{\varepsilon^4 + 6V^2\varepsilon^2 + 54V^4}}. \quad (4.244)$$

$k = 3$, $m_k^{\text{`odd'}} = +2, 0, -2$,

$$\begin{pmatrix} 2\varepsilon & \sqrt{30}\,V & 0 \\ \sqrt{30}\,V & 0 & \sqrt{30}\,V \\ 0 & \sqrt{30}\,V & -2\varepsilon \end{pmatrix} \leftrightarrow H, \quad (4.245)$$

$$\lambda = 0, \pm 2\sqrt{\varepsilon^2 + 15V^2}. \quad (4.246)$$

(g) $n = 8$, $k^{\max}=4$,

$$\begin{aligned} \lambda = 0, &\pm\sqrt{10\varepsilon^2 + 118V^2 \pm \sqrt{\varepsilon^4 - 2V^2\varepsilon^2 + 225V^4}}, \\ &\pm\sqrt{5\varepsilon^2 + 113V^2 \pm 4\sqrt{\varepsilon^4 + 38V^2\varepsilon^2 + 550V^4}}. \end{aligned} \quad (4.247)$$

Reference

[1] Lipkin H J, Meshkov N and Glick A J 1965 *Nucl. Phys.* **62** 188

Chapter 5

Group theory and quantum mechanics

Following the definition of a group, the role of group theory and symmetry in quantum mechanics is introduced. This is handled by way of presenting a synoptic view of quantum mechanics, 'seen through a group theoretical lens', especially the view of angular momentum theory. Many of the concepts developed in Volume 1 and the preceding chapters, herein, have been shaped to arrive at this exposition of group theory. The distinction between discrete and continuous groups is clarified. Details of Lie groups and their associated algebras, suitable for quantum mechanical applications, are presented. The fundamental concept of a group generator is emphasized. The use of Lie algebras for model building is explained, especially spectrum generating groups and their associated algebras (also known as dynamical algebras). Representation of groups is introduced, including matrices and polynomial functions. The distinction between compact and non-compact groups is explained. The groups $SO(\mathrm{n})$ and $SU(\mathrm{n})$ are introduced, together with their Casimir operators.

Concepts: group definition; Abelian group; translation group; rotation group; space–time transformations; symmetry transformations in quantum mechanics; discrete groups; continuous groups; matrix groups; Lie groups; Lie algebras; group generator; dynamical group; dynamical algebra; spectrum generating algebra (SGA); compact groups; non-compact groups; $SO(3)$; $SU(2)$; $U(n)$; $SO(n)$; commutator bracket tables; Casimir operators.

Group theory plays a fundamental role in the processes of the world around us: objects get translated and rotated. If a book on a closely-packed shelf has its title upside down, one takes the book out (translation, T) turns the title right-side up (rotation, R) and returns the book to the shelf (inverse translation, T^{-1}): $T^{-1}RT$—a *similarity transformation*. In the quantum world we are especially concerned with rotations in the space around us, $(3, \mathbb{R})$ and in Hilbert space, $(n, \mathbb{C})$, i.e. the complex linear space in n dimensions, where n is the number of distinct observable results of measurements defined by a maximal set of commuting operators.

doi:10.1088/978-0-7503-2171-6ch5

In this chapter, a basic formal outline is given of the essential elements of group theory, and its role in quantum mechanics. Much of this material is a restatement of forgoing subject matter. The intent is rather akin to a first formal presentation of language, when one is introduced to the fact that one is using ‘prose’ and that prose can be broken down into rules of grammar, etc. One does not need to know the rules of grammar to converse; but to write, it helps to know a few rules. To take the adage ‘the pen is mightier than the sword’, at some risk, we pose: ‘group theory is mightier than blind numerical computation’.

The role of group theory in quantum mechanics enters at two levels: its role in transformation theory, and its role in the algebra of infinitesimal group generators. The algebras of group generators are introduced here, and this is formally developed in the next chapter (chapter 6).

5.1 Definition of a group

A group is a mathematical structure defined by the following: It is a set of elements $\{G_i\}$, where i may be discrete or continuous, such that

I.

$$G_j \circ G_k = G_l; \tag{5.1}$$

where the symbol ‘$\circ$’ denotes the group ‘product’ which may be multiplication or addition or some other law of mathematical combination; and G_j, G_k, $G_l \in \{G_i\}$. This is called *closure*.

II.

$$G_j \circ (G_k \circ G_l) = (G_j \circ G_k) \circ G_l. \tag{5.2}$$

This is called *associativity*.

III.

$$G_j \circ I = G_j. \tag{5.3}$$

I is called the *identity*. $I \in \{G_i\}$.

IV.

$$G_j \circ G_j^{-1} = I. \tag{5.4}$$

G_j^{-1} is called the *inverse* of G_j. $G_j^{-1} \in \{G_i\}$.

It is not necessary that $G_i \circ G_j = G_j \circ G_i$, i.e. that the group product be commutative.

Definition of an Abelian group

An *Abelian* group is a group with a commutative product.

5.2 Groups and transformation

Transformations of physical structures (or mathematical structures) naturally give rise to realisations of groups.

5.2.1 Translations

Consider a quantity, $f(x)$, which depends on a position coordinate x. Then, for a translation along the x-axis through a distance α described by the operation $T_x(\alpha)$:

$$T_x(\alpha)f(x) = f(x - \alpha). \tag{5.5}$$

Thus,

$$T_x(\beta)T_x(\alpha)f(x) = T_x(\beta)f(x - \alpha) = f(x - \alpha - \beta) = T_x(\alpha + \beta)f(x), \tag{5.6}$$

$$\therefore\ T_x(\beta)T_x(\alpha) = T_x(\alpha + \beta), \tag{5.7}$$

which manifestly expresses the group closure axiom. The other group axioms follow directly for translations.

- The elements of translation groups, so expressed, form a continuous set.
- Translation groups with discrete sets of elements arise, e.g. when considering translations that leave crystalline lattices unchanged.
- Translation groups are directly extended to any number of spatial dimensions, e.g.

$$T_y(b)T_x(a) = T_n(c), \tag{5.8}$$

  where n and c are uniquely defined.
- Translation groups are Abelian.

5.2.2 Rotations

Rotations can be defined in two or more dimensions.

(a) Rotations in a plane: e.g.

$$R_z(\phi_2)R_z(\phi_1) = R_z(\phi_2 + \phi_1), \tag{5.9}$$

where z is an axis perpendicular to the plane and ϕ_2 and ϕ_1 are angles. Such rotations form an Abelian group. Moreover, there is a '*compactness*' to rotations, i.e.

$$R_z(\phi + 2\pi) = R_z(\phi). \tag{5.10}$$

(b) Rotations in three dimensions: e.g.

$$R_y(\chi)R_x(\phi) = R_n(\psi), \tag{5.11}$$

where n and ψ are uniquely defined, form a group. They are compact, but they are *non-Abelian* (see figure 5.1).

5.2.3 Space–time transformation

Translations and rotations in three dimensions play a vital role in the discussion of *physical space*. Physical space together with time constitutes *space–time* or *Minkowski* space. Minkowski space possesses a rich variety of transformations:

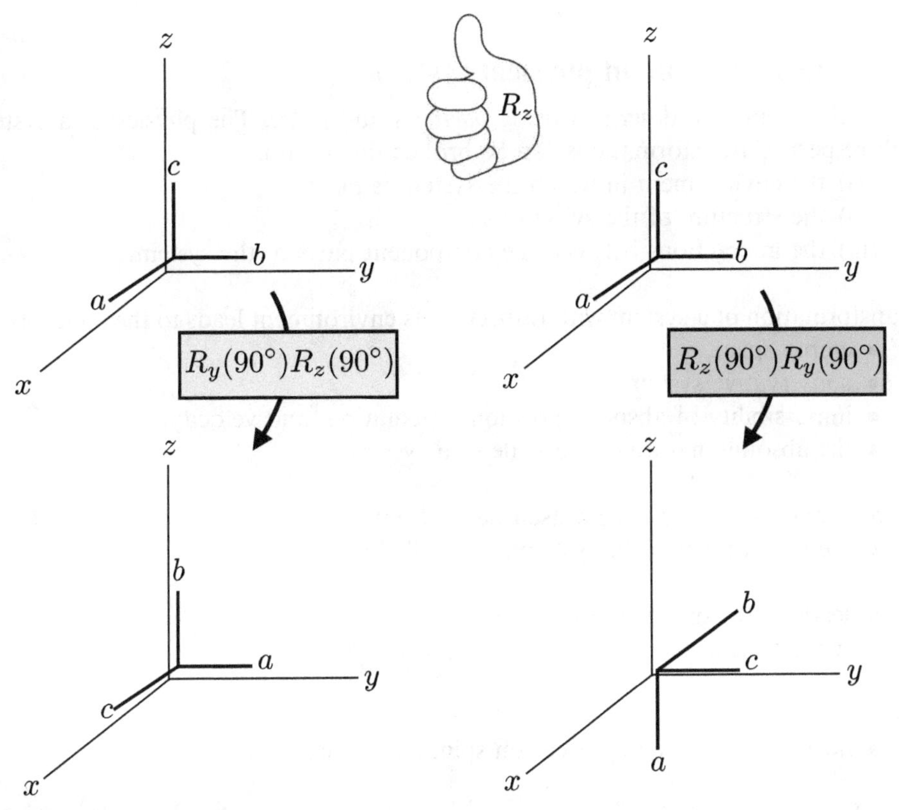

Figure 5.1. Rotation of a moving frame (a, b, c) with respect to a fixed frame (x, y, z). On the left the sequence 'rotate counterclockwise through 90° about the z-axis followed by rotate through 90° about the y-axis' is illustrated. On the right the sequence 'rotate counterclockwise through 90° about the y-axis followed by rotate through 90° about the z-axis' is illustrated. The sequential operations do not commute. The 'right-hand' rule is depicted to help visualise the rotations.

Translations (T)
Rotations (R) Space inversion (π)
Time translations (τ)
Lorentz boosts[1] (Λ) Time reversal ($\mathcal{T}$)
The following groupings of the above have special names:
Poincaré or *inhomogeneous Lorentz* group
$T, R, \tau, \Lambda, \pi, \mathcal{T}$
(Homogeneous) *Lorentz* group
R, Λ, π, τ
Proper orthochronous Lorentz group
R, Λ
Galilean group
$\Lambda_{\text{non-relativistic}}$ (non-relativistic: $v \ll c$, i.e. $\gamma = 1$)
Euclidean group in three dimensions
R, T.

[1] $x' = \gamma(x - v_x t)$, $\gamma = \left(1 - \frac{v_x^2}{c^2}\right)^{-\frac{1}{2}}$, $t' = \gamma\left(t - \frac{v_x}{c^2}x\right)$ for an x-direction boost.

5.3 Transformation on physical systems

Physical systems are described using *particles* and *fields*. The physics of a system with respect to transformations can be broken down into:

(i) the environment in which the system is placed,
(ii) the structure of the system itself,
(iii) the interactions between the component parts of the system.

Transformation of a system with respect to its environment leads to the concepts of:

- empty space,
- conservative systems,
- impossibility of absolute position, orientation, and velocity,
- the absolute nature of the scale of a system.

Transformations of the system itself depend on:

- the level at which the system is modelled;

which leads to the distinction between

- space–time structure;

and

- intrinsic structure (e.g. electron spin, electric charge).

Transformations of the space–time structure of a system lead to the concepts of:

- phase space,
- equations of motion,
- permutations of particles,
- shape,
- normal coordinates,
- constants of motion.

Transformations of the intrinsic structure of a system lead to the concept of:

- gauge.

Transformation of the interactions between the component parts of the system reveals:

- the equations of motion,
- the constants of motion,

and in complex many-body systems leads to

- the collective modes of motion.

5.4 Quantum mechanics: a synoptic view

- Quantum mechanics is a highly successful theory of the (mechanical) behaviour of small physical systems.
- It is complete, theoretically, i.e. it is able to describe all known phenomena observed in small physical systems (and it is essential to the understanding of certain macroscopic physical systems, e.g. superconductors, superfluids, quantum Hall effect devices).

The axioms on which quantum mechanics is founded can be stated in a number of ways. The following is neither unique nor, in its entirety, necessary, but it is sufficient:

- The *state* of a quantum system is completely described by an *element* (*state vector*) of a *Hilbert*[2] *space*.
- If the number of distinguishable states of the system is n, the Hilbert space is n dimensional.
- The *physical quantities* (*dynamical variables*) associated with a system are represented by *Hermitian operators*. These operators act on the elements of the Hilbert space describing the system, to yield other elements of the space.
- If an operator $\hat{A}$ acting on an element ψ_α of the Hilbert space obeys

$$\hat{A}\psi_\alpha = \alpha\psi_\alpha,$$

  then ψ_α represents a state of the system with the value α for the physical quantity represented by $\hat{A}$, where α is a real number. This equation is an *eigenvalue equation* with eigenvalue α and *eigenvector* ψ_α. Hermitian operators have real eigenvalues.
- If two *distinguishable* states of a system have the same value α for a physical quantity A then there is a second physical quantity B with respect to which the states ψ_{α_1}, ψ_{α_2} obey

$$\hat{B}\psi_{\alpha_1} = \beta_1\psi_{\alpha_1},\ \hat{B}\psi_{\alpha_2} = \beta_2\psi_{\alpha_2},\ \beta_1 \neq \beta_2;$$

  - The states ψ_{α_1} and ψ_{α_2} are said to be *degenerate* with respect to the physical quantity A.
  - The physical quantities A and B are said to be *compatible*.
  - The operators $\hat{A}$ and $\hat{B}$ *commute*.
- Different physical quantities, for a given physical system, are not necessarily compatible: if this occurs, then the operators representing the incompatible quantities *do not commute*.
- The number of physical quantities that provide the distinction between the different states of the system must be determined by experiment; and these

[2] A complex linear space with finite inner products.

physical quantities are represented by a set of commuting operators called *a complete set of commuting operators.*

- The *scale* in which our world is quantized must be determined by experiment and is described by a single parameter, commonly chosen to be

$$1.054\,571\,817(\text{exact}) \times 10^{-34}\ \text{J s} \equiv \hbar.$$

- The manner in which the scale factor is incorporated into the axioms of quantum mechanics must also be determined by experiment. However, the exact manner of this incorporation is extremely abstract.
  - The most commonly used axioms for incorporating $\hbar$ into quantum mechanics are:

$$[\hat{r}_\alpha, \hat{p}_\beta] := \hat{r}_\alpha \hat{p}_\beta - \hat{p}_\beta \hat{r}_\alpha = i\hbar \hat{I} \delta_{\alpha\beta};\ \alpha, \beta = x, y, z, \tag{5.12}$$

where x, y and z span the three spatial dimensions of Minkowski space and $\delta_{\alpha\beta} = 1$ if $\alpha = \beta$, $\delta_{\alpha\beta} = 0$ if $\alpha \neq \beta$; and

$$[\hat{S}_\alpha, \hat{S}_\beta] := i\hbar \varepsilon_{\alpha\beta\gamma} \hat{S}_\gamma;\ \alpha, \beta, \gamma = x, y, z, \tag{5.13}$$

$$\varepsilon_{xyz} = \varepsilon_{yzx} = \varepsilon_{zxy} = +1,$$
$$\varepsilon_{yxz} = \varepsilon_{xzy} = \varepsilon_{zyx} = -1,$$

where $\hat{S}_x$, $\hat{S}_y$, and $\hat{S}_z$ are the three components of intrinsic spin.

- The operators of quantum mechanics are formed from the dynamical variables of classical mechanics as follows:

$$r^n_{\alpha,\text{class}} \to \hat{r}^n_\alpha,\ p_{\alpha,\text{class}} \to \hat{p}^n_\alpha,\ \alpha = x, y, z, \tag{5.14}$$

$$r^n_{\alpha,\text{class}} p^m_{\alpha,\text{class}} \to \hat{r}^n_\alpha \hat{p}^m_\alpha\Big)_{\text{symmetrized}}, \tag{5.15}$$

$$\text{e.g. } r_\alpha p_\alpha \to \frac{1}{2}(\hat{r}_\alpha \hat{p}_\alpha + \hat{p}_\alpha \hat{r}_\alpha). \tag{5.16}$$

The states defined by the preceding axioms are *stationary states*, i.e. they do not change with time.

Quantum mechanical systems may also evolve in time. There are a variety of ways of incorporating time dependence into the description of quantum mechanical systems. The two commonly used methods are the *Schrödinger picture* and the *Heisenberg picture*:

- In the Schrödinger picture, time dependence is incorporated into the state vectors according to

$$i\hbar \frac{\partial}{\partial t} \psi(t) = \hat{H} \psi(t), \tag{5.17}$$

where $\hat{H}$ is the Hamiltonian operator.

- In the Heisenberg picture, time dependence is incorporated into the operators according to

$$\frac{d\hat{A}(t)}{dt} = \frac{1}{i\hbar}[\hat{A}(t), \hat{H}] + \frac{\partial \hat{A}(t)}{\partial t}, \tag{5.18}$$

where $\hat{A}$ is an operator representing a dynamical variable, and the term $\frac{\partial \hat{A}(t)}{\partial t}$ is only non-zero if A has a classical time dependence.

There is one aspect of quantum mechanics, which has a time dependence, for which no description is possible: the act of measurement! Herein lies the inherent uncertainty in quantum mechanics. Specifically, one is dealing with probabilities and this necessitates accumulating statistical distributions of quantities: these distributions may be 'sharp' or 'diffuse'. Great caution is needed in attempting to qualify uncertainty, in the sense that paradoxes arise if one supposes that one can talk or think in terms of a single measurement on a given system. See below and see the footnote on p. 8-1 for some further comments.

The probabilistic nature and consequent uncertainties of quantum mechanics are incorporated using the following axioms:

- For a system in the state ψ_α, the *probability of observing* the system to be in the state ψ_β is $|(\psi_\alpha, \psi_\beta)|^2$, where $(\psi_\alpha, \psi_\alpha) = (\psi_\beta, \psi_\beta) = 1$ and '(,)' denotes an inner product.
- For a system in the state ψ, the *expectation value* of the physical quantity A, $\langle A \rangle$ is given by

$$\langle A \rangle = (\psi, \hat{A}\psi), \tag{5.19}$$

where $(\psi, \psi) = 1$.
- Although the language used suggests that it is applicable to individual quantum systems, the probabilistic nature of quantum mechanics only makes sense when comparison with experiment involves many measurements either repeatedly on an individual system or on many identical copies of that system. To allow for experimental uncertainty (i.e. non-quantum mechanical uncertainty) the formalism of density matrices must be used (cf. Volume 1, section 7.4).
- A useful expression for uncertainty in a dynamical variable A with respect to a state represented by the vector ψ is $(\psi, \hat{A}^2\psi) - (\psi, \hat{A}\psi)^2$. It is variously called the *dispersion* of A or the *variance* of A or the *mean-square deviation* in A.

For a system consisting of many identical particles:

- The operation of permutation of the ith and jth particles, $\hat{P}_{ij}$ results in

$$\hat{P}_{ij}\psi = +\psi \text{ for bosons}, \tag{5.20}$$

$$\hat{P}_{ij}\psi = -\psi \text{ for fermions}, \tag{5.21}$$

where ψ is a many-particle state vector.

- The quantum mechanics of many-particle systems is conveniently formulated using the *occupation number representation.* A state with n_1 identical particles in state 1, n_2 identical particles in state 2, etc., is written

$$(a_1^\dagger)^{n_1}(a_2^\dagger)^{n_2}\ldots|0\rangle,$$

where the operator $a_1^\dagger$ creates a particle in state 1, etc., and $|0\rangle$ is the *particle vacuum*, i.e. the state corresponding to no particles; $\langle 0|0\rangle = 1$.

- The particle creation operators, $a_i^\dagger$ and their corresponding annihilation operators, a_i obey

$$\left[a_i, a_j^\dagger\right] = \delta_{ij}, [a_i, a_j] = 0, \left[a_i^\dagger, a_j^\dagger\right] = 0, \forall\, i, j \tag{5.22}$$

for bosons; and

$$\left\{a_i, a_j^\dagger\right\} = \delta_{ij}, \{a_i, a_j\} = 0, \left\{a_i^\dagger, a_j^\dagger\right\} = 0, \forall\, i, j \tag{5.23}$$

for fermions, where

$$\{a_i, a_j\} := a_i a_j + a_j a_i \tag{5.24}$$

is called an *anticommutator bracket.*

5.5 Symmetry transformations in quantum mechanics

Transformations of quantum mechanical systems are of great interest when there is a symmetry, i.e. *when the transformation leaves the system unchanged.*

Because the state of a quantum mechanical system is completely described by an element of a Hilbert space, there is a *one-to-one correspondence* between distinct physical transformations of a system and transformations of its Hilbert space.

Both the transformations in physical space (on the physical system) and the transformations in Hilbert space (on the state vectors describing the physical system) are *fully-equivalent realisations* of a group.

In quantum mechanics the observable quantities are probabilities and expectation values:

- A *probability* is invariant under a transformation, $\hat{U}$ of a state vector, $\psi' = \hat{U}\psi$ if

$$(\psi', \psi') = (\hat{U}\psi, \hat{U}\psi) = (\psi, \hat{U}^\dagger\hat{U}\psi) = (\psi, \psi), \tag{5.25}$$

i.e. if

$$\hat{U}^\dagger\hat{U} = \hat{I}. \tag{5.26}$$

Thus, the symmetry operators of quantum mechanics are *unitary*. In quantum mechanics, we are only interested in *unitary representations* of the groups that act in Hilbert space.

- An *expectation value* is invariant under $\hat{U}$ if

$$(\psi', \hat{A}\psi') = (\hat{U}\psi, \hat{A}\hat{U}\psi) = (\psi, \hat{U}^{\dagger}\hat{A}\hat{U}\psi) = (\psi, \hat{A}\psi), \tag{5.27}$$

i.e. if

$$\hat{U}^{\dagger}\hat{A}\hat{U} = \hat{A}. \tag{5.28}$$

This is a *similarity transformation* that obeys

$$[\hat{A}, \hat{U}] = 0. \tag{5.29}$$

5.5.1 The unitary transformations for translations, rotations, and time evolution in quantum mechanics

- Translations[3] of state vectors are described by

$$\hat{T}(\vec{a}) = \exp\left(\frac{-i\vec{p}_{\text{op}} \cdot \vec{a}}{\hbar}\right), \tag{5.30}$$

where $\vec{p}_{\text{op}} = (\hat{p}_x, \hat{p}_y, \hat{p}_z)$ and the displacement is $\vec{a}$.

- Rotations of state vectors are described by

$$\hat{R}(\vec{n}, \phi) = \exp\left(\frac{-i\hat{J}_{\text{op}} \cdot \vec{n}\phi}{\hbar}\right), \tag{5.31}$$

where $\vec{J}_{op} = (\hat{J}_x, \hat{J}_y, \hat{J}_z)$, $\vec{n}$ is a unit vector along the axis of rotation and ϕ is the angle of rotation. The operators, $\hat{J}_x$, $\hat{J}_y$, $\hat{J}_z$ may be components of angular momentum, e.g. $\hat{J}_x = \hat{L}_x = \hat{y}\hat{p}_z - \hat{z}\hat{p}_y$; or they may be components of spin, e.g. $\hat{J}_x = \hat{S}_x$; or they may be the sum of angular momentum and spin, e.g. $\hat{J}_x = \hat{L}_x + \hat{S}_x$. Spin and angular momentum obey $[\hat{S}_i, \hat{L}_j] = 0$, $i, j = x, y, z$.

- Time evolution of state vectors is described by

$$\hat{U}(t) = \exp\left(\frac{-i\hat{H}t}{\hbar}\right), \tag{5.32}$$

where $\hat{H}$ is the Hamiltonian and t is the time change.

[3] For $\hat{T}_x(a)\psi(x) = \psi(x - a)$, consider a Taylor series expansion of $\psi(x - a)$:

$$\begin{aligned}\psi(x - a) &= \left(1 - a\frac{\mathrm{d}}{\mathrm{d}x} + \frac{a^2}{2!}\frac{\mathrm{d}^2}{\mathrm{d}x^2} + \cdots\right)\psi(x) \\ &= \exp\left(-a\frac{\mathrm{d}}{\mathrm{d}x}\right)\psi(x) \rightarrow \frac{i\hat{p}_x}{\hbar} = \frac{\mathrm{d}}{\mathrm{d}x}.\end{aligned}$$

5.5.2 Consequences of symmetry in quantum mechanics

If a quantum mechanical system possesses a symmetry, the system will exhibit the following properties:

- The system will possess *degeneracy*. This is seen to follow from $(\psi', \hat{A}\psi') = (\psi, \hat{A}\psi)$ and $\psi' = \hat{U}\psi$, i.e. there is more than one state with the same expectation value of a particular physical quantity.
- The system may possess *constants of motion*. These occur when the Hamiltonian is invariant under $\hat{U} = \exp(\frac{-i\hat{K}k}{\hbar})$; whence

$$[\hat{H}, \hat{K}] = 0$$

and from

$$\frac{\mathrm{d}\hat{K}}{\mathrm{d}t} = \frac{1}{i\hbar}[\hat{K}, \hat{H}] + \frac{\partial \hat{K}}{\partial t} \tag{5.33}$$

if $\frac{\partial \hat{K}}{\partial t} = 0$ then $\frac{\mathrm{d}\hat{K}}{\mathrm{d}t} = 0$ and K is a constant with respect to time.

The practical consequences of the above are:

- The Hilbert space will be reduced to degenerate subspaces. The characterising eigenvalue of this subspace can be obtained by solving a single eigenvalue problem for any one of the state vectors within the subspace. This will solve the particular eigenvalue problem for all the other state vectors in the subspace. Judicious choice of a state in the subspace may facilitate the solution. The constants of motion will provide labelling quantum numbers. These can be found by searching for dynamical variables that commute with the Hamiltonian. (A subspace may possess a sub-subspace, and so on: this will lead to a set of characterising eigenvalues.)

There are more subtle consequences of symmetry in quantum mechanics. These consequences are briefly outlined here. Their details require considerable discussion and constitute a major theme in the application of quantum mechanics to physical systems.

Symmetry in quantum mechanics leads to simple properties of matrix elements. The embodiment of this is contained in the so-called *Wigner–Eckart theorem*:

- If the Hilbert space describing a quantum mechanical system possesses degenerate subspaces then:
    (i) Matrix elements between pairs of states, both of which are in the same subspace, can be expressed as a common factor (which is characteristic of the operator and the subspace) multiplied by a number which is a *coupling coefficient* corresponding to a well-defined *Kronecker product*. All such matrix elements are reducible to simple multiples of each other, where the multiples are ratios of coupling coefficients.

(ii) A similar property to (i) is possessed by matrix elements between states that lie in different subspaces (the initial state in one subspace, the final state in the other).
(iii) For a given operator, some pairs of subspaces are connected by non-zero matrix elements, other pairs of subspaces are connected only by zero matrix elements: such features of quantum mechanical systems are referred to as *selection rules*.

Besides quantum mechanical systems possessing symmetry, there are quantum mechanical systems that possess *approximate symmetry*. Such systems have Hamiltonians of the form

$$H = H_0 + V,$$

where H_0 possesses symmetry and V breaks this symmetry, but $V \ll H_0$. The consequences of this are that:

- The system possesses approximately degenerate subspaces, approximately good quantum numbers, and approximate selection rules.
- The symmetry breaking effects of V can be calculated in a straightforward manner using the *Wigner–Eckart theorem* and *perturbation theory*.

There is one symmetry, exhibited by many quantum systems, that heavily dominates the subject of symmetry in quantum mechanics: rotational symmetry and the conservation of angular momentum. This topic contains many of the paradigms of symmetry in quantum mechanics and its consequences. There is one symmetry, exhibited by *many-body quantum systems*, that overwhelmingly dominates the subject of symmetry in quantum mechanics: permutational symmetry. All many-fermion states must be *antisymmetric*.

5.6 Models with symmetry in quantum mechanics

For many-body quantum systems, we generally resort to *model descriptions*. These descriptions are simpler than the actual physical systems, and they are *solvable*. Some examples are:

- In the quantum mechanics of the *hydrogen atom*, the quarks in the nucleus are modelled as a single particle, the proton, to which certain intrinsic properties are ascribed—spin, charge, magnetic moment, mass, and charge radius.
- In the quantum mechanics of *molecular rotations and vibrations*, the electrons in the atoms are ignored and the molecule is modelled as an array of masses (atoms) with inter-atomic potentials (chemical bonds). Intrinsic properties include bond lengths, bond angles, inter-atomic potential parameters. Many of the consequences of specified bond lengths and angles are succinctly described by a *molecular point group*—the group of transformations which keep at least one point in the molecule fixed and which leave the molecule unchanged.

- In the quantum mechanics of *crystalline lattice vibrations*, a similar approach to molecular vibrations is taken. The *crystallographic groups* are groups of rotations, reflections, translations and combinations of these that leave crystal lattices invariant.
- In the quantum mechanics of *nuclear rotations and vibrations*, the nucleons can be ignored and the nucleus is then modelled as a droplet of fluid with a well-defined surface and shape. It is also possible to model the inertial flows of the fluid, e.g. rigid flow or irrotational flow. In such a system, the transformations that leave the shape and current flows of the nucleus invariant constitute the *intrinsic symmetry group* of the nucleus.

The above examples of modelling and symmetry can be regarded as '*geometric*' in content. But *algebraic* modelling is also possible and, indeed, is the area of greatest activity nowadays.

Permutational symmetry manifestly lacks a geometrical content: but it can be given a geometrical representation, e.g. by putting objects in boxes.

5.7 Groups and algebras

Consider a Hamiltonian which is invariant with respect to the unitary transformations

$$\hat{U}_1 = \exp\left(\frac{-i\hat{K}_1 k_1}{\hbar}\right),\ \hat{U}_2 = \exp\left(\frac{-i\hat{K}_2 k_2}{\hbar}\right), \tag{5.34}$$

then

$$[\hat{H}, \hat{K}_1] = 0,\ [\hat{H}, \hat{K}_2] = 0, \tag{5.35}$$

and if $\frac{\partial \hat{K}_1}{\partial t} = 0, \frac{\partial \hat{K}_2}{\partial t} = 0$, then $\frac{\mathrm{d}\hat{K}_1}{\mathrm{d}t} = 0, \frac{\mathrm{d}\hat{K}_2}{\mathrm{d}t} = 0$ and K_1 and K_2 are constants of motion. Then, from

$$[\hat{H}, [\hat{K}_1, \hat{K}_2]] = [\hat{H}, \hat{K}_1\hat{K}_2] - [\hat{H}, \hat{K}_2\hat{K}_1] = 0, \tag{5.36}$$

$[\hat{K}_1, \hat{K}_2]$ is also a constant of motion. If the set of all operators $\{\hat{K}_i\}$ which are constants of motion obey

$$[\hat{K}_j, \hat{K}_k] = \sum_l c^l_{jk}\hat{K}_l,\ \hat{K}_l \in \{\hat{K}_i\}\ \forall j, k, \tag{5.37}$$

and

$$[\hat{K}_j, [\hat{K}_k, \hat{K}_l]] + [\hat{K}_l, [\hat{K}_j, \hat{K}_k]] + [\hat{K}_k, [\hat{K}_l, \hat{K}_j]] = 0 \tag{5.38}$$

(called the *Jacobi identity*), where the c^l_{jk} (called *structure constants*) are complex numbers, then the set of operators $\{\hat{K}_i\}$ generates the symmetry *Lie algebra* of the system with the Hamiltonian $\hat{H}$.

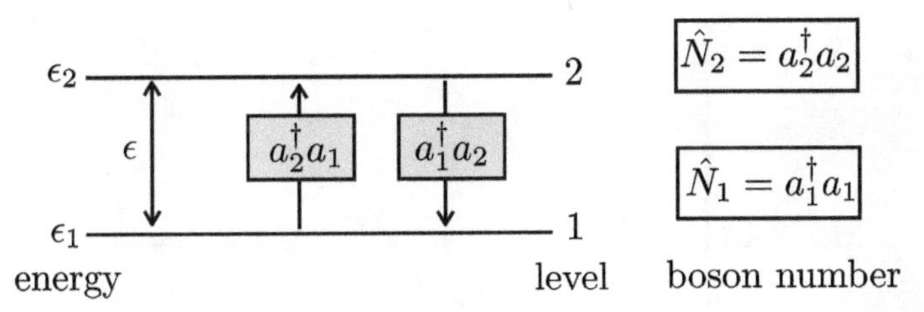

Figure 5.2. Schematic view of a two-level 'boson' model. The operator $a_2^\dagger a_1$ excites a boson from the lower level to the upper level and $a_1^\dagger a_2$ de-excites such a boson.

The pre-eminent example of a symmetry Lie algebra in quantum mechanics is rotational symmetry and angular momentum. The generators of rotations, $\hat{L}_x$, $\hat{L}_y$, and $\hat{L}_z$ possess the Lie algebra given by

$$[\hat{L}_\alpha, \hat{L}_\beta] = i\hbar\varepsilon_{\alpha\beta\gamma}\hat{L}_\gamma; \quad \alpha, \beta, \gamma = x, y, z, \tag{5.39}$$

called *SO(3) or SU(2)*.

Commonly, in quantum mechanics, the focus is (almost) entirely on the symmetry Lie algebra, not the symmetry group, of a physical system. Indeed, sometimes a system possesses an algebraic structure for which it makes little or no sense to seek a geometrical type of symmetry.

A practical consequence of the emphasis on algebraic structures in quantum mechanics is that modelling of systems is often done algebraically and there may not be a corresponding symmetry associated with the system geometry.

An example of a simple model system with a well-defined algebraic structure, but no specific geometrical structure, is a two-level system for identical bosons (see figure 5.2). For a model Hamiltonian

$$\hat{H} = \varepsilon_1 a_1^\dagger a_1 + \varepsilon_2 a_2^\dagger a_2; \tag{5.40}$$

if $\varepsilon_1 = \varepsilon_2$ (i.e. the two levels are degenerate in energy), then $\hat{H}$ commutes with $a_2^\dagger a_1$ and $a_1^\dagger a_2$ and a (trivial) algebraic structure results; for a fixed boson number, an *so*(3) algebra results; and for $\varepsilon_1 \neq \varepsilon_2$ this is an *so*(3) dynamical algebra (see next section).

5.8 Dynamical or spectrum generating algebras

Consider a model Hamiltonian of the form

$$\hat{H} = \alpha\hat{L}^2 + \beta L_z^2, \tag{5.41}$$

where $\hat{L}^2 = \hat{L}_x^2 + \hat{L}_y^2 + \hat{L}_z^2$ and $\{\hat{L}_x, \hat{L}_y, \hat{L}_z\}$ are the generators of the Lie algebra *so*(3). From the theory of *so*(3) (the theory of angular momentum in quantum mechanics),

$$\begin{aligned} \hat{L}^2|lm\rangle &= l(l+1)|lm\rangle, \qquad & \hat{L}_z|lm\rangle &= m|lm\rangle \qquad (\hbar \equiv 1), \\ l &= 0, 1, 2, \ldots, & m &= \pm l, \pm(l-1), \ldots, \pm 2, \pm 1, 0, \end{aligned} \tag{5.42}$$

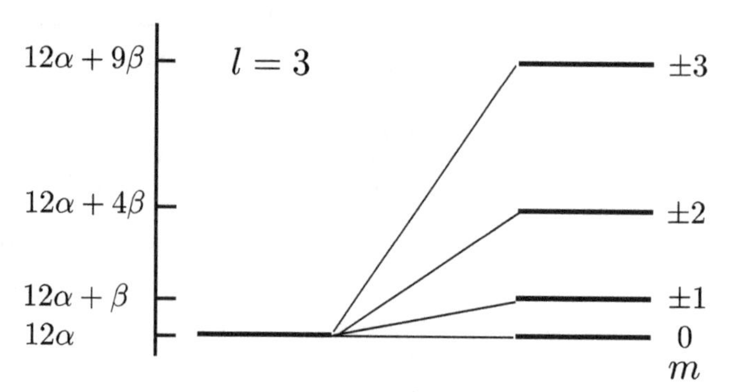

Figure 5.3. Energy spectrum for a system with the Hamiltonian, equation (5.41) for $l = 3$. Note the two-fold degeneracy in the quantum number, m.

$$\hat{H}|lm\rangle = (\alpha l(l+1) + \beta m^2)|lm\rangle, \tag{5.43}$$

and the energy eigenvalues can be depicted as shown in figure 5.3. *Note*: *$\hat{H}$ splits but does not mix* the $|lm\rangle$ states. The Hamiltonian $\hat{H}$ is said to possess an *so*(3) dynamical algebra and the structure of the algebra generates a spectrum. The algebra is called a *spectrum-generating algebra, SGA*.

The two-level system described on the previous page also has an *so*(3) dynamical algebra; but it has a different spectrum to the above.

5.9 Matrix groups

Matrix algebra naturally has a group structure, i.e. products of matrices possess closure and associativity; and the identity matrix and the condition for an inverse are defined. Within these structures there are continuous matrix groups and discrete matrix groups. Continuous matrix groups have an infinite number of group elements because their matrix elements depend on continuous parameters, e.g.

$$\begin{pmatrix} a_{11} & a_{12} \\ a_{21} & a_{22} \end{pmatrix}$$

form an infinite set because a_{11}, a_{12}, a_{21}, a_{22} are continuous. Discrete groups have a finite number of elements.

5.9.1 Discrete matrix groups

Examples of discrete groups are molecular point groups, crystallographic groups, and permutation groups. Matrix representations are realised by introducing geometric manifolds or permutational diagrams. For molecules and crystals, suitable geometric manifolds are the systems themselves. Symmetry transformations are made up of translations (for crystals), rotations, reflections, inversions and combinations of these. Matrix representations are straightforwardly formulated. Permutation groups may be applied to systems with or without an inherent geometry. Groups of permutations of identical objects are called symmetric groups.

Example 5-1. Matrix representations for the symmetric group, S_3.

The following operators for three identical objects leave the system unchanged:

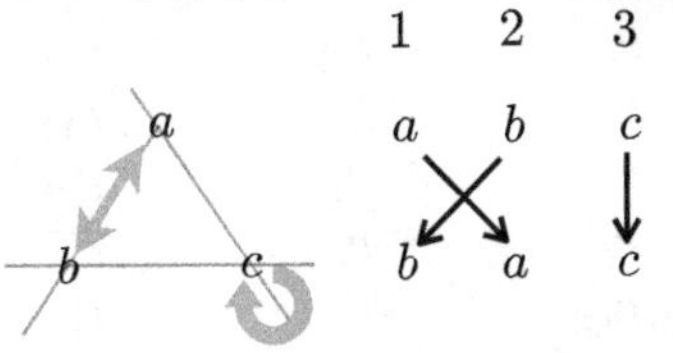

$$P_{12}: \begin{pmatrix} b \\ a \\ c \end{pmatrix} = \begin{pmatrix} 0 & 1 & 0 \\ 1 & 0 & 0 \\ 0 & 0 & 1 \end{pmatrix} \begin{pmatrix} a \\ b \\ c \end{pmatrix}$$

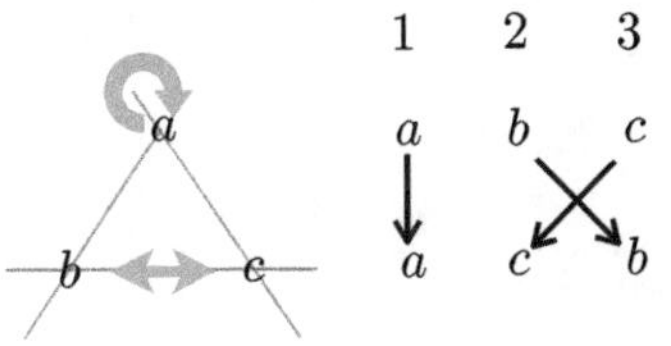

$$P_{23}: \begin{pmatrix} a \\ c \\ b \end{pmatrix} = \begin{pmatrix} 1 & 0 & 0 \\ 0 & 0 & 1 \\ 0 & 1 & 0 \end{pmatrix} \begin{pmatrix} a \\ b \\ c \end{pmatrix}$$

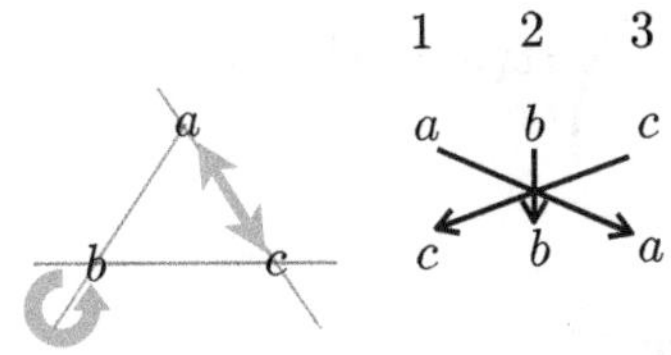

$$P_{13}: \begin{pmatrix} c \\ b \\ a \end{pmatrix} = \begin{pmatrix} 0 & 0 & 1 \\ 0 & 1 & 0 \\ 1 & 0 & 0 \end{pmatrix} \begin{pmatrix} a \\ b \\ c \end{pmatrix}$$

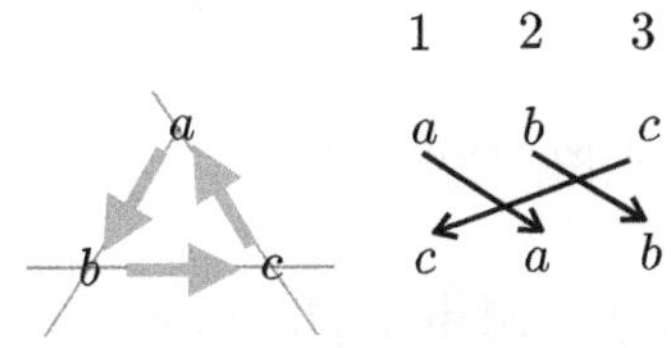

$$P_{+}: \begin{pmatrix} c \\ a \\ b \end{pmatrix} = \begin{pmatrix} 0 & 0 & 1 \\ 1 & 0 & 0 \\ 0 & 1 & 0 \end{pmatrix} \begin{pmatrix} a \\ b \\ c \end{pmatrix}$$

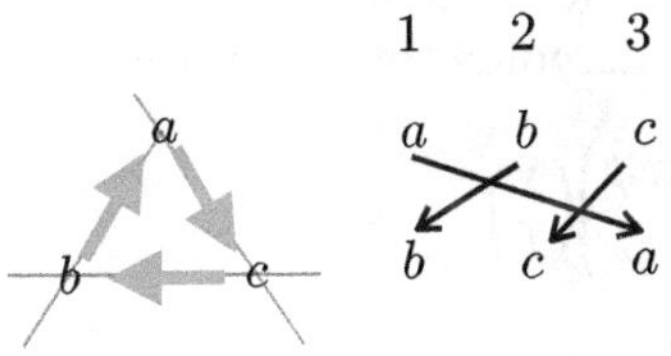

$$P_{-}: \begin{pmatrix} b \\ c \\ a \end{pmatrix} = \begin{pmatrix} 0 & 1 & 0 \\ 0 & 0 & 1 \\ 1 & 0 & 0 \end{pmatrix} \begin{pmatrix} a \\ b \\ c \end{pmatrix}$$

They can be arranged in a multiplication table (see table 5.1). Note 1: the order of multiplication is $P_{12}P_{23} = P_+$, $P_{12}P_{13} = P_-$, i.e. left-hand column first. Note 2: I is the identity, i.e. 'no permutation'. (The inverses are $P_{12}^{-1} = P_{12}$, $P_+^{-1} = P_-$, etc.; $P_{ij} = P_{ji}$, $i = 1, 2, 3$.)

5.9.2 Continuous matrix groups

Continuous matrix groups possess the remarkable property that they can be expressed in terms of *infinitesimal* steps. Thus, the essence of rotations in a plane can be expressed as $R(\phi + \mathrm{d}\phi, \phi)$ where the essential process is moving a vector making an angle ϕ with respect to a reference axis through an infinitesimal angle $\mathrm{d}\phi$. It can be depicted as shown in figure 5.4.

Table 5.1. Multiplication table for the permutation group, S_3.

	P_{12}	P_{23}	P_{13}	P_+	P_-
P_{12}	I	P_+	P_-	P_{23}	P_{13}
P_{23}	P_-	I	P_+	P_{13}	P_{12}
P_{13}	P_+	P_-	I	P_{12}	P_{23}
P_+	P_{13}	P_{12}	P_{23}	P_-	I
P_-	P_{23}	P_{13}	P_{12}	I	P_+

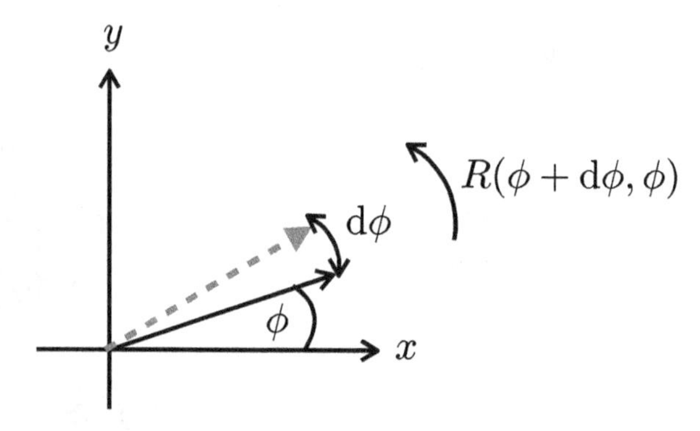

Figure 5.4. Infinitesimal rotation of a vector in a plane from the position defined by ϕ (x-axis: reference axis) to the position $\phi + \mathrm{d}\phi$ by the rotation operator $R(\phi + \mathrm{d}\phi, \phi)$.

All rotations in a plane can be achieved by repeated multiplication of infinitesimal rotations. This can be expressed using matrices as follows. From the basic representation of a point (x, y) in the plane, undergoing rotation through an angle ϕ,

$$\begin{pmatrix} x' \\ y' \end{pmatrix} = \begin{pmatrix} \cos\phi & -\sin\phi \\ \sin\phi & \cos\phi \end{pmatrix}\begin{pmatrix} x \\ y \end{pmatrix}, \tag{5.44}$$

for counterclockwise rotations, consider the limit

$$\lim_{\phi\to 0}\begin{pmatrix} \cos\phi & -\sin\phi \\ \sin\phi & \cos\phi \end{pmatrix} \to \begin{pmatrix} 1 & -\phi \\ \phi & 1 \end{pmatrix}, \tag{5.45}$$

where the small angle approximation, $\sin\phi \approx \phi$ (ϕ measured in radians) has been made. Then from

$$\begin{pmatrix} 1 & -\phi \\ \phi & 1 \end{pmatrix} = \begin{pmatrix} 1 & 0 \\ 0 & 1 \end{pmatrix} + \begin{pmatrix} 0 & -\phi \\ \phi & 0 \end{pmatrix}, \tag{5.46}$$

$$\lim_{\phi\to 0}\begin{pmatrix} \cos\phi & -\sin\phi \\ \sin\phi & \cos\phi \end{pmatrix} = I + g\phi, \tag{5.47}$$

where I is the identity matrix for the plane and

$$g := \begin{pmatrix} 0 & -1 \\ 1 & 0 \end{pmatrix}. \tag{5.48}$$

Consider now repeating an infinitesimal rotation $\frac{\phi}{N}$, where N is a very large number, N times, i.e. $(I + \frac{g\phi}{N})^N$. The $N \to \infty$ limit of this process is $\exp(g\phi)$. This can be explored on a calculator by evaluating $(1 + \frac{1}{N})^N$ for $N = 10, 10^3, 10^6, \ldots$ and is seen to approach 2.7183…. Here, expanding,

$$\exp(g\phi) = I + g\phi + \frac{1}{2!}g^2\phi^2 + \frac{1}{3!}g^3\phi^3 + \cdots, \tag{5.49}$$

but

$$g^2 = \begin{pmatrix} 0 & -1 \\ 1 & 0 \end{pmatrix}\begin{pmatrix} 0 & -1 \\ 1 & 0 \end{pmatrix} = \begin{pmatrix} -1 & 0 \\ 0 & -1 \end{pmatrix} = -I, \tag{5.50}$$

whence

$$\begin{aligned} \exp(g\phi) &= I\left(1 - \frac{1}{2!}\phi^2 + \cdots\right) + g\left(\phi - \frac{1}{3!}\phi^3 + \cdots\right) \\ &= I\cos\phi + g\sin\phi \\ &= \begin{pmatrix} 1 & 0 \\ 0 & 1 \end{pmatrix}\cos\phi + \begin{pmatrix} 0 & -1 \\ 1 & 0 \end{pmatrix}\sin\phi \\ &= \begin{pmatrix} \cos\phi & -\sin\phi \\ \sin\phi & \cos\phi \end{pmatrix}. \end{aligned} \tag{5.51}$$

One notices that g and the complex number[4] i possess an isomorphic algebra, recall $e^{i\phi} = \cos\phi + i\sin\phi$.

5.9.3 Compact and non-compact groups

Rotations in a plane are described by a single parameter, the angle ϕ, and the parameter is *compact*, i.e. $0 \leqslant \phi \leqslant 2\pi$. With the operation of reflection, represented by the matrix $\begin{pmatrix} 1 & 0 \\ 0 & -1 \end{pmatrix}$, the groups so defined are $SO(2, \mathbb{R})$ or $SO(2)$ and $O(2, \mathbb{R})$ or $O(2)$; where 'O' stands for orthogonal, and 'S' stands for 'special' which implies that the determinant of the 2×2 matrix (group elements) have the value $+1$. Manifestly, these groups preserve a length in the plane, i.e.

$$(x')^2 + (y')^2 = x^2 + y^2. \tag{5.52}$$

[4] Indeed, complex numbers can be used to represent both operands and operators in a plane, viz. $e^{i\phi}(re^{i\theta}) = re^{i(\theta+\phi)}$, cf. $Rv = v'$.

One can also define the group that ensures

$$(x')^2 - (y')^2 = x^2 - y^2. \tag{5.53}$$

The matrix representation of this group is parameterised as

$$R(\chi) = \begin{pmatrix} \cosh\chi & \sinh\chi \\ \sinh\chi & \cosh\chi \end{pmatrix}, \tag{5.54}$$

where $0 \leqslant \chi$, i.e. it is non-compact. It is encountered in physics, e.g. for space–time in one spatial dimension, where $x^2 - (ct)^2$ is conserved. The group is designated $SO(1,1)$ or $O(1,1)$.

In three dimensions one encounters, e.g. $SO(2,1)$ which would preserve the quantity $x_1^2 + x_2^2 - x_3^2$. Compactness is a property of the *manifold* on which the group elements act. The group $SO(3)$ describes moving a point on the surface of a sphere; the group $SO(2,1)$ describes moving a point on the surface of a hyperboloid. Thus, manifolds are geometric surfaces.

5.9.4 Polynomial representation of groups

Consider

$$\begin{pmatrix} x_1' \\ x_2' \end{pmatrix} = \begin{pmatrix} \cos\phi & -\sin\phi \\ \sin\phi & \cos\phi \end{pmatrix} \begin{pmatrix} x_1 \\ x_2 \end{pmatrix}, \tag{5.55}$$

i.e.

$$x_1' = x_1\cos\phi - x_2\sin\phi, \tag{5.56}$$

$$x_2' = x_1\sin\phi + x_2\cos\phi \tag{5.57}$$

and the binomials

$$t_1 := x_1^2, \quad t_2 := \sqrt{2}x_1x_2, \quad t_3 := x_2^2. \tag{5.58}$$

These binomials transform under $SO(2)$, equation (5.55), as

$$\begin{pmatrix} t_1' \\ t_2' \\ t_3' \end{pmatrix} = \begin{pmatrix} \cos^2\phi & -\sqrt{2}\cos\phi\sin\phi & \sin^2\phi \\ \sqrt{2}\cos\phi\sin\phi & \cos^2\phi - \sin^2\phi & -\sqrt{2}\cos\phi\sin\phi \\ \sin^2\phi & \sqrt{2}\cos\phi\sin\phi & \cos^2\phi \end{pmatrix} \begin{pmatrix} t_1 \\ t_2 \\ t_3 \end{pmatrix}. \tag{5.59}$$

Equation (5.59) describes the transformation of a three-component structure under $SO(2)$. The structure is a rank-2 polynomial constructed on the space $(2, \mathbb{R})$. There is only one parameter, the angle ϕ, which is a characteristic property of the group $SO(2)$. This process can be extended by defining

$$w_1 := x_1^3, \quad w_2 := \sqrt{3}x_1^2x_2, \quad w_3 := \sqrt{3}x_1x_2^2, \quad w_4 := x_2^3. \tag{5.60}$$

The factors '$\sqrt{2}$' in equation (5.58) and '$\sqrt{3}$' in equation (5.60) are not necessary, but yield a more symmetric pattern. Polynomials of any degree are permissible.

The matrix in equation (5.59) is *reducible*. Calling this matrix M, the similarity transformation

$$C := \frac{1}{2}\begin{pmatrix} 1 & 0 & -1 \\ 0 & \sqrt{2} & 0 \\ 1 & 0 & 1 \end{pmatrix} \tag{5.61}$$

effects

$$C^T M C = \begin{pmatrix} 1 & 0 & 0 \\ 0 & \cos^2\phi - \sin^2\phi & -2\cos\phi\sin\phi \\ 0 & 2\cos\phi\sin\phi & \cos^2\phi - \sin^2\phi \end{pmatrix}, \tag{5.62}$$

where C^T is the transpose of C and $C^T = C^{-1}$, i.e. $C^T C = I$. This result is straightforwardly understood when it is recognised that $t_1 + t_3 = x_1^2 + x_2^2$ is invariant under rotations in a plane: it is a scalar. The linear combination $t_1 - t_3$ together with t_2 span an *irreducible* two-dimensional subspace. The space spanned by (t_1, t_2, t_3) is said to be reducible under $SO(2)$ transformation.

5.10 Generators of continuous groups and Lie algebras

Rotation through an angle ϕ in a counterclockwise direction in a plane is described by (cf. equation (5.44))

$$R(\phi) = \begin{pmatrix} \cos\phi & -\sin\phi \\ \sin\phi & \cos\phi \end{pmatrix}. \tag{5.63}$$

The essential operation that underlies this is the recognition that it is defined infinitesimally by (cf. equation (5.47))

$$R(\phi + \mathrm{d}\phi, \phi) = I + g\mathrm{d}\phi, \tag{5.64}$$

where

$$I = \begin{pmatrix} 1 & 0 \\ 0 & 1 \end{pmatrix}, \quad g = \begin{pmatrix} 0 & -1 \\ 1 & 0 \end{pmatrix}. \tag{5.65}$$

Any rotation in a plane (there are an infinite number of possibilities) is achieved by repeating this process

$$R(\phi) = \lim_{N\to\infty}\left(I + \frac{g\phi}{N}\right)^N. \tag{5.66}$$

The process is described by a single *generator* g. Rotations in a plane have one generator and one parameter (an angle, ϕ).

5.10.1 The matrix group $SO(3)$ and its generators

Rotations in three dimensions, in the space $(3, \mathbb{R})$ (the space where we live) can be described by three generators, say g_1, g_2, g_3, and three parameters (three angles). There are various ways to define these angles: they could be with respect to three Cartesian axes, in which case $\{g_1, g_2, g_3\} = \{g_x, g_y, g_z\}$; they could be with respect to an axis and a single angle, in which case two angles are needed to specify the axis, $\hat{n}$ and $\{g_1, g_2, g_3\} = \{g_n, g_a, g_b\}$, where a and b are orthogonal axes[5] with respect to $\hat{n}$. The possibilities are infinite. For three Cartesian axes

$$R_x(\alpha) = \begin{pmatrix} 1 & 0 & 0 \\ 0 & \cos\alpha & -\sin\alpha \\ 0 & \sin\alpha & \cos\alpha \end{pmatrix}, \tag{5.67}$$

$$R_y(\beta) = \begin{pmatrix} \cos\beta & 0 & \sin\beta \\ 0 & 1 & 0 \\ -\sin\beta & 0 & \cos\beta \end{pmatrix}, \tag{5.68}$$

$$R_z(\gamma) = \begin{pmatrix} \cos\gamma & -\sin\gamma & 0 \\ \sin\gamma & \cos\gamma & 0 \\ 0 & 0 & 1 \end{pmatrix}. \tag{5.69}$$

The generators corresponding to these rotations are

$$g_x = \begin{pmatrix} 0 & 0 & 0 \\ 0 & 0 & -1 \\ 0 & 1 & 0 \end{pmatrix}, \tag{5.70}$$

$$g_y = \begin{pmatrix} 0 & 0 & 1 \\ 0 & 0 & 0 \\ -1 & 0 & 0 \end{pmatrix}, \tag{5.71}$$

$$g_z = \begin{pmatrix} 0 & -1 & 0 \\ 1 & 0 & 0 \\ 0 & 0 & 0 \end{pmatrix}. \tag{5.72}$$

The generators $\{g_x, g_y, g_z\}$ do not commute with each other, viz.

$$[g_x, g_y] = g_z, \tag{5.73}$$

$$[g_y, g_z] = g_x, \tag{5.74}$$

$$[g_z, g_x] = g_y. \tag{5.75}$$

[5] But no rotations take place about these axes.

They define an algebra with three elements. It is a Lie algebra, as first communicated by the Norwegian mathematician, Sophus Lie, ca. 1890. They also define a group under the group product, $[A, B] = AB - BA = C$, i.e. they obey closure. It is worth observing that, by adopting equations (5.73)–(5.75) as the essential structure of $(3, \mathbb{R})$ with respect to rotations, by invoking a scale factor $\hbar$ (which has dimensions of angular momentum), and by requiring that operators be Hermitian, one arrives axiomatically at

$$[\hat{L}_x, \hat{L}_y] = i\hbar\hat{L}_z, \tag{5.76}$$

$$[\hat{L}_y, \hat{L}_z] = i\hbar\hat{L}_x, \tag{5.77}$$

$$[\hat{L}_z, \hat{L}_x] = i\hbar\hat{L}_y. \tag{5.78}$$

It would be fair to say that the mathematical structures that we use to represent and conduct quantum mechanics work because of the underlying Lie algebra structures. Lie algebras have undergone an elegant and sophisticated evolution since Sophus Lie's work, notably by Élie Cartan. Here and in the next chapter (chapter 6) we provide basic details.

5.10.2 Unitary groups and $SU(2)$

The mathematical structure of quantum mechanics 'resides' in Hilbert space. Indeed, the operators $\{\hat{L}_x, \hat{L}_y, \hat{L}_z\}$ given in equations (5.76)–(5.78) act on elements of Hilbert space, not on vectors in $(3, \mathbb{R})$. Hilbert space is a complex linear space with a finite norm. As such, we are concerned with unitary transformations (as well as orthogonal transformations in $(3, \mathbb{R})$, cf. equations (5.67)–(5.69)).

The simplest, non-trivial[6] group associated with Hilbert space is $SU(2)$. The group $SU(2)$ is not amenable to a simple pictorial view, such as presented in section 5.9.2 and figure 5.4, for $SO(2)$. The simplest realisation of an $SU(2)$ transformation can be expressed as

$$\begin{pmatrix} z_1' \\ z_2' \end{pmatrix} := \begin{pmatrix} a & b \\ -b^* & a^* \end{pmatrix} \begin{pmatrix} z_1 \\ z_2 \end{pmatrix}, \tag{5.79}$$

$$|z_1'|^2 + |z_2'|^2 = |z_1|^2 + |z_2|^2, \tag{5.80}$$

where z_1, z_2 span the space $(2, \mathbb{C})$ and a, b are parameters that are complex. As such, a and b represent four parameters (each has a real and an imaginary part); but the transformation is unitary, i.e.

$$\begin{pmatrix} a^* & -b \\ b^* & a \end{pmatrix} \begin{pmatrix} a & b \\ -b^* & a^* \end{pmatrix} = \begin{pmatrix} |a|^2 + |b|^2 & 0 \\ 0 & |a|^2 + |b|^2 \end{pmatrix} = I \tag{5.81}$$

imposes the constraint $|a|^2 + |b|^2 = 1$, yielding a three-parameter group.

[6] There is the trivial group $U(1) \equiv SU(1)$, which is just multiplication by a phase factor, $e^{-i\phi}$.

As with $SO(3)$, there are various ways to define these three parameters. If one chooses to match the rotation angles used in $SO(3)$, one obtains the expressions

$$a = e^{\frac{-i(\alpha+\gamma)}{2}} \cos\frac{\beta}{2}, \quad b = -e^{\frac{-i(\alpha-\gamma)}{2}} \sin\frac{\beta}{2}, \tag{5.82}$$

when the Euler angle representation is used, and

$$a = \cos\frac{\phi}{2} - in_z \sin\frac{\phi}{2}, \quad b = (-in_x - n_y)\sin\frac{\phi}{2}, \tag{5.83}$$

when the axis-angle representation is used, cf. equations (1.53) and (1.56), respectively.

5.11 The unitary and orthogonal groups in *n* dimensions, $U(n)$ and $SO(n)$

To see how group structures, such as $SO(2)$, $SO(3)$, and $SU(2)$ can be explored and how they might apply to quantum mechanics, it is useful to introduce the groups $U(n)$ and $SO(n)$.

The $U(n)$ group is defined by transformations in an n-dimensional field of complex numbers $(z_1, z_2, \dots, z_n)$,

$$z_i' := \sum_{j=1}^{n} U_{ij} z_j, \qquad i = 1, 2, \dots, n, \tag{5.84}$$

where the matrix U satisfies

$$\sum_{j=1}^{n} U_{ij} U_{kj}^* = \delta_{ik}; \quad i, k = 1, 2, \dots, n, \tag{5.85}$$

i.e.

$$U^\dagger U = I. \tag{5.86}$$

Consider the infinitesimal unitary transformations

$$U = I + i\varepsilon\hat{S} + \cdots, \tag{5.87}$$

where $\hat{S}$ is Hermitian and ε is an infinitesimal real number. Then, the action of $\hat{S}$ on an arbitrary function $f(z_1, z_2, \dots, z_n) := f(z_i)$ is

$$\begin{aligned} \hat{S}f(z_i) = f(z_i') &= f\left(z_i + i\varepsilon\sum_{j=1}^{n} S_{ij} z_j\right) \\ &= f(z_i) + i\varepsilon\sum_{i=1}^{n}\sum_{j=1}^{n} S_{ij} z_j \frac{\partial f}{\partial z_i} + \cdots, \end{aligned} \tag{5.88}$$

i.e. to first order in ε, the infinitesimal unitary transformation is expressed by the increment $i\varepsilon \sum_{i,j=1}^{n} S_{ij} G_j^i$, where

$$G_j^i := z_j \frac{\partial}{\partial z_i}. \tag{5.89}$$

There are n^2 such operators, and since

$$\left[\frac{\partial}{\partial z_i}, z_j\right] = \delta_{ij}, \tag{5.90}$$

$$[G_i^j, G_k^l] = G_i^l \delta_{jk} - G_k^j \delta_{il}; \quad i, j, kl = 1, 2, \ldots, n, \tag{5.91}$$

which defines a $u(n)$ Lie algebra[7].

Example 5-2. Derivation of the Lie algebra u*(2) from the Lie group* U*(2).*
The Lie group $U(2)$ can be expressed as

$$\begin{pmatrix} c_{11} & c_{12} \\ c_{21} & c_{22} \end{pmatrix} \begin{pmatrix} z_1 \\ z_2 \end{pmatrix} = \begin{pmatrix} z_1' \\ z_2' \end{pmatrix}. \tag{5.92}$$

The generators of infinitesimal transformations can be expressed as

$$G_1^1 = z_1 \frac{\partial}{\partial z_1}, \quad G_1^2 = z_1 \frac{\partial}{\partial z_2}, \quad G_2^1 = z_2 \frac{\partial}{\partial z_1}, \quad G_2^2 = z_2 \frac{\partial}{\partial z_2}. \tag{5.93}$$

Commutator brackets for the generators can be evaluated using the identity

$$[AB, CD] = AC[B, D] + A[B, C]D + C[A, D]B + [A, C]DB, \tag{5.94}$$

whence, e.g.

$$[G_1^1, G_1^2] = \left[z_1 \frac{\partial}{\partial z_1}, z_1 \frac{\partial}{\partial z_2}\right] = z_1 \left[\frac{\partial}{\partial z_1}, z_1\right] \frac{\partial}{\partial z_2} = G_1^2. \tag{5.95}$$

The 16 commutator brackets that result can be depicted as in table 5.2. Note 1: the order of the commutator brackets is, e.g. $[G_1^1, G_1^2] = G_1^2$, $[G_1^1, G_2^1] = -G_2^1$. Note 2: only the upper 'triangle' is needed because $[A, B] = -[B, A]$. The group $SU(2)$ follows directly from $U(2)$ by imposing the constraint

$$G_1^1 + G_2^2 = I, \tag{5.96}$$

whence, $SU(2)$ has three generators. (Note: adding entries in the columns, for G_1^1 and G_2^2, always gives zero.) Then, defining

[7] Note, upper case letters are used to denote Lie groups, e.g. $U(n)$, and lower case letters denote Lie algebras, viz. $u(n)$.

Table 5.2. Commutator bracket table for $u(2)$.

	G_1^1	G_1^2	G_2^1	G_2^2
G_1^1	0	G_1^2	$-G_2^1$	0
G_1^2	$-G_1^2$	0	$G_1^1 - G_2^2$	G_1^2
G_2^1	G_2^1	$G_2^2 - G_1^1$	0	$-G_2^1$
G_2^2	0	$-G_1^2$	G_2^1	0

$$g_0 := \frac{i}{2}(G_1^1 - G_2^2), \quad g_1 := \frac{1}{2}(G_1^2 - G_2^1), \quad g_2 := \frac{i}{2}(G_1^2 + G_2^1), \tag{5.97}$$

$$[g_0, g_1] = g_2, \quad [g_1, g_2] = g_0, \quad [g_2, g_0] = g_1, \tag{5.98}$$

i.e. the Lie algebra of $\{g_0, g_1, g_2\}$ is closed and it is isomorphic to equations (5.73)–(5.75): the algebra of $so(3)$. The Lie algebras $su(2)$ and $so(3)$ are isomorphic.

The $SO(n)$ group is defined by transformations in an n-dimensional field of real numbers $(x_1, x_2, \dots, x_n)$,

$$x_i' = \sum_{j=1}^{n} R_{ij} x_j, \quad i = 1, 2, \dots, n, \tag{5.99}$$

where the matrix R satisfies

$$\sum_{j=1}^{n} R_{ij} R_{kj} = \delta_{ik}, \quad i, k = 1, 2, \dots, n; \tag{5.100}$$

i.e.

$$R\tilde{R} = I, \tag{5.101}$$

where $\tilde{R}$ is the transpose of R.

Now, consider the infinitesimal orthogonal transformations

$$R = I + \varepsilon T + \cdots, \tag{5.102}$$

where T is a real asymmetric matrix since

$$R\tilde{R} \approx (I + \varepsilon T)(1 + \varepsilon \tilde{T}) \approx I + \varepsilon(T + \tilde{T}) = I, \tag{5.103}$$

i.e.

$$T_{ji} = -T_{ij}. \tag{5.104}$$

Then consider the action of T on an arbitrary function, $g(x_i)$,

$$\begin{aligned} Tg(x_i) &= g(x_i') \\ &= g\left(x_i + \varepsilon \sum_{j=1}^{n} T_{ij}x_j\right) \\ &= g(x_i) + \varepsilon \sum_{i,j=1}^{n} T_{ij}x_j \frac{\partial g}{\partial x_i} + \cdots, \\ &= g(x_i) + \varepsilon \sum_{i,j=1}^{n} T_{ij}\left(x_j \frac{\partial g}{\partial x_i} - x_i \frac{\partial g}{\partial x_j}\right) + \cdots. \end{aligned} \tag{5.105}$$

We define (note: $i \neq j$)

$$\Lambda_{ij} := G_i^j - G_j^i, \tag{5.106}$$

where

$$G_j^i := x_j \frac{\partial}{\partial x_i}, \text{ etc.} \tag{5.107}$$

There are $\frac{n(n-1)}{2}$ such generators and

$$[\Lambda_{ij}, \Lambda_{kl}] = \Lambda_{il}\delta_{jk} + \Lambda_{jk}\delta_{il} + \Lambda_{lj}\delta_{ik} + \Lambda_{ki}\delta_{jl}, \tag{5.108}$$

where, $i, j, k, l = 1, 2, \dots, n$, which defines the Lie algebra $so(n)$.

The commutator brackets for, e.g. $so(4)$ follow directly from equation (5.108) (see table 5.3). Note: $\Lambda_{12} = -\Lambda_{21}$, etc.

5.12 Casimir invariants and commuting operators

Casimir invariants are operators that commute with all of the generators of a particular Lie algebra.

Table 5.3. Commutator bracket table for $so(4)$.

	Λ_{12}	Λ_{13}	Λ_{14}	Λ_{23}	Λ_{24}	Λ_{34}
Λ_{12}	0	$-\Lambda_{23}$	$-\Lambda_{24}$	Λ_{13}	Λ_{14}	0
Λ_{13}	Λ_{23}	0	$-\Lambda_{34}$	$-\Lambda_{12}$	0	Λ_{14}
Λ_{14}	Λ_{24}	Λ_{34}	0	0	$-\Lambda_{12}$	$-\Lambda_{13}$
Λ_{23}	$-\Lambda_{13}$	Λ_{12}	0	0	Λ_{34}	Λ_{24}
Λ_{24}	$-\Lambda_{14}$	0	Λ_{12}	$-\Lambda_{34}$	0	Λ_{23}
Λ_{34}	0	$-\Lambda_{14}$	Λ_{13}	$-\Lambda_{24}$	$-\Lambda_{23}$	0

5.12.1 The Casimir invariants of *u(n)*

$$C_{u(n)}^{(1)} := \sum_{i=1}^{n} G_i^i, \tag{5.109}$$

viz.

$$\begin{aligned}\left[C_{u(n)}^{(1)}, G_k^l\right] &= \sum_{i=1}^{n}[G_i^i, G_k^l] \\ &= \sum_{i=1}^{n}(G_i^l\delta_{ik} - G_k^i\delta_{il}) \\ &= 0.\end{aligned} \tag{5.110}$$

$$C_{u(n)}^{(2)} := \sum_{i,j=1}^{n} G_i^j G_j^i. \tag{5.111}$$

$$C_{u(n)}^{(3)} := \sum_{i,j,k=1}^{n} G_i^j G_j^k G_k^i, \tag{5.112}$$

etc. Note the index 'contraction'.

Example 5-3. The Casimir invariants of u*(2) and* su*(2).*
From equation (5.109)

$$C_{u(2)}^{(1)} := G_1^1 + G_2^2, \tag{5.113}$$

cf. equation (5.96) and table 5.2.

From equation (5.111),

$$C_{u(2)}^{(2)} := G_1^1G_1^1 + G_1^2G_2^1 + G_2^1G_1^2 + G_2^2G_2^2; \tag{5.114}$$

then, using $g_0 := \frac{i}{2}(G_1^1 - G_2^2)$, $g_1 := \frac{1}{2}(G_1^2 - G_2^1)$, $g_2 := \frac{i}{2}(G_1^2 + G_2^1)$, cf. equations (5.97) and (5.113), it follows that

$$C_{u(2)}^{(2)} = -8\left(g_0^2 + g_1^2 + g_2^2\right) + 2\left[C_{u(2)}^{(1)}\right]^2. \tag{5.115}$$

We define

$$C_{su(2)}^{(2)} := g_0^2 + g_1^2 + g_2^2. \tag{5.116}$$

5.12.2 The Casimir invariants of *so(n)*

There is no first-order Casimir invariant for *so*(*n*). The second-order invariant is

$$C_{so(n)}^{(2)} := \frac{1}{2}\sum_{i,j=1}^{n} \Lambda_{ij}\Lambda_{ji}. \tag{5.117}$$

Example 5-4. The Casimir invariant of SO*(3).*
From equation (5.117)

$$C^{(2)}_{so(3)} := \Lambda_{12}\Lambda_{21} + \Lambda_{13}\Lambda_{31} + \Lambda_{23}\Lambda_{32}, \tag{5.118}$$

cf. table 5.3, defining

$$g_1 := \Lambda_{23}, \ \ g_2 := \Lambda_{31}, \ \ g_3 := -\Lambda_{12}, \tag{5.119}$$

from which

$$C^{(2)}_{so(3)} = -\left(g_1^2 + g_2^2 + g_3^2\right) \tag{5.120}$$

follows. (Note:

$$[g_1, g_2] = g_3, \ \ [g_2, g_3] = g_1, \ \ [g_3, g_1] = g_2. \ ) \tag{5.121}$$

IOP Publishing

Quantum Mechanics for Nuclear Structure, Volume 2
An intermediate level view
Kris Heyde and John L Wood

Chapter 6

Algebraic structure of quantum mechanics

The algebraic structure of quantum mechanics is emphasized. This is illustrated by reiterating the *so*(3) ~ *su*(2) treatment of angular momentum and spin. The algebra *su*(1,1) ~ *so*(2,1) ~ *sl*(2,R) is sketched. The main focus is on rank-2 Lie algebras. Details of *su*(3) and the isotropic harmonic oscillator in three dimensions are introduced. The *so*(4) structure of the Kepler problem (hydrogen-like systems) is presented. Uses of *so*(5) and *sp*(3,R), as applied to nuclei are briefly outlined. Root and weight diagrams are explained. Young diagram techniques for elucidating *SU*(3) group irreps are built from first principles. First steps into Cartan theory of Lie algebras are made, using *su*(3) to illustrate this language.

Concepts: commutator bracket algebra; ladder operators; Casimir operators; *so*(3) ~ *su*(2); *su*(1,1) ~ *so*(2,1) ~ *sl*(2,R); su(3); three-dimensional harmonic oscillator; *hw*(3); root diagram; weight diagram; *so*(4); Kepler problem and hydrogen atom; *so*(5); *sp*(3,R); Young diagrams; Young tableaux; highest-weight state; multiplicity of weights; dimensions of irreps; Robinson hook-length method; sub-irreps; Kronecker products of irreps; Cartan structure; Cartan subalgebra; Weyl reflection theorem; *g*(2).

At the heart of quantum mechanics is the relationship

$$[\hat{x}_i, \hat{p}_j] := i\hbar\delta_{ij} = i\hbar\delta_{ij}\hat{I}, \quad i, j = 1, 2, 3, \tag{6.1}$$

where $\hat{x}_i$ and $\hat{p}_i$ are *operators* representing position and momentum with respect to the x_i axes. The equation involves a commutator bracket viz. $[\hat{A}, \hat{B}] := \hat{A}\hat{B} - \hat{B}\hat{A}$, $i = \sqrt{-1}$, $\hbar$ is the fundamental constant describing the scale of quantum phenomena, δ_{ij} is the Kronecker delta, ($\delta_{ij} = 1$, $i = j$, $\delta_{ij} = 0$, $i \neq j$), and $\hat{I}$ is the identity operator. Equation (6.1) defines an algebra.

It is specified that operators in quantum mechanics act on elements of a Hilbert space (a complex linear space possessing an inner product) and that outcomes of observations in quantum mechanics are the eigenvalues of equations of the form

$\hat{A}|\alpha_i\rangle = \alpha_i|\alpha_i\rangle$, where α_i is a real number, $\hat{A}$ is an operator corresponding to a dynamical variable, and $|\alpha_i\rangle$ is an eigenvector. The set $\{|\alpha_i\rangle\}$, $i = 1, \ldots, n$, where n possible outcomes of the measurement of A are observed, span a Hilbert space which serves to describe the physics of a system with respect to the observable A. Probabilities are given by $|\langle\alpha|\beta\rangle|^2$, where $|\alpha\rangle$ and $|\beta\rangle$ are any two elements of the Hilbert space and the interpretation is 'for a system in the state $|\beta\rangle$, what is the probability that it is observed in the state $|\alpha\rangle$?' The condition imposed on $|\langle\alpha|\beta\rangle|^2$ in order that it corresponds to probability is that the representation $\{|\alpha\rangle\}$ must be *unitary*.

One proceeds to 'solve' the quantum mechanics of a system by finding the eigenvalues of the Hamiltonian operator and of all the operators that commute with the Hamiltonian. The condition of unitarity is imposed on the eigenvectors $\{|\alpha\rangle\}$ to complete this task. Then, the processes that can occur in the system are elucidated by computing matrix elements for all operators of interest that have an action in the Hilbert space of the system. In this chapter the algebras associated with some quantum mechanical systems of especial interest will be explored and an introduction to Young diagram techniques and Cartan theory of Lie algebras will be made.

6.1 Angular momentum theory as an application of a Lie algebra

The theory of angular momentum (and spin) in quantum mechanics is instantly recognised by most physicists via the set of algebraic equations

$$[\hat{J}_x, \hat{J}_y] = i\hbar\hat{J}_z, \tag{6.2}$$

$$[\hat{J}_y, \hat{J}_z] = i\hbar\hat{J}_x, \tag{6.3}$$

$$[\hat{J}_z, \hat{J}_x] = i\hbar\hat{J}_y, \tag{6.4}$$

or

$$[\hat{J}_i, \hat{J}_k] = i\hbar\varepsilon_{ijk}\hat{J}_k, \quad i, j, k = (1, 2, 3) \equiv (x, y, z) \tag{6.5}$$

and ε_{ijk} is the Levi-Civita symbol. Often $J_i \equiv L_i$ when angular momentum is involved, $J_i \equiv S_i$ when spin is involved, and $J_i \equiv L_i + S_i$ when a system has contributions to total spin from both intrinsic spin and angular momentum.

Remarkably, for angular momentum, the relationships,

$$[\hat{L}_i, \hat{L}_j] = i\hbar\varepsilon_{ijk}\hat{L}_k, \tag{6.6}$$

can be derived from equation (6.1). However, for intrinsic spin,

$$[\hat{S}_i, \hat{S}_j] := i\hbar\varepsilon_{ijk}\hat{S}_k \tag{6.7}$$

must be stated as an axiom[1]. These three equations define the Lie algebra *su*(2); it is isomorphic to the Lie algebra *so*(3). (Recall the groups *SU*(2) and *SO*(3) are homomorphic: there is a 1:2 correspondence in group elements. This is dramatised by the rotation of spinors, i.e. spin-$\frac{1}{2}$ states in Hilbert space which acquire a quantum phase factor of −1 when rotated through 2π, cf. section 1.16.)

A standard procedure for 'solving' spin-angular problems is to define the operators (e.g. $J_i \equiv L_i$):

$$\hat{L}^2 := \hat{L}_x^2 + \hat{L}_y^2 + \hat{L}_z^2, \tag{6.8}$$

$$\hat{L}_\pm := \hat{L}_x \pm i\hat{L}_y, \tag{6.9}$$

to arrive at the algebraic relations

$$[\hat{L}^2, \hat{L}_x] = 0, \ \ [\hat{L}^2, \hat{L}_y] = 0, \ \ [\hat{L}^2, \hat{L}_z] = 0, \tag{6.10}$$

$$[\hat{L}_z, \hat{L}_\pm] = \pm\hbar\hat{L}_z, \ \ [\hat{L}_+, \hat{L}_-] = 2\hbar\hat{L}_0. \tag{6.11}$$

The operator $\hat{L}^2$ is termed a *Casimir operator* or *Casimir invariant* with respect to *su*(2). It is the only such operator for *su*(2): it commutes with all of the elements $\{\hat{L}_x, \hat{L}_y, \hat{L}_z\}$ of the algebra and any functions of these elements, such as equation (6.9). The operators $\hat{L}_+$ and $\hat{L}_-$ are termed *raising and lowering operators*, respectively, or *'ladder' operators*. One then arrives at simultaneous eigenkets, $|lm\rangle$ which obey

$$\hat{L}^2|lm\rangle = l(l+1)\hbar^2|lm\rangle, \tag{6.12}$$

$$\hat{L}_z|lm\rangle = m\hbar|lm\rangle, \tag{6.13}$$

$$\hat{L}_\pm|lm\rangle = \sqrt{(l \mp m)(l \pm m + 1)}\,\hbar|l,m \pm 1\rangle, \tag{6.14}$$

$l = 0, 1, 2, \cdots, m = +l, +l-1, \ldots, +1, 0, -1, \ldots, -l$ for angular momentum (and $m = +j, +j-1, \ldots, +\frac{1}{2}, -\frac{1}{2}, \ldots, -j$ if a spin-$\frac{1}{2}$ particle is present). The quantization necessitates that only *unitary* representations are permitted, i.e.

$$(\hat{L}_+)^\dagger = (\hat{L}_-). \tag{6.15}$$

The full importance of equation (6.15) is realised, e.g. when considering the state '$\hat{L}_+|lm\rangle$': in order for an inner product with a finite norm to exist, we require

$$(\hat{L}_+|lm\rangle)^\dagger(\hat{L}_+|lm\rangle) = C_{lm}^*C_{lm}, \tag{6.16}$$

$$\therefore \langle lm|(\hat{L}_+)^\dagger\hat{L}_+|lm\rangle = |C_{lm}|^2, \tag{6.17}$$

[1] This is true in non-relativistic quantum mechanics; but in relativistic quantum mechanics, as shown by Dirac, spin emerges from the so-called Dirac equation.

$$\therefore \langle lm|\hat{L}_-\hat{L}_+|lm\rangle = |C_{lm}|^2, \tag{6.18}$$

$$\therefore \langle lm|\left(\hat{L}^2 - \hat{L}_z^2 - \hbar\hat{L}_z\right)|lm\rangle = |C_{lm}|^2, \tag{6.19}$$

$$\therefore \{l(l+1)\hbar^2 - m^2\hbar^2 + m\hbar\}\langle lm|lm\rangle = |C_{lm}|^2, \tag{6.20}$$

$$\therefore C_{lm} = e^{i\phi}\sqrt{l(l+1) - m(m+1)}\,\hbar, \tag{6.21}$$

or

$$C_{lm} = e^{i\phi}\sqrt{(l-m)(l+m+1)}\,\hbar, \tag{6.22}$$

usually making the choice $\phi := 0$.

The entire spectrum of spin-angular momentum follows from equation (6.14) (i.e. from the result, equation (6.22)) by repeated application of $\hat{L}_+$ to the 'lowest-weight' state $|lm\rangle = |l, -l\rangle$ (or of $\hat{L}_-$ to the 'highest-weight' state $|lm\rangle = |l, +l\rangle$) for $l = 1, 2, 3, \ldots$ ($l = 0$ is trivial) or similarly for $j = \frac{1}{2}, \frac{3}{2}, \frac{5}{2}, \ldots$ with $\hat{J}_\pm$ acting on $|j, m_j\rangle = |j, \pm j\rangle$. This process is known technically as *inducing* representations.

From the above basic algebraic relations, irreducible representations are defined: the term 'irreducible' refers to the fact that $\hat{L}_\pm$ cannot change l (as also, rotations cannot change the magnitude of a vector such as $\vec{L}$). Further, irreducible tensor operators are defined: these are called spherical tensor operators with respect to $SO(3)$ and $SU(2)$ tensor operators with respect to $SU(2)$. Then follows the construction of new irreducible representations (irreps from here on) by taking Kronecker products of (generally simpler) pairs of irreps. This process involves Clebsch–Gordan coefficients, and applies to combining kets with kets, bras with bras, outer products of kets with bras, tensor operators with tensor operators, tensor operators with kets, and (adjoints of) tensor operators with bras, e.g.

$$|JM\rangle = \sum_{m_1}\langle j_1 m_1 j_2 m_2|JM\rangle|j_1 m_1\rangle|j_2 m_2\rangle, \tag{6.23}$$

$$M = m_1 + m_2, \quad j_1 + j_2 \geqslant J \geqslant |j_1 - j_2|, \tag{6.24}$$

and (cf. equation (3.72))

$$\hat{T}_M^{(J)} = \sum_{m_1}\langle j_1 m_1 j_2 m_2|JM\rangle \hat{T}_{m_1}^{(j_1)}\hat{T}_{m_2}^{(j_2)}. \tag{6.25}$$

The powerful Wigner–Eckart theorem follows (cf. section 3.3), viz.

$$\langle j_1 m_1|\hat{T}_q^{(k)}|j_2 m_2\rangle = \frac{1}{(2k+1)}\sum_{m_1}\langle j_1 m_1 j_2 m_2|kq\rangle\langle j_1||\hat{T}^{(k)}||j_2\rangle. \tag{6.26}$$

All of the above stems from the relationship

$$\hat{L} := \vec{r} \times \vec{p}, \tag{6.27}$$

and equation (6.1), for angular momentum and the algebraic isomorphism of spin in quantum mechanics, i.e. equations (6.2)–(6.4), with $\vec{J} \equiv \vec{S}$. Thus, the algebra of angular momentum and (by algebraic isomorphism) of spin is the result of equation (6.27) which is the 'synthesis' of a new algebraic structure, manifest in equations (6.2)–(6.4), from the original (axiomatic) algebraic structure, equation (6.1).

Beyond the widespread applicability of the algebraic structure, defined above, to finite many-body quantum systems, the structure leads to an enormous reduction of the Hilbert space into subspaces labelled by l (or j), i.e. the Hilbert space for finite many-body quantum systems is *reducible*. Then, at the hands of the Wigner–Eckart theorem, computations are 'reduced', often to simple ratios of products and quotients of integers (given by Clebsch–Gordan coefficients).

6.2 The Lie algebra $su(1,1) \sim sp(1,R)$

In the preceding section, the algebraic consequences of $\hat{L} := \vec{r} \times \vec{p}$, i.e. equation (6.27) were outlined, stemming from $[\hat{x}_i, \hat{p}_j] = i\hbar\delta_{ij}\hat{I}$, i.e. equation (6.1). One can enquire: what other functions of $\hat{x}_i$, $\hat{p}_i$ (and by implication, their corresponding classical counterparts) have algebras that are realised in physical systems?

Consider the following three operators,

$$\hat{T}_1 := \hat{\mathbf{r}} \cdot \hat{\mathbf{r}}, \tag{6.28}$$

$$\hat{T}_2 := \hat{\mathbf{r}} \cdot \hat{\mathbf{p}} + \hat{\mathbf{p}} \cdot \hat{\mathbf{r}}, \tag{6.29}$$

$$\hat{T}_3 := \hat{\mathbf{p}} \cdot \hat{\mathbf{p}}. \tag{6.30}$$

These operators obey the closed set of commutator bracket relations,

$$[\hat{T}_1, \hat{T}_2] = 4i\hbar\hat{T}_1, \tag{6.31}$$

$$[\hat{T}_2, \hat{T}_3] = 4i\hbar\hat{T}_3, \tag{6.32}$$

$$[\hat{T}_3, \hat{T}_1] = -2i\hbar\hat{T}_2, \tag{6.33}$$

which follows directly from equation (6.1). Taking the linear combinations, viz.

$$\hat{K}_1 := \frac{i}{4}(\hat{T}_1 - \hat{T}_3), \quad \hat{K}_2 := \frac{-1}{4}\hat{T}_2, \quad \hat{K}_3 := \frac{i}{4}(\hat{T}_1 + \hat{T}_3), \tag{6.34}$$

yields the commutator bracket relations

$$[\hat{K}_1, \hat{K}_2] = -i\hbar\hat{K}_3, \tag{6.35}$$

$$[\hat{K}_2, \hat{K}_3] = i\hbar\hat{K}_1, \tag{6.36}$$

$$[\hat{K}_3, \hat{K}_1] = i\hbar\hat{K}_2. \tag{6.37}$$

Equations (6.35)–(6.37) bear a close resemblance to equations (6.2)–(6.4). They can be placed in a single set of algebraic relations, viz.

$$[\hat{P}_1, \hat{P}_2] := \gamma i\hbar\hat{P}_3, \tag{6.38}$$

$$[\hat{P}_2, \hat{P}_3] := i\hbar\hat{P}_1, \tag{6.39}$$

$$[\hat{P}_3, \hat{P}_1] := i\hbar\hat{P}_2; \tag{6.40}$$

where for $\gamma = +1$, $\hat{P}_i := \hat{J}_i$, and for $\gamma = -1$, $\hat{P}_i := K_i$, $i = 1, 2, 3$. The algebra defined by equations (6.35)–(6.37) is called $su(1,1)$ and the algebra defined by equations (6.2)–(6.4) (as already detailed) is called $su(2)$. As $su(2)$ and $so(3)$ are isomorphic algebras, so $su(1,1)$ is isomorphic to $so(2,1)$. Recall, the group $SO(2,1)$ describes the invariance

$$-x_1^2 + x_2^2 + x_3^2 = -(x_1')^2 + (x_2')^2 + (x_3')^2, \tag{6.41}$$

which defines a hyperboloidal surface (much as $x_1^2 + x_2^2 + x_3^2 = \text{const.}$ defines a spherical surface). The algebra $su(1,1)$ is also isomorphic to the so-called *symplectic algebra* in one dimension, $sp(1, \mathbb{R})$. Symplectic transformations, and their associated group elements are well known from how they act on *phase space* (phase space describes the domain of the dynamical variables $\{x_{in}, p_{in}\}$, $i = 1, 2, 3$, and n designates that there are n particles) which is a real space. The scalar form of equations (6.28)–(6.30) means that they are one dimensional, e.g. they could be used to describe radial distributions in the phase space associated with $(3, \mathbb{R})$ for a single particle. Indeed, in Volume 1, chapter 12, this algebra is used to solve central force problems, specifically the isotropic three-dimensional harmonic oscillator and the hydrogen atom: Therein, one can find more details of the algebras $su(1,1) \sim sp(1, \mathbb{R}) \sim so(2,1)$.

The algebra defined by equations (6.28)–(6.30) can be seen to connect directly to the isotropic three-dimensional harmonic oscillator, via its Hamiltonian,

$$\hat{H} = \frac{\hat{p}^2}{2m} + \frac{1}{2}k\hat{r}^2, \tag{6.42}$$

when the scale factor $\alpha := \sqrt{\frac{m\omega}{\hbar}}$, $\omega := \sqrt{\frac{k}{m}}$ is introduced to yield

$$\hat{H} = \hat{P}^2 + \hat{Q}^2, \tag{6.43}$$

$$\hat{P} := \frac{\hat{p}}{\alpha\hbar}, \quad \hat{Q} := \alpha x, \tag{6.44}$$

whence, via equation (6.34),

$$\hat{H} = \hat{T}_3 + \hat{T}_1 = -4i\hat{K}_3. \tag{6.45}$$

Then, defining

$$\hat{K}_{\pm} := \hat{K}_1 \pm i\hat{K}_2, \tag{6.46}$$

$$\therefore [\hat{H}, \hat{K}_{\pm}] = \mp 4i\hbar\hat{K}_{\pm}. \tag{6.47}$$

The algebraic structure originating in equations (6.28)–(6.30), together with the definition of $\hat{H}$, equation (6.42), constitutes a dynamical algebra, where the Hamiltonian is a function of the generators of the algebra and can be solved using the ladder operators, equations (6.46) and (6.47), applied to equation (6.45) to obtain energy eigenvalues and energy eigenvectors.

An insightful procedure for understanding the algebraic content of equations (6.45)–(6.47) is to define the operators

$$a_i^{\dagger} := \frac{-i\hat{p}_i}{\alpha\hbar} + \alpha\hat{x}_i, \tag{6.48}$$

$$a_i := \frac{i\hat{p}_i}{\alpha\hbar} + \alpha\hat{x}_i, \quad i = 1, 2, 3, \tag{6.49}$$

whence

$$\hat{H} = \sum_{i=1}^{3}\left(a_i^{\dagger}a_i + \frac{1}{2}\right)\hbar\omega, \tag{6.50}$$

and, via

$$\hat{x}_i = \frac{1}{2\alpha}(a_i^{\dagger} + a_i), \tag{6.51}$$

$$\hat{p}_i = i\alpha\hbar(a_i^{\dagger} - a_i), \tag{6.52}$$

we obtain the dependence of $\hat{K}_{\pm}$ on $a_i^{\dagger}$, a_i, viz.

$$\hat{K}_{+} \sim \mathbf{a}^{\dagger} \cdot \mathbf{a}^{\dagger} = \sum_{i=1}^{3} a_i^{\dagger}a_i^{\dagger} \tag{6.53}$$

and

$$\hat{K}_{-} \sim \mathbf{a} \cdot \mathbf{a} = \sum_{i=1}^{3} a_i a_i, \tag{6.54}$$

i.e. they are 'double' raising and lowering operators. In terms of the simple harmonic oscillator, Volume 1, chapter 5, they produce two ladders, viz.

$$E_n^{\text{even}} = \frac{1}{2}\hbar\omega, \frac{5}{2}\hbar\omega, \frac{9}{2}\hbar\omega, \ldots \tag{6.55}$$

and

$$E_n^{\text{odd}} = \frac{3}{2}\hbar\omega, \frac{7}{2}\hbar\omega, \frac{11}{2}\hbar\omega, \ldots. \tag{6.56}$$

The algebra defined by $\{a^\dagger, a, \hat{I}\}$, recall $[a, a^\dagger] = \hat{I}$, is called the Heisenberg–Weyl algebra in one dimension, $hw(1)$; and the algebra defined by $\{a_i^\dagger, a_i, \hat{I};\ i = 1, 2, 3\}$ is $hw(3)$. These algebras are examples of 'factorization' algebras[2]: note the Hamiltonian is factorized, manifestly $\hat{H} \sim a^\dagger a$ into the product of two operators. The algebra $hw(3)$ is another way in which the basic commutator algebra, equation (6.1) can be 'mapped' into a new algebraic structure, albeit with scaling of x_i and p_i as manifest in equations (6.48) and (6.49). (The origin of this algebra is due to Dirac who based the definitions of $a^\dagger$ and a on *phasors*, a method by which simple harmonic oscillations can be analysed.)

6.3 Rank-2 Lie algebras

The algebras introduced in sections 6.1 and 6.2 are rank-1 Lie algebras. What this means is that within the algebra there are no pairs of commuting operators. The algebra $hw(3)$ has three independent copies of $hw(1)$ and so has rank-3: $hw(3)$ will reappear in this section. Rank-2 Lie algebras possess one pair of commuting generators. By choosing various linear combinations of generators, this pair can be adapted to address various requirements: some pairs have a physics interpretation, some pairs facilitate the 'solving' of the structure of the algebra, especially the labelling of irreps. In quantum mechanics, rank-2 algebras which are functions of $\{x_i, p_i;\ i = 1, 2, 3\}$ are of especial interest. In sections 6.1 and 6.2 we exhausted the simplest of such functions—a vector cross product, $so(3) \sim su(2)$, a vector scalar product, $so(2,1) \sim su(1,1) \sim sp(1, \mathbb{R})$, and $hw(1)$. These are all rank-1 Lie algebras. Thus, we turn to a 'higher level' of functions of x_i, p_i.

6.3.1 *su*(3) and the isotropic harmonic oscillator in three dimensions

The isotropic harmonic oscillator in three dimensions has already received considerable attention. We summarise first what has been learned, and then the $su(3)$ algebraic structure is introduced.

The three-dimensional isotropic harmonic oscillator has the Hamiltonian

$$\hat{H} = \frac{\hat{p}^2}{2m} + \frac{1}{2}k\hat{r}^2. \tag{6.57}$$

Much can be learned about this system by treating it as three uncoupled one-dimensional harmonic oscillators:

$$\hat{H} = \sum_{j=x,y,z,}\left(\frac{\hat{p}_j^2}{2m} + \frac{1}{2}k\hat{r}_j^2\right). \tag{6.58}$$

[2] There are many useful factorization algebras encountered in quantum mechanics. Some examples were developed for central force problems in Volume 1, chapter 12.

Thus, defining

$$a_j^\dagger := \frac{1}{\sqrt{2}}\left(\frac{-i\hat{p}_j}{\alpha\hbar} + \alpha\hat{r}_j\right), \quad a_j := \frac{1}{\sqrt{2}}\left(\frac{i\hat{p}_j}{\alpha\hbar} + \alpha\hat{r}_j\right), \tag{6.59}$$

where $\alpha := \left(\frac{m\omega}{\hbar}\right)^{\frac{1}{2}}, \ \omega := \left(\frac{k}{m}\right)^{\frac{1}{2}}$,

$$\left[a_i, a_j^\dagger\right] = \delta_{ij}, \ [a_i, a_j] = \left[a_i^\dagger, a_j^\dagger\right] = 0, \tag{6.60}$$

$$\hat{H} = \sum_j \left(a_j^\dagger a_j + \frac{1}{2}\right)\hbar\omega, \tag{6.61}$$

$$\hat{H}|n_x n_y n_z\rangle = \left(n_x + n_y + n_z + \frac{3}{2}\right)\hbar\omega|n_x n_y n_z\rangle, \tag{6.62}$$

$$a_x^\dagger|n_x n_y n_z\rangle = \sqrt{n_x + 1}\,|n_x + 1, n_x n_y\rangle, \text{ etc.}, \tag{6.63}$$

$$a_x|n_x n_y n_z\rangle = \sqrt{n_x}\,|n_x - 1, n_x n_y\rangle, \text{ etc.}, \tag{6.64}$$

$$n_x, n_y, n_z = 0, 1, 2, \ldots, \quad n = n_x + n_y + n_z. \tag{6.65}$$

The foregoing treatment of the 3D harmonic oscillator leads immediately to the identification of $SU(3)$ as the symmetry group of the system by considering the nine operators

$$a_i^\dagger a_j; \ i, j = x, y, z; \ \left[\hat{H}, a_i^\dagger a_j\right] = 0. \tag{6.66}$$

From this set, the linear combination $a_x^\dagger a_x + a_y^\dagger a_y + a_z^\dagger a_z$ is removed because it is related to the total energy of the system which is conserved. This leaves eight operators which define an $su(3)$ algebra. The $su(3)$ irreps are labelled by $n = n_x + n_y + n_z$ (or by the total energy of the system). The degeneracies of the irreps are directly computed in table 6.1, i.e. $n[d] = 0[1], 1[3], 2[6], 3[10], 4[15], \cdots, n[\frac{1}{2}(n+1)(n+2)]$.

Although the 3D harmonic oscillator possesses $SU(3)$ symmetry, the foregoing solution uses only the algebraic structure[3]

$$u(3) = hw(1)_x \otimes hw(1)_y \otimes hw(1)_z, \tag{6.67}$$

where the 'Heisenberg–Weyl' algebras are defined by $\{a_i, a_i^\dagger, \hat{N}_i \equiv a_i^\dagger a_i, \hat{I}; i = x, y, x\}$.

$$[a_i, a_i^\dagger] = \hat{I}, \qquad [a_i, \hat{N}_i] = a_i, \qquad [a_i^\dagger, \hat{N}_i] = -a_i^\dagger, \tag{6.68}$$

[3] $u(3)$ is a rank-3 Lie algebra, e.g. the three operators $a_1^\dagger a_1$, $a_2^\dagger a_2$, and $a_3^\dagger a_3$ are mutually commuting.

Table 6.1. Degeneracies of the three-dimensional harmonic oscillator elucidated using the partitioning of n over n_x, n_y, n_z.

n	n_x	n_y	n_z	Degeneracy, d
0	0	0	0	1
1	1	0	0	
	0	1	0	3
	0	0	1	
2	2	0	0	
	0	2	0	
	0	0	2	
	1	1	0	6
	1	0	1	
	0	1	1	

$$[a_i, \hat{I}] = 0, \qquad [a_i^\dagger, \hat{I}] = 0, \qquad [\hat{N}_i, \hat{I}] = 0. \tag{6.69}$$

The algebra $hw(1)$ is non-compact.

The solution of the 3D harmonic oscillator using $u(3) = \prod_{i=x,y,z} \otimes\, hw(1)_i$ reveals none of the symmetry of the system, although the degeneracies reveal that it must be present.

The 3D harmonic oscillator Hamiltonian commutes with $\hat{L}_x$, $\hat{L}_y$, $\hat{L}_z$ because of its rotational symmetry, and thus the 3D harmonic oscillator eigenstates can be labelled with the quantum numbers l and m of the simultaneous observables $\hat{L}^2$ and $\hat{L}_z$. The $(2l + 1)$ degeneracy of angular momentum irreps is less than the degeneracy of the 3D harmonic oscillator energy shells, indicating that there must be another conserved dynamical quantity.

The 3D harmonic oscillator Hamiltonian also commutes with the components of the tensor operator, $\hat{\mathbf{A}}$,

$$\hat{A}_{ij} := \frac{1}{2}(\hat{p}_i\hat{p}_j + mk\hat{r}_i\hat{r}_j), \quad i, j = x, y, z. \tag{6.70}$$

It is useful to define (note: the symbol 'k' is serving two independent standard, familiar roles here)

$$\hat{M}_i := \hat{A}_{jk} = \hat{A}_{kj} = \frac{1}{2}(\hat{p}_j\hat{p}_k + mk\hat{r}_j\hat{r}_k), \tag{6.71}$$

$$\hat{N}_i := \hat{A}_{ii} = \frac{1}{2}\left(\hat{p}_i^2 + mk\hat{r}_i^2\right), \tag{6.72}$$

$$\hat{L}_i := \hat{r}_j\hat{p}_k - \hat{r}_k\hat{p}_j. \tag{6.73}$$

Table 6.2. Tabulation of the 36 commutator bracket relations for the operators defined in equations (6.71)–(6.73), where $\zeta = \frac{mk}{4}$.

	$\hat{L}_x$	$\hat{L}_y$	$\hat{L}_z$	$\hat{M}_x$	$\hat{M}_y$	$\hat{M}_z$	$\hat{N}_x$	$\hat{N}_y$	$\hat{N}_z$
$\hat{L}_x$		$i\hbar\hat{L}_z$	$-i\hbar\hat{L}_y$	$i\hbar(\hat{N}_z - \hat{N}_y)$	$-i\hbar\hat{M}_z$	$i\hbar\hat{M}_y$	0	$2i\hbar\hat{M}_x$	$-2i\hbar\hat{M}_x$
$\hat{L}_y$			$i\hbar\hat{L}_x$	$i\hbar\hat{M}_z$	$i\hbar(\hat{N}_x - \hat{N}_z)$	$-i\hbar\hat{M}_x$	$-2i\hbar\hat{M}_y$	0	$2i\hbar\hat{M}_y$
$\hat{L}_z$				$-i\hbar\hat{M}_y$	$i\hbar\hat{M}_x$	$i\hbar(\hat{N}_y - \hat{N}_x)$	$2i\hbar\hat{M}_z$	$-2i\hbar\hat{M}_z$	0
$\hat{M}_x$					$-i\hbar\zeta\hat{L}_z$	$i\hbar\zeta\hat{L}_y$	0	$-2i\hbar\zeta\hat{L}_x$	$2i\hbar\zeta\hat{L}_x$
$\hat{M}_y$						$-i\hbar\zeta\hat{L}_x$	$2i\hbar\zeta\hat{L}_y$	0	$-2i\hbar\zeta\hat{L}_y$
$\hat{M}_z$							$-2i\hbar\zeta\hat{L}_z$	$2i\hbar\zeta\hat{L}_z$	0
$\hat{N}_x$								0	0
$\hat{N}_y$									0
$\hat{N}_z$									

The nine operators $\{\hat{L}_x, \hat{L}_y, \hat{L}_z, \hat{M}_x, \hat{M}_y, \hat{M}_z, \hat{N}_x, \hat{N}_y, \hat{N}_z\}$ obey 36 commutator brackets (see table 6.2):

- The commutation properties of these operators with $\hat{H}$ follow from $m\hat{H} = \hat{N}_x + \hat{N}_y + \hat{N}_z$ (remove this from set).
- Cannot disentangle the nine operators into two subsets of operators $\{A_i\}$ and $\{B_j\}$ such that $[A_i, B_j] = 0$, for all i, j.
- Maximum mutually commuting subset, start with $\hat{L}_z$: commutes with $\hat{N}_z$ and $\hat{N}_x + \hat{N}_y$. Choose $\frac{1}{\sqrt{6}}(2\hat{N}_z - \hat{N}_x - \hat{N}_y) := \hat{Q}_0^{(2)}$.
- Choice of $\hat{Q}_0^{(2)}$ because with

$$\hat{Q}_{\pm 1}^{(2)} := -i\hat{M}_x \mp \hat{M}_y, \quad \hat{Q}_{\pm 2}^{(2)} := \frac{1}{2}(\hat{N}_x - \hat{N}_y \pm 2i\hat{M}_z), \tag{6.74}$$

$\{\hat{Q}_0^{(2)}, \hat{Q}_{\pm 1}^{(2)}, \hat{Q}_{\pm 2}^{(2)}\}$ form a spherical tensor of rank-2, recall (cf. equations (3.45) and (3.46))

$$\left[\hat{L}_z, \hat{T}_q^{(k)}\right] = q\hbar\hat{T}_q^{(k)}, \tag{6.75}$$

$$\left[\hat{L}_\pm, \hat{T}_q^{(k)}\right] = \sqrt{(k \mp q)(k \pm q + 1)}\,\hbar\hat{T}_{q\pm 1}^{(k)}. \tag{6.76}$$

- Need raising and lowering operators:

From the maximal mutually commuting subset

$$\hat{L}_z(=\hat{L}_0), \quad \hat{Q}_0 = \frac{1}{\sqrt{6}}(2\hat{N}_z - \hat{N}_x - \hat{N}_y), \tag{6.77}$$

and $\hat{L}_{\pm}$

$$[\hat{L}_0, \hat{L}_{\pm}] = \pm\hbar\hat{L}_{\pm}; \tag{6.78}$$

we immediately obtain from $[\hat{L}_z, \hat{T}_q^{(k)}] = q\hbar T_q^{(k)}$ (equation (6.75)):

$$[\hat{L}_0, \hat{Q}_{\pm 1}] = \pm\hbar\hat{Q}_{\pm 1}, \tag{6.79}$$

$$[\hat{L}_0, \hat{Q}_{\pm 2}] = \pm 2\hbar\hat{Q}_{\pm 2}. \tag{6.80}$$

For the commutator brackets of $\hat{L}_{\pm}$, $\hat{Q}_{\pm 1}$, $\hat{Q}_{\pm 2}$ with $\hat{Q}_0$ (cf. equation (6.75)):

$$[\hat{Q}_0, \hat{L}_{\pm}] = -\sqrt{6}\hbar\hat{Q}_{\pm 1} \tag{6.81}$$

and

$$[\hat{Q}_0, \hat{Q}_{\pm 1}] = -\frac{1}{2}\sqrt{\frac{3}{2}}\hbar mk\hat{L}_{\pm}, \tag{6.82}$$

$$[\hat{Q}_0, \hat{Q}_{\pm 2}] = 0. \tag{6.83}$$

Evidently,

$$\begin{aligned}\left[\hat{Q}_0, \left(\hat{L}_{\pm} \mp \frac{2}{\sqrt{mk}}\hat{Q}_{\pm 1}\right)\right] &= [\hat{Q}_0, \hat{L}_{\pm}] \mp \frac{2}{\sqrt{mk}}[\hat{Q}_0, \hat{Q}_{\pm 1}] \\ &= -\sqrt{6}\hbar\hat{Q}_{\pm 1} \mp \frac{2}{\sqrt{mk}}\left(-\frac{1}{2}\sqrt{\frac{3}{2}}\right)\hbar mk\hat{L}_{\pm} \\ &= \pm\sqrt{3mk}\,\hbar\left(\hat{L}_{\pm} \mp \frac{2}{\sqrt{mk}}\hat{Q}_{\pm 1}\right).\end{aligned} \tag{6.84}$$

Define

$$\hat{P}_{\pm} := \hat{L}_{\pm} \mp \frac{2}{\sqrt{mk}}\hat{Q}_{\pm 1}, \quad \hat{R}_{\pm} := \hat{L}_{\pm} \pm \frac{2}{\sqrt{mk}}\hat{Q}_{\pm 1}, \tag{6.85}$$

whence ($\kappa := \sqrt{\frac{3mk}{2}}$)

$$[\hat{Q}_0, \hat{P}_{\pm}] = \pm\kappa\hbar\hat{P}_{\pm}, \quad [\hat{Q}_0, \hat{R}_{\pm}] = \mp\kappa\hbar\hat{R}_{\pm}, \quad [\hat{Q}_0, \hat{Q}_{\pm 2}] = 0, \tag{6.86}$$

$$[\hat{L}_0, \hat{P}_{\pm}] = \pm\hbar\hat{P}_{\pm}, \quad [\hat{L}_0, \hat{R}_{\pm}] = \pm\hbar\hat{R}_{\pm}, \quad [\hat{L}_0, \hat{Q}_{\pm 2}] = \pm 2\hbar\hat{Q}_{\pm 2}, \tag{6.87}$$

$$[\hat{L}_0, \hat{Q}_0] = 0. \tag{6.88}$$

This is the 'Cartan' representation for *su*(3) in terms of $\{\hat{L}_0, \hat{Q}_0, \hat{P}_{\pm}, \hat{R}_{\pm}, \hat{Q}_{\pm 2}\}$. The action of the raising and lowering operators can be depicted in a *root diagram* (see figure 6.1).

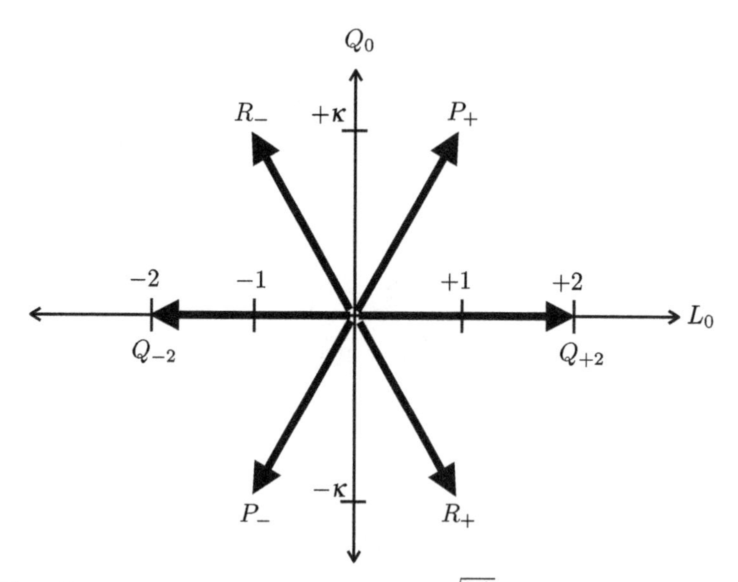

Figure 6.1. The $su(3)$ root diagram for $P_\pm$, $R_\pm$, $Q_{\pm 2}$, where $\kappa = \sqrt{\frac{3mk}{2}}$ (and $\hbar \equiv 1$). The action of the arrows is interpreted in terms of the 'raising' and 'lowering' manifested in equations (6.86)–(6.88). Thus, e.g. P_+ raises Q_0 by $+\kappa\hbar$ and L_0 by $\hbar$.

To find the irreducible representations of $su(3)$ (and subsequently Kronecker products and $su(3)$ tensor operators) we could proceed by analogy with $su(2)$ to map out the weight space. However, it is very tedious. The elegant procedure is using Cartan theory which will be developed in section 6.7. For the present, a few simple irreps in $su(3)$ weight space are shown in figure 6.2. From the weight diagrams, figure 6.2 it is evident where the 'extra' degeneracy comes from, i.e. the degeneracy due to the multiple angular momentum values within a given oscillator shell.

The algebraic structure embodied in the tensor, $\mathbf{A}$, equation (6.70) is also a key to its role as an invariant quantity for the classical mechanics of the three-dimensional isotropic harmonic oscillator. Thus,

$$Q_{ij}^{\text{class}} = \frac{1}{2}\sum_{n=1}^{A}(p_{ni}p_{nj} + mkx_{ni}x_{nj}), \quad i, j = 1, 2, 3, \tag{6.89}$$

$$\therefore \frac{\mathrm{d}Q_{ij}^{\text{class}}}{\mathrm{d}t} = \frac{1}{2}\sum_{n=1}^{A}\left(\frac{\mathrm{d}p_{ni}}{\mathrm{d}t}p_{nj} + p_{ni}\frac{\mathrm{d}p_{nj}}{\mathrm{d}t} + mk\frac{\mathrm{d}x_{ni}}{\mathrm{d}t}x_{nj} + mkx_{ni}\frac{\mathrm{d}x_{nj}}{\mathrm{d}t}\right). \tag{6.90}$$

Now, $p_{ni} = m\frac{\mathrm{d}x_{ni}}{\mathrm{d}t}$, etc.,

$$\therefore \frac{\mathrm{d}Q_{ij}^{\text{class}}}{\mathrm{d}t} = \frac{1}{2}\sum_{n=1}^{A}\left\{p_{ni}\left(\frac{\mathrm{d}p_{nj}}{\mathrm{d}t} + kx_{nj}\right) + p_{nj}\left(\frac{\mathrm{d}p_{ni}}{\mathrm{d}t} + kx_{ni}\right)\right\}. \tag{6.91}$$

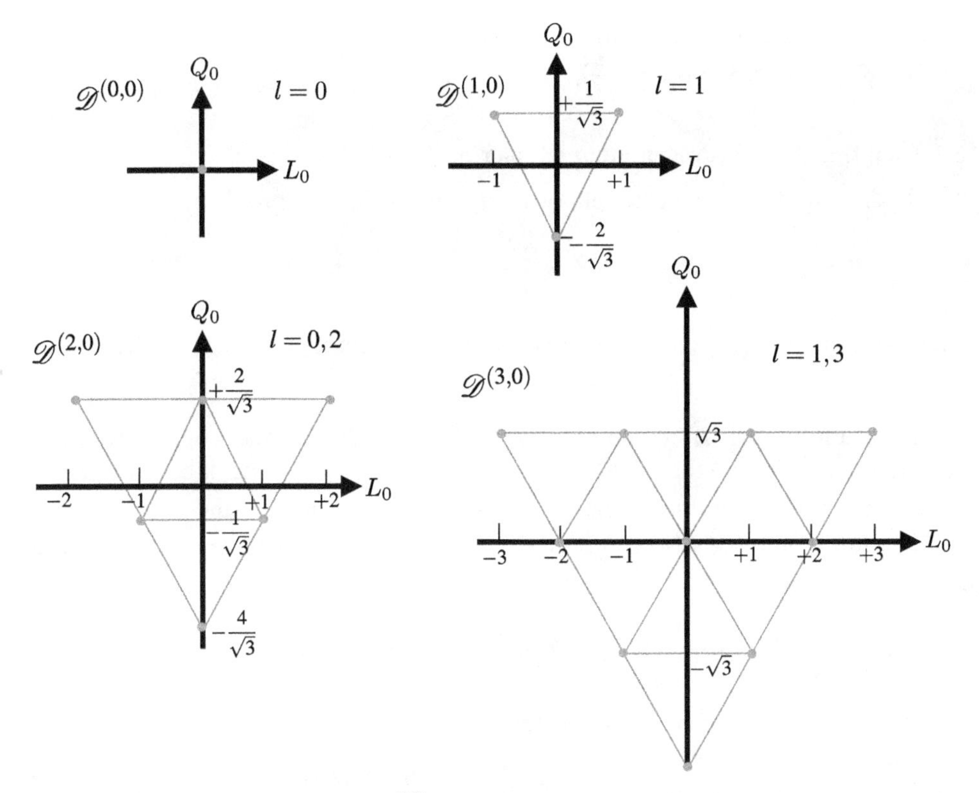

Figure 6.2. $su(3)$ weight diagrams for the $\mathcal{D}^{(\lambda,0)}$ irreps with $\lambda = 0$, 1, 2, and 3. The angular momentum values are deduced from the L_0 eigenvalues.

But,

$$m\frac{\mathrm{d}^2x_{ni}}{\mathrm{d}t^2} = F(x_{ni}) = -kx_{ni}, \text{ etc.,} \tag{6.92}$$

$$\therefore m\frac{\mathrm{d}}{\mathrm{d}t}\left(\frac{\mathrm{d}x_{ni}}{\mathrm{d}t}\right) = -kx_{ni}, \text{ etc.,} \tag{6.93}$$

$$\therefore \left(\frac{\mathrm{d}p_{ni}}{\mathrm{d}t} + kx_{ni}\right) = 0, \text{ etc.,} \tag{6.94}$$

$$\therefore \frac{\mathrm{d}Q_{ij}^{\text{class}}}{\mathrm{d}t} = 0, \tag{6.95}$$

i.e. Q_{ij}^{class} does not change with time.

The dynamical variable Q_{ij}^{class} describes the elliptical nature of the orbits under the influence of a central force, $F(r) = -kr$. The shape (major and minor axes) and location (centre of ellipse at the origin) are fixed, but the orientation is not fixed (unlike Kepler orbits, $F(r) = -\frac{k}{r}$).

6.3.2 *so*(4) and the hydrogen atom (Kepler problem)

The hydrogen atom is an example of the quantum Kepler problem, i.e.

$$\hat{H} = \frac{\hat{p}^2}{2\mu} + \frac{k}{\hat{r}}, \tag{6.96}$$

where μ is the reduced mass of the two-body system and, in the case of the H atom, $k = \frac{-e^2}{4\pi\varepsilon_0}$ (in SI units).

The Hamiltonian $\hat{H}$ commutes with $\hat{L}_x$, $\hat{L}_y$, $\hat{L}_z$ because of its rotational symmetry, and thus, the H atom energy eigenstates can be labelled with the quantum numbers l and m of the simultaneous observables $\hat{L}^2$ and $\hat{L}_z$.

The Hamiltonian $\hat{H}$ also commutes with the components of the vector operator $\vec{A}$,

$$\vec{A} := \frac{1}{2\mu}(\vec{L} \times \vec{p} - \vec{p} \times \vec{L}) - k\frac{\vec{r}}{r}, \ (\vec{L}, \vec{p} \text{ and } \vec{r} \text{ are also operators}) \tag{6.97}$$

called the Laplace–Runge–Lenz[4] vector, i.e.

$$[\hat{H}, \hat{A}_i] = 0, \ \forall\, i = x, y, z, \ [\hat{H}, \hat{A}^2] = 0. \tag{6.98}$$

Because, $\vec{A}$ is a vector operator, it obeys

$$[\hat{L}_i, \hat{A}_j] = i\hbar\varepsilon_{ijk}\hat{A}_k. \tag{6.99}$$

Further, the components of the operator $\vec{A}$ obey

$$[\hat{A}_i, \hat{A}_j] = i\hbar\varepsilon_{ijk}\left(\frac{-2\hat{H}}{\mu}\right)\hat{L}_k. \tag{6.100}$$

For a given (bound-state) energy shell, $\hat{H}$ becomes a constant, $-E$ and defining

$$\vec{K} := \sqrt{\frac{\mu}{2|E|}}\vec{A}, \tag{6.101}$$

then

$$[\hat{K}_i, \hat{K}_j] = i\hbar\varepsilon_{ijk}\hat{L}_k. \tag{6.102}$$

The operators $\{\hat{L}_x, \hat{L}_y, \hat{L}_z, \hat{K}_x, \hat{K}_y, \hat{K}_z\}$ define an *so*(4) Lie algebra and the H atom is said to possess an *SO*(4) symmetry or invariance group.

[4] Sometimes Pauli's name is associated with this vector because he was the first person to use it to quantize the H atom.

This $so(4)$ Lie algebra possesses a quadratic Casimir invariant,

$$C^{(2)}_{so(4)} = \hat{L}^2 + \hat{K}^2; \tag{6.103}$$

and $\vec{L}$ and $\vec{A}$ are orthogonal

$$\vec{L} \cdot \vec{A} = \vec{A} \cdot \vec{L} = 0. \tag{6.104}$$

The Hamiltonian can be expressed:

$$\hat{H} = \frac{-k^2\mu}{2} \frac{1}{\hat{K}^2 + \hat{L}^2 + \hbar^2}, \tag{6.105}$$

and thus its connection to the $so(4)$ algebra is manifest—it is a simple linear function of $C^{(2)}_{so(4)}$.

Unfortunately, deriving most of the above details involves extremely heavy commutator bracket manipulations. There appears to be no simple derivation.

There are also hidden subtleties:

- the $so(4)$ algebra is only obtained for $E < 0$, i.e. for bound states;
- the $so(4)$ algebraic structure is different for each energy shell, i.e. the K are different for different shells;
- it appears that replacing $\hat{H}$ by $-E$ to get $\vec{K}$ and hence the $so(4)$ algebra assumes the result that we are setting out to derive; however, we are only replacing an operator ($\hat{H}$) with one that commutes with the Hamiltonian (trivially) and thus can be replaced with a constant.

An $so(4)$ Lie algebra can be directly defined using the representation introduced in chapter 5:

$$\Lambda_{12} = x_1\frac{\partial}{\partial x_2} - x_2\frac{\partial}{\partial x_1}, \quad \Lambda_{14} = x_1\frac{\partial}{\partial x_4} - x_4\frac{\partial}{\partial x_1}, \tag{6.106}$$

$$\Lambda_{23} = x_2\frac{\partial}{\partial x_3} - x_3\frac{\partial}{\partial x_2}, \quad \Lambda_{24} = x_2\frac{\partial}{\partial x_4} - x_4\frac{\partial}{\partial x_2}, \tag{6.107}$$

$$\Lambda_{31} = x_3\frac{\partial}{\partial x_1} - x_1\frac{\partial}{\partial x_3}, \quad \Lambda_{34} = x_3\frac{\partial}{\partial x_4} - x_4\frac{\partial}{\partial x_3}, \tag{6.108}$$

whence 15 commutator brackets are obtained:

$$[\Lambda_{12}, \Lambda_{23}] = -\Lambda_{31}, \; [\Lambda_{23}, \Lambda_{31}] = -\Lambda_{12}, \; [\Lambda_{31}, \Lambda_{12}] = -\Lambda_{23}, \tag{6.109}$$

$$[\Lambda_{12}, \Lambda_{14}] = -\Lambda_{24}, \ldots, \text{ i. e. } [\Lambda_{ij}, \Lambda_{kl}] = \Lambda_{il}\delta_{jk} + \Lambda_{jk}\delta_{il} + \Lambda_{lj}\delta_{ik} + \Lambda_{ki}\delta_{jl}. \tag{6.110}$$

Two of these operators can be chosen as commuting operators, e.g. Λ_{12} and Λ_{34}: $so(4)$ is a rank-2 Lie algebra.

The operators $\{\Lambda_{12}, \Lambda_{23}, \Lambda_{31}\}$ form an $so(3)$ subalgebra.

The $so(4)$ algebra introduced for the H atom in terms of $\{\hat{L}_x, \hat{L}_y, \hat{L}_z, \hat{K}_x, \hat{K}_y, \hat{K}_z\}$ obeys the commutator bracket algebra ($i, j, k = x, y, z$):

$$[\hat{L}_i, \hat{L}_j] = i\hbar\varepsilon_{ijk}\hat{L}_k \quad \text{(3 non-zero Lie products)}, \tag{6.111}$$

$$[\hat{K}_i, \hat{K}_j] = i\hbar\varepsilon_{ijk}\hat{L}_k \quad \text{(6 non-zero Lie products)}, \tag{6.112}$$

$$[\hat{L}_i, \hat{K}_j] = i\hbar\varepsilon_{ijk}\hat{K}_k \quad \text{(6 non-zero Lie products)}; \tag{6.113}$$

and three zero Lie products, $[\hat{L}_i, \hat{K}_i] = 0, \ i = x, y, z$.

The association (cf. $\{x, y, z\} \equiv \{1, 2, 3\}$)

$$\hat{L}_x = -i\hbar\Lambda_{23}, \quad \hat{L}_y = -i\hbar\Lambda_{31}, \quad \hat{L}_z = -i\hbar\Lambda_{12}, \tag{6.114}$$

$$\hat{K}_x = -i\hbar\Lambda_{14}, \quad \hat{K}_y = -i\hbar\Lambda_{24}, \quad \hat{K}_z = -i\hbar\Lambda_{34}, \tag{6.115}$$

can be made.

The operators $\{\hat{L}_x, \hat{L}_y, \hat{L}_z\}$ define the angular momentum subalgebra of the H atom.

The operators $\{\hat{K}_x, \hat{K}_y, \hat{K}_z\}$ are 'entangled' with the angular momentum subalgebra and do not form a separate subalgebra.

Two unentangled subalgebras can be formed from

$$\vec{M} := \frac{1}{2}(\vec{L} + \vec{K}), \quad \vec{N} := \frac{1}{2}(\vec{L} - \vec{K}), \tag{6.116}$$

whence

$$[\hat{M}_i, \hat{M}_j] = i\hbar\varepsilon_{ijk}\hat{M}_k, \quad [\hat{N}_i, \hat{N}_j] = i\hbar\varepsilon_{ijk}\hat{N}_k, \tag{6.117}$$

$$[\hat{M}_i, \hat{N}_j] = 0, \quad \forall\, i, j = x, y, z. \tag{6.118}$$

Further, $\hat{M}^2$ and $\hat{N}^2$ are all-commuting.

These two subalgebras are both $su(2)$. We indicate this decomposition by

$$so(4) = su(2) \times su(2). \tag{6.119}$$

This decomposition provides the way to solve for the irreps and eigenvalues of a system with $SO(4)$ symmetry. (However, note that neither of these $su(2)$ subalgebras is the angular momentum subalgebra, or any other 'physical' subalgebra in the case of the H atom.)

The two $su(2)$ algebras, say $su_M(2)$ and $su_N(2)$, have the standard solutions

$$\hat{M}_z|j_m m_m\rangle = m_m\hbar|j_m m_m\rangle, \quad \hat{N}_z|j_n m_n\rangle = m_n\hbar|j_n m_n\rangle, \tag{6.120}$$

$$\hat{M}^2|j_m m_m\rangle = j_m(j_m + 1)\hbar^2|j_m m_m\rangle, \ \hat{N}^2|j_n m_n\rangle = j_n(j_n + 1)\hbar^2|j_n m_n\rangle. \tag{6.121}$$

The H atom Hamiltonian can now be rewritten as

$$\hat{H} = \frac{-\mu k^2}{2}\frac{1}{2\hat{M}^2 + 2\hat{N}^2 + \hbar^2}, \tag{6.122}$$

whence the energy eigenvectors are $|j_m m_m j_n m_n\rangle$ and the energy eigenvalues are

$$E(j_m j_n) = \frac{-\mu k^2}{4\hbar^2}\frac{1}{\left\{j_m(j_m + 1) + j_n(j_n + 1) + \frac{1}{2}\right\}}. \tag{6.123}$$

The eigenvalue spectra of j_m and j_n, viz. $0, \frac{1}{2}, 1, \frac{3}{2}, \ldots$ are not independent because $\vec{L} \cdot \vec{A} = \vec{A} \cdot \vec{L} = 0$, whence

$$(\vec{M} + \vec{N}) \cdot (\vec{M} - \vec{N}) = (\vec{M} - \vec{N}) \cdot (\vec{M} + \vec{N}) = 0, \tag{6.124}$$

$$\therefore M^2 - \vec{M} \cdot \vec{N} + \vec{N} \cdot \vec{M} - N^2 = M^2 + \vec{M} \cdot \vec{N} - \vec{N} \cdot \vec{M} - N^2 = 0, \tag{6.125}$$

$$\therefore \vec{M} \cdot \vec{N} = \vec{N} \cdot \vec{M}, \tag{6.126}$$

and

$$M^2 = N^2. \tag{6.127}$$

Thus,

$$j_m = j_n \equiv \nu, \quad \nu = 0, \frac{1}{2}, 1, \frac{3}{2}, \cdots \tag{6.128}$$

and

$$E_\nu = \frac{-\mu k^2}{2\hbar^2}\frac{1}{(2\nu + 1)^2}; \tag{6.129}$$

and for

$$k^2 = \left(\frac{-e^2}{4\pi\varepsilon_0}\right)^2, \quad 2\nu + 1 = n, \tag{6.130}$$

$$E_n = \frac{-R_y}{n^2}, \quad n = 1, 2, 3, \ldots, \tag{6.131}$$

where

$$R_y = \frac{\mu e^4}{8\varepsilon_0^2\hbar^2} = 13.606\,\text{eV}. \tag{6.132}$$

The degeneracies of the problem emerge from the eigenvalue spectra of $j_m m_m$, $j_n m_n$ and

$$M_z = \frac{1}{2}(L_z + K_z), \quad N_z = \frac{1}{2}(L_z - K_z), \tag{6.133}$$

whence

j_m	j_n	m_m	m_n	(m_l, m_k)
0	0	0	0	(0, 0)
$\frac{1}{2}$	$\frac{1}{2}$	$\pm\frac{1}{2}$	$\pm\frac{1}{2}$	(1, 0), (0, 1), (0, −1), (−1, 0)
1	1	0, ±1	0, ±1	(2, 0), (1, 1), (0, 2), (1, −1), (0, 0)
				(−1, 1), (0, −2), (−1, −1), (−2, 0)

Thus, using the m-scheme for m_l:

$$\begin{aligned} j_m = j_n &= 0, \quad l = 0, \\ j_m = j_n &= \frac{1}{2}, \quad l = 0, 1, \\ j_m = j_n &= 1, \quad l = 0, 1, 2, \end{aligned} \tag{6.134}$$

etc., i.e. the values of l occurring in each $so(4)$ irrep are given by

$$l = l_{\max}, l_{\max} - 1, \dots, 1, 0, \quad l_{\max} = j_m + j_n = n - 1. \tag{6.135}$$

These results can be depicted in *weight space* using *weight diagrams* (see figure 6.3). The action of the raising and lowering operators can be depicted using *root diagrams* (see figure 6.4).

The $so(4)$ Lie algebra is said to have rank two because its weight space (and root space) are two-dimensional. This reflects the two commuting $so(4)$ generators, L_z and K_z.

The $|j_m m_m\rangle$, $|j_n m_n\rangle$ states are coupled to states of good l and m using Clebsch–Gordan coefficients:

$$|lm_l\rangle = \sum_{m_m (m_n = m_l - m_m)} |j_m m_m j_n m_n\rangle \langle j_m m_m j_n m_n | l m_l\rangle. \tag{6.136}$$

6.4 *so*(5) and models with 'quadrupole' degrees of freedom (Bohr model)

As an illustration of algebraic modelling, a few details of the algebra that lies behind the Bohr model of nuclear quadrupole collective structure are given.

The Bohr model introduces five (collective) coordinates via an expression of the quadrupole shape of the nucleus, treated as a liquid drop, viz.

$$R(\theta, \phi) = R_0 \left\{ 1 + \sum_{2,\mu} \alpha^*_{2,\mu} Y_{2,\mu}(\theta, \phi) \right\}, \tag{6.137}$$

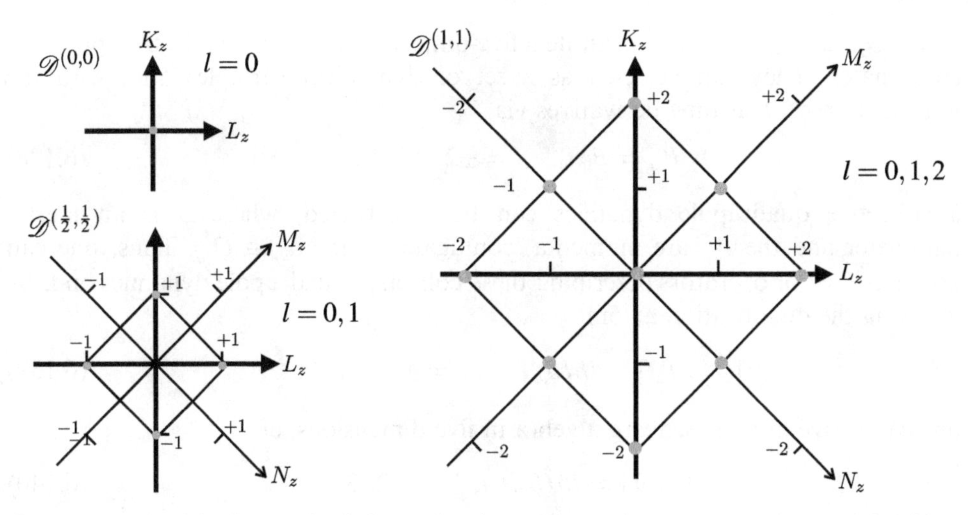

Figure 6.3. $so(4)$ weight diagrams for the lowest three irreps. The angular momentum values are deduced from the L_z eigenvalues.

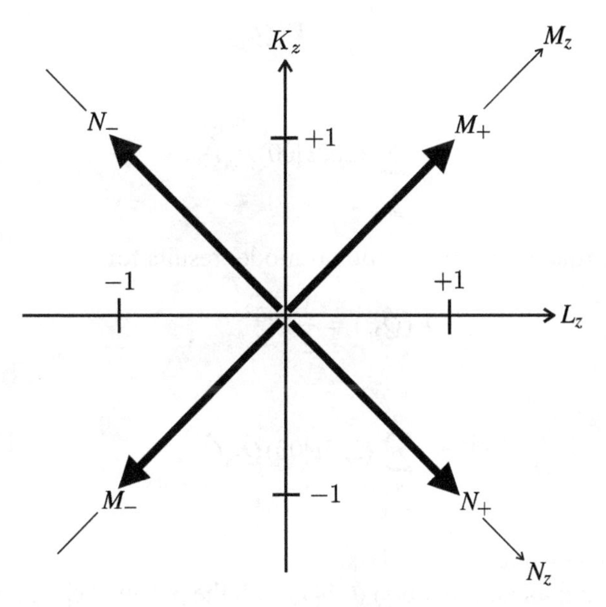

Figure 6.4. The $so(4)$ root diagram for $M_\pm$, $N_\pm$. The action of the arrows is interpreted in terms of the raising and lowering of K_z and L_z. Thus, e.g. M_+ raises K_z by +1 and L_z by +1, cf. figure 6.3.

$\mu = +2, +1, 0, -1, -2$, where R describes a sharp surface for the liquid drop, R_0 is the equivalent spherical drop radius, $Y_{2\mu}(\theta, \phi)$ are the spherical harmonics of rank-2, and the $\alpha_{2,\mu}$ are 'collective' coordinates for the model. (The appearance of the complex conjugates, $\alpha^*_{2,\mu}$ is to ensure that $R(\theta, \phi)$ is real, recall the $Y_{2\mu}(\theta, \phi)$ are, in general, complex.)

The $\alpha_{2\mu}$, $\mu = \pm 2, \pm 1, 0$, constitute a five-dimensional space of collective 'position' coordinates. They can be used as a set of dynamical variables and, with the introduction of their time derivatives via

$$P_{2\mu} := B\dot{\alpha}_{2\mu}, \quad \mu = \pm 2, \pm 1, 0, \tag{6.138}$$

a collective quadrupole dynamics can be constructed, where B is an inertia parameter and the $P_{2\mu}$ are momenta 'conjugate' to the $\alpha_{2\mu} \equiv Q_{2\mu}$. Thus, one can arrive at a set of operators describing these collective quadrupole dynamics and, by invoking the quantization axiom

$$[\hat{Q}_{2\mu}, \hat{P}_{2\mu}] := i\hbar\hat{I}\delta_{\mu\nu}, \quad \mu, \nu = \pm 2, \pm 1, 0, \tag{6.139}$$

one can arrive at a Heisenberg algebra in five dimensions, cf.

$$[\hat{x}_i, \hat{p}_j] := i\hbar\hat{I}\delta_{ij}, \quad i, j = 1, 2, 3 \tag{6.140}$$

for physical space (equation (6.1)).

Model Hamiltonians can be constructed, viz.

$$\hat{H} := \frac{\hat{P}^2}{2B} + V(\hat{Q}_{2\mu}), \tag{6.141}$$

where

$$\hat{P}^2 = \sum_{\substack{\mu \\ \nu=-\mu}} \langle 2\mu 2\nu | 00 \rangle \hat{P}_{2\mu}\hat{P}_{2\nu}. \tag{6.142}$$

For example, a (quadrupole) vibrational model results for

$$V(\hat{Q}_{2\mu}) = \frac{1}{2}C\hat{Q}^2, \tag{6.143}$$

where

$$\hat{Q}^2 = \sum_{\substack{\mu \\ \nu=-\mu}} \langle 2\mu 2\nu | 00 \rangle \hat{Q}_{2\mu}\hat{Q}_{2\nu}. \tag{6.144}$$

Recall Hamiltonians are $so(3)$ scalars.

The model Hamiltonian, equation (6.141) with the potential choice, equation (6.143) can be solved by treating it as five copies of the Heisenberg–Weyl algebra $hw(1)$, viz.

$$hw(5) = \prod_{i=1}^{5} \otimes\, hw(1)_i, \tag{6.145}$$

by defining

$$B_{2\mu}^{\dagger} := \frac{1}{\sqrt{2}}\left(\alpha\hat{Q}_{2\mu} - \frac{i}{\alpha\hbar}\hat{P}_{2\mu}\right), \tag{6.146}$$

$$B_{2\mu} := \frac{1}{\sqrt{2}}\left(\alpha\hat{Q}_{2\mu} + \frac{i}{\alpha\hbar}\hat{P}_{2\mu}\right), \tag{6.147}$$

where $\mu = \pm 2, \pm 1, 0$, and

$$\alpha := \sqrt{\frac{B\omega}{\hbar}}, \tag{6.148}$$

$$\omega := \sqrt{\frac{C}{B}}. \tag{6.149}$$

This leads directly to the energy eigenvalues

$$E = \left(N + \frac{5}{2}\right)\hbar\omega, \tag{6.150}$$

$N = 0, 1, 2....$ The quantum number N is partitioned over the n_μ as shown in table 6.3. The model is termed the five-dimensional isotropic harmonic vibrator or quadrupole vibrator. The degeneracies can be classified in terms of angular momentum quantum numbers. Recall, for a finite system such as a liquid drop, angular momentum is conserved. Thus, with the implied association of spin 2 with

Table 6.3. The first few states of the quadrupole vibrator model. The dots denote that the '2' and the two '1' entries must be distributed over all possible combinations of columns.

N	n_{+2}	n_{-1}	n_0	n_{-1}	n_{-2}	degeneracy	L
0	0	0	0	0	0	1	0
1	1						
		1					
			1			5	2
				1			
					1		
2	2					5	
	...					15	0,2,4
	1	1				10	
	...						

each quantum, the m-scheme yields the angular momentum values given in table 6.3, e.g. $L = 4$ yields nine states with directional components $M = \pm4, \pm3, \pm2, \pm1, 0$. The degeneracy in L values already for $N = 2$, raises the question 'is there a richer dynamics at work behind the scene?'. Such a dynamics emerges for $su(3)$ and $so(4)$ as detailed in sections 6.3.1 and 6.3.2; and, indeed, there is further dynamics behind the quadrupole vibrator model and it emerges from an $so(5)$ symmetry possessed by the model. A few details are given in the following.

The five model coordinates $Q_{2\mu}$ can be viewed as defining an isotropic five-dimensional space, $(5, \mathbb{R})$. Just as $(3, \mathbb{R})$ can be expressed in terms of radial and angle coordinates, $(5, \mathbb{R})$ can be similarly expressed: there is one radial coordinate and four angle coordinates. This can be viewed, much as one views a spherical surface using (r, θ, ϕ), as a hyperspherical surface using $(\beta, \gamma, \theta_1, \theta_2, \theta_3)$, where β is the radial coordinate, γ describes the axial asymmetry of the quadrupole surface, and $\theta_1, \theta_2, \theta_3$ describe the orientation Euler angles of the ellipsoidal shape of the surface: the coordinates '1, 2, 3' are a set of body-fixed coordinates $\bar{Q}_{2\mu}$, such that

$$\bar{Q}_{2,0} := \beta \cos\gamma, \tag{6.151}$$

$$\bar{Q}_{2,\pm1} := 0, \tag{6.152}$$

$$\bar{Q}_{2,\pm2} := \frac{1}{\sqrt{2}}\beta \sin\gamma, \tag{6.153}$$

and they are connected to the laboratory coordinates $Q_{2\mu}$ by

$$Q_{2m} = \sum_{\mu} \bar{Q}_{2\mu} \mathcal{D}^{(2)}_{\mu m}(\theta, \phi), \tag{6.154}$$

where $\mathcal{D}^{(2)}_{\mu,m}(\theta, \phi)$ is a rotation matrix. The rotational invariance in five dimensions, i.e. the $SO(5)$ invariance, is manifest in the parameterisation used, equations (6.151)–(6.154), viz.

$$\sum_{\mu} \bar{Q}^2_{2\mu} = \beta^2. \tag{6.155}$$

The $SO(5)$ invariance of the model can be developed using the algebra introduced in section 5.10.3. Thus, the algebra $so(5)$ has 10 generators, $\Lambda_{ij}(=-\Lambda_{ji})$, $i \neq j$, $i, j = 1, \dots, 5$. A table similar to table 5.2 can be constructed and one finds that the algebra has rank-2 (among the 10 generators, there are no triples of generators that are mutually commuting). One can proceed to define the Casimir invariant $C^{(2)}_{so(5)}$, and thus arrive at an explanation of the angular momentum degeneracy for a given value of N, the number of oscillator quanta.

A fully algebraic treatment of the Bohr model, with accommodation of deformed potentials, follows from the above details by using the algebra so(5) ⊗ su(1,1), where the su(1,1) algebra describes the 'radial' dynamics manifest in the model coordinate

β: the structure closely matches that of central force problems in (3, ℝ), as developed in Volume 1, section 12.7, using $so(3) \otimes su(1,1)$.[5]

6.5 The Lie algebra $sp(3, \mathbb{R})$ and microscopic models of nuclear collectivity

Using the approach adopted at the outset in this chapter, viz. the construction of algebras from x_i, p_i, $i = 1, 2, 3$ via polynomials, there is one further algebra which is simply defined. This involves the quadratic polynomials $x_i x_j$. To express $x_i x_j$ in familiar terms,

$$Q_{ij} := x_i x_j \leftrightarrow \begin{matrix} xx & xy & xz \\ yx & yy & yz \\ zx & zy & zz \end{matrix} \tag{6.156}$$

is a rank-2 symmetric Cartesian tensor. The symmetry is manifest in $xy = yx$, $xz = zx$, $yz = zy$. Thus, there are six independent components to $x_i x_j$; further, they are reducible by recognising that

$$x^2 + y^2 + z^2 := r^2 \tag{6.157}$$

is invariant under $SO(3)$ transformations. This reduction yields the scalar, r^2 and a rank-2 spherical tensor with components, $Q_\mu^{(2)}$, viz.

$$Q_0^{(2)} := 3z^2 - r^2, \tag{6.158}$$

$$Q_{\pm 1}^{(2)} := -2z(x \pm iy), \tag{6.159}$$

$$Q_{\pm 2}^{(2)} := (x \pm iy)^2, \tag{6.160}$$

where the linear combinations of xy, etc., are chosen to match spherical harmonics, cf. section 1.9.

From the $Q_{ij} = x_i x_j$, 'quadrupole' coordinate operators

$$\hat{Q}_{ij} := \hat{x}_i \hat{x}_j \tag{6.161}$$

can be defined; and via time derivatives, $\dot{x}_i$, etc., which lead to canonical momenta $\hat{p}_i := m\dot{x}_i$, etc., the operators

$$\hat{S}_{ij} := \hat{x}_i \hat{p}_j + \hat{p}_i \hat{x}_i, \tag{6.162}$$

$$\hat{L}_{ij} := \hat{x}_i \hat{p}_j - \hat{x}_j \hat{p}_i, \tag{6.163}$$

[5] Details are given in [1].

and

$$\hat{K}_{ij} := \hat{p}_i \hat{p}_j, \tag{6.164}$$

are obtained.

The commutator brackets for $\{\hat{Q}_{ij}, \hat{S}_{ij}, \hat{L}_{ij}, \hat{K}_{ij}\}$ follow from $[\hat{x}, \hat{p}_j] = i\hbar\hat{I}\delta_{ij}$ and define the algebra $sp(3, \mathbb{R})$, $sp \equiv$ symplectic [Greek: 'to fold together']. This is expressed most concisely using the Heisenberg–Weyl operators

$$a_i^\dagger = \frac{1}{\sqrt{2}}\left(\alpha\hat{x}_i - \frac{i}{\alpha\hbar}\hat{p}_i\right), \tag{6.165}$$

$$a_i = \frac{1}{\sqrt{2}}\left(\alpha\hat{x}_i + \frac{i}{\alpha\hbar}\hat{p}_i\right), \tag{6.166}$$

where $\alpha := \sqrt{\frac{m\omega}{\hbar}}$, $\omega := \sqrt{\frac{k}{m}}$, and thus via their inverses

$$\hat{x}_i = \frac{1}{\sqrt{2}\alpha}(a_i^\dagger + a_i), \tag{6.167}$$

$$\hat{p}_i = \frac{i}{\sqrt{2}}\alpha\hbar(a_i^\dagger - a_i), \tag{6.168}$$

we obtain

$$\hat{Q}_{ij} = \frac{1}{2\alpha^2}\left\{a_i^\dagger a_j^\dagger + a_i^\dagger a_j + a_i a_j^\dagger + a_i a_j\right\}. \tag{6.169}$$

Defining

$$\hat{A}_{ij} := a_i^\dagger a_j^\dagger, \tag{6.170}$$

$$\hat{B}_{ij} := a_i a_j, \tag{6.171}$$

$$\hat{C}_{ij} := \left(a_i^\dagger a_j + a_j a_i^\dagger\right) = a_i^\dagger a_j + \frac{1}{2}\delta_{ij}, \tag{6.172}$$

then

$$\hat{Q}_{ij} = \frac{1}{2\alpha^2}(\hat{A}_{ij} + \hat{B}_{ij} + \hat{C}_{ij} + \hat{C}_{ji}), \tag{6.173}$$

$$\hat{S}_{ij} = i\hbar(\hat{A}_{ij} - \hat{B}_{ij}), \tag{6.174}$$

$$\hat{L}_{ij} = -i\hbar(\hat{C}_{ij} - \hat{C}_{ji}), \tag{6.175}$$

$$\hat{K}_{ij} = -\frac{\alpha^2\hbar^2}{2}(\hat{A}_{ij} + \hat{B}_{ij} - \hat{C}_{ij} - \hat{C}_{ji}). \tag{6.176}$$

Then, commutator brackets can be evaluated via

$$[a_i, a_j^\dagger] = \hat{I}\delta_{ij}, \quad i, j = 1, 2, 3, \tag{6.177}$$

$$[a_i, a_j] = 0 \ \left[a_i^\dagger, a_j^\dagger\right] = 0, \tag{6.178}$$

e.g.

$$[\hat{C}_{ij}, \hat{C}_{kl}] = \left[\left(a_i^\dagger a_j + \frac{1}{2}\delta_{ij}\right), \left(a_k^\dagger a_l + \frac{1}{2}\delta_{kl}\right)\right], \tag{6.179}$$

and, recalling $[AB, CD] = AC[B, D] + A[B, C]D + C[A, D]B + [A, C]DB$ yields

$$[\hat{C}_{ij}, \hat{C}_{kl}] = \hat{C}_{il}\delta_{jk} - \hat{C}_{jk}\delta_{il}. \tag{6.180}$$

The generators of the $sp(3, \mathbb{R})$ algebra, $\{\hat{Q}_{ij}, \hat{S}_{ij}, \hat{L}_{ij}, \hat{K}_{ij}\}$ total 21—6 each for $\hat{Q}_{ij}$, $\hat{S}_{ij}$, and $\hat{K}_{ij}$, and 3 for $\hat{L}_{ij}$. Thus, there are 210 commutator brackets. This algebraic structure is greatly simplified by defining

$$\hat{Q}_{ij} := \frac{1}{2}\left(\alpha^2\hat{x}_i\hat{x}_j + \frac{1}{\alpha\hbar^2}\hat{p}_i\hat{p}_j\right), \tag{6.181}$$

cf. equation (6.70). It follows from equations (6.167) and (6.168) that

$$\hat{Q}_{ij} = \frac{1}{2}(\hat{C}_{ij} + \hat{C}_{ji}) = \hat{C}_{ij} + \frac{1}{2}\delta_{ij}. \tag{6.182}$$

The operators $\{\hat{Q}_{ij}, \hat{L}_{ij}\}$ define an $su(3)$ subalgebra for $sp(3, \mathbb{R})$, and $\hat{A}_{ij}$, $\hat{B}_{ij}$ act as double raising and lowering operators, respectively, when acting on oscillator 'shells' defined by

$$\hat{H} := \sum_{i=1}^{3}\hat{Q}_{ii} = \left(N + \frac{3}{2}\right)\hbar\omega, \tag{6.183}$$

$N = 0, 1, 2, \ldots$. A further simplification is achieved by expressing the raising and lowering operators in $so(3)$ spherical tensor form, $\hat{A}_{2\mu}$, viz.

$$\hat{A}_{2,0} := 2\hat{A}_{11} - \hat{A}_{22} - \hat{A}_{33}, \tag{6.184}$$

$$\hat{A}_{2,\pm 1} := \mp(\hat{A}_{12} \pm i\hat{A}_{13}), \tag{6.185}$$

$$\hat{A}_{2,\pm 2} := \hat{A}_{22} - \hat{A}_{33} \pm 2i\hat{A}_{23}, \tag{6.186}$$

with similar expressions for $\hat{B}_{2\mu}$. These raising and lowering operators describe giant quadrupole resonances in nuclei. (It is important to emphasize here that these modes have nothing to do with low-energy quadrupole vibrations such as can be modelled using the algebra in section 6.4.) Giant monopole resonances are described by

$$\hat{A}_{0,0} := \sum_{i=1}^{3} \hat{A}_{ii}, \tag{6.187}$$

and

$$\hat{B}_{0,0} := \sum_{i=1}^{3} \hat{B}_{ii}, \tag{6.188}$$

cf. section 6.2.

The adaptation to nuclei is via

$$Q_{ij} := \sum_{n=1}^{A} x_{ni} x_{nj}, \tag{6.189}$$

$i, j \equiv x, y, z$, for a nucleus with A nucleons (recall protons and neutrons possess near identical masses and so do not need to be distinguished herein).

6.6 Young tableaux

Starting from the fundamental representation of a group ($SU(2)$—spinor representation, $SU(3)$—quark representations, …, etc.), all the irreducible representations of a group can be constructed.

The method has been introduced for $SU(2)$ where two spin-$\frac{1}{2}$ or spinor irreps were combined in a direct or tensor or Kronecker product to yield a spin-1 and a spin-0 irrep (section 2.1). (The method is implicit in the $SU(2)$ extensions to the generation of the Clebsch–Gordan series (section 2.2) and the Schwinger representation (section 1.7).) There exists a very powerful language for manipulating irrep products which was introduced by Alfred Young in a study of the permutational symmetries of tensors. The language is variously referred to as Young tableaux or Young diagrams or Young frames or Weyl tableaux or tensor tableaux.

Consider first a single box and the labels 1 and 2:

$$\Box \begin{cases} \boxed{1} \\ \boxed{2} \end{cases}$$

This is just a way of representing an $SU(2)$ spinor. Now consider:

$$\boxed{1\,|\,1} \equiv |11\rangle \tag{6.190}$$

$$\Box\Box \quad \boxed{1\,|\,2} \equiv \frac{1}{\sqrt{2}}(|12\rangle + |21\rangle) \tag{6.191}$$

$$\boxed{2\,|\,2} \equiv |22\rangle \tag{6.192}$$

and

$$\begin{array}{|c|}\hline \ \\ \hline \ \\ \hline\end{array} \to \begin{array}{|c|}\hline 1\\ \hline 2\\ \hline\end{array} \equiv \frac{1}{\sqrt{2}}(|12\rangle - |21\rangle) \tag{6.193}$$

These are just ways of representing the possible couplings of two spinors ($1 \equiv u$ or $\uparrow$ or $+\frac{1}{2}$, $2 \equiv d$ or $\downarrow$ or $-\frac{1}{2}$).

Some rules for these tableaux immediately are evident:

(a) For row tableaux, when moving from left to right, the numbers entered in each successive box must not decrease.

Reason: $\begin{array}{|c|c|}\hline 2&1\\\hline\end{array}$ is indistinguishable from $\begin{array}{|c|c|}\hline 1&2\\\hline\end{array}$ to allow such a tableau would result in 'double counting'.

(b) For column tableaux, when moving from top to bottom, the numbers entered in each successive box must increase.

Reason: $\begin{array}{|c|}\hline 1\\\hline 1\\\hline\end{array}$ and $\begin{array}{|c|}\hline 2\\\hline 2\\\hline\end{array}$ are no states at all; $\begin{array}{|c|}\hline 2\\\hline 1\\\hline\end{array}$ is indistinguishable from $\begin{array}{|c|}\hline 1\\\hline 2\\\hline\end{array}$—it differs only by an overall phase of (-1).

To continue, for $SU(2)$, following rules (1) and (2), a third box (containing a (1) or a (2)) can be combined with the various two-box states as follows:

$$\begin{array}{|c|c|}\hline \ & \ \\\hline\end{array} \otimes \begin{array}{|c|}\hline \ \\\hline\end{array} = \left.\begin{array}{c} \begin{array}{|c|c|c|}\hline 1&1&1\\\hline\end{array} \\ \begin{array}{|c|c|c|}\hline 1&1&2\\\hline\end{array} \\ \begin{array}{|c|c|c|}\hline 1&2&2\\\hline\end{array} \\ \begin{array}{|c|c|c|}\hline 2&2&2\\\hline\end{array} \end{array}\right\} j = 1 \otimes j = \frac{1}{2} \Rightarrow j = \frac{3}{2}$$

$$\left.\begin{array}{c} \begin{array}{|c|c|}\hline 1&1\\\hline 2&\multicolumn{1}{c}{}\\\cline{1-1}\end{array} \\ \begin{array}{|c|c|}\hline 1&2\\\hline 2&\multicolumn{1}{c}{}\\\cline{1-1}\end{array} \end{array}\right\} j = 1 \otimes j = \frac{1}{2} \Rightarrow j = \frac{1}{2}$$

$$\begin{array}{|c|}\hline 1\\\hline 2\\\hline\end{array} \otimes \begin{array}{|c|}\hline \ \\\hline\end{array} = \left.\begin{array}{c} \begin{array}{|c|c|}\hline 1&1\\\hline 2&\multicolumn{1}{c}{}\\\cline{1-1}\end{array} \\ \begin{array}{|c|c|}\hline 1&2\\\hline 2&\multicolumn{1}{c}{}\\\cline{1-1}\end{array} \end{array}\right\} j = 0 \otimes j = \frac{1}{2} \Rightarrow j = \frac{1}{2}$$

A third rule for tableaux has been used in the above:

(c) The size of columns cannot increase when moving from left to right.

Reason: e.g. $\begin{array}{|c|c|}\hline 1&1\\\hline \multicolumn{1}{c|}{}&2\\\cline{2-2}\end{array}$ is indistinguishable from $\begin{array}{|c|c|}\hline 1&1\\\hline 2&\multicolumn{1}{c}{}\\\cline{1-1}\end{array}$ because in either case the '2' is antisymmetrized w.r.t. the '1's'; to allow such a tableau would, again, result in 'double counting'.

A fourth rule emerges from the above:

(d) Wherever $\begin{array}{|c|}\hline 1\\\hline 2\\\hline\end{array}$ appears in an $SU(2)$ tableau it can be removed:

Reason: It corresponds to $j = 0$ and so adds nothing to irrep building.

The procedure for building $SU(2)$ irreps using tensor tableaux is now clear: assemble all possible row tableaux, adhering to rule (1). Irreps are distinguished by the number of boxes in the row.

The preceding development, standing alone, is a rather laborious path to a simple algorithm stated in the previous paragraph. The purpose was to develop the tensor tableau calculus as a language. Its usefulness emerges when groups of rank higher than one are encountered.

6.6.1 *SU*(3) tensor tableau calculus

The fundamental or 'quark' representation of *SU*(3) is depicted

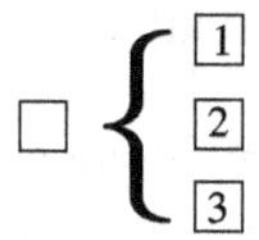

in tensor tableau form.

Higher *SU*(3) irreps are 'assembled' using quark 'building blocks' in tensor tableaux, following the rules developed in the discussion of *SU*(2). This is shown in figure 6.5.

These tableaux can be attached to weight diagrams as shown in figure 6.6.

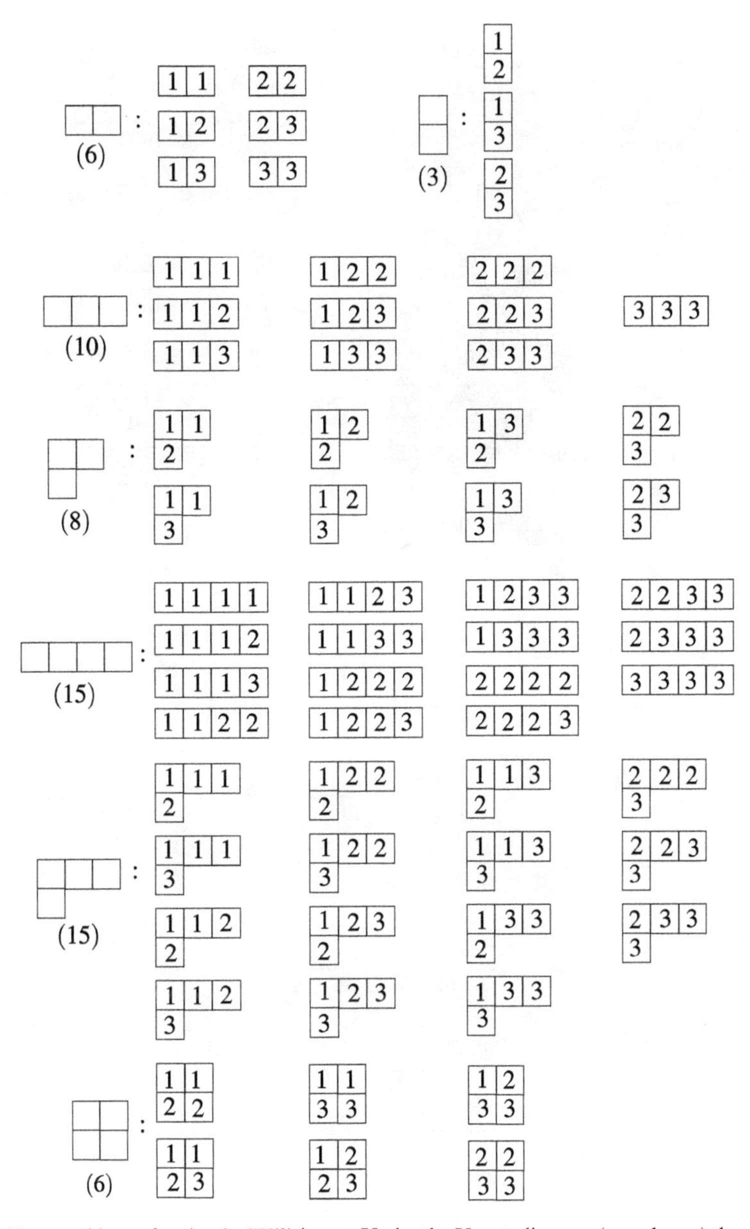

Figure 6.5. Young tableaux for simple $SU(3)$ irreps. Under the Young diagram (open boxes) the dimension of the irrep is given.

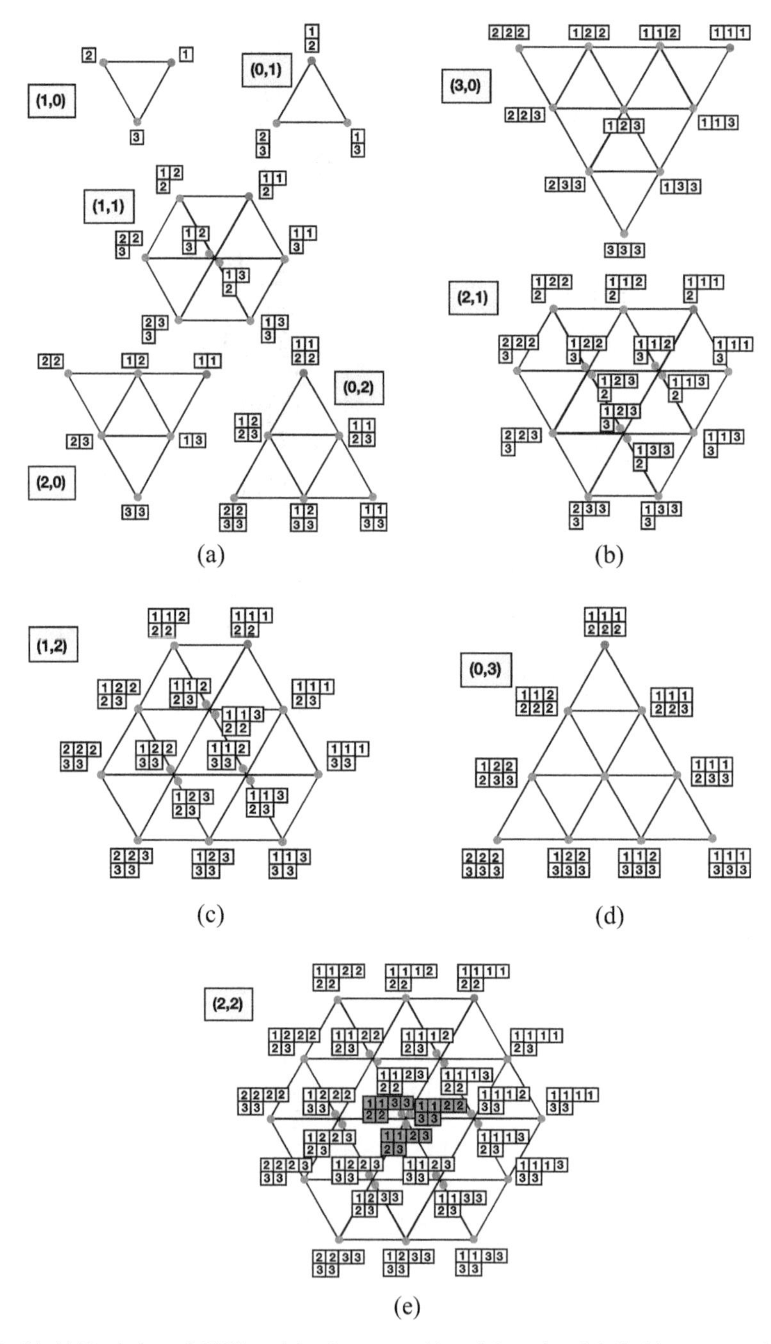

Figure 6.6. (a)–(e) Depiction of $SU(3)$ weight diagrams with weight points labelled by tensor tableaux. The diagrams present all of the low-dimension $SU(3)$ irreps, designated by the labelling quantum numbers, (λ, μ): (1,0), (0,1), (1,1), (2,0), (0,2), (3,0), (2,1), (1,2), (2,2), and (0,3). The dots designate the number of weight points at each location in the weight diagram. The tableaux for the multiplicity-3 weight point associated with the (2,2) diagram (figure 6.6(e)) are shaded pink. The total number of dots associated with each weight diagram is the dimension of the irrep. The 'highest-weight' point in each diagram is indicated by a red dot.

6.6.2 Multiplicity of a weight state in an *SU*(3) irrep

Consider the tableaux

$$\begin{array}{|c|c|c|c|}\hline 1&1&2&2\\ \hline 3&3 \\ \cline{1-2}\end{array}\ ,\quad \begin{array}{|c|c|c|c|}\hline 1&1&2&3\\ \hline 2&3 \\ \cline{1-2}\end{array}\ ,\quad \begin{array}{|c|c|c|c|}\hline 1&1&3&3\\ \hline 2&2 \\ \cline{1-2}\end{array}\ ,$$

these are the permutations of the labels that obey the rules for the box entries and they directly give the multiplicity (three) for the central weights in (2,2), cf. figure 6.6(e).

The general pattern of weight multiplicities for *SU*(3) is indicated in figure 6.7:

(a) the outermost weights have multiplicity one;
(b) for a hexagonal perimeter, the next 'ring' inwards has weights of multiplicity two;
(c) for each ring inwards the multiplicity increases by one until a triangular ring is reached: thereafter the multiplicity of each weight point remains constant and equal to that of the outermost triangular ring.

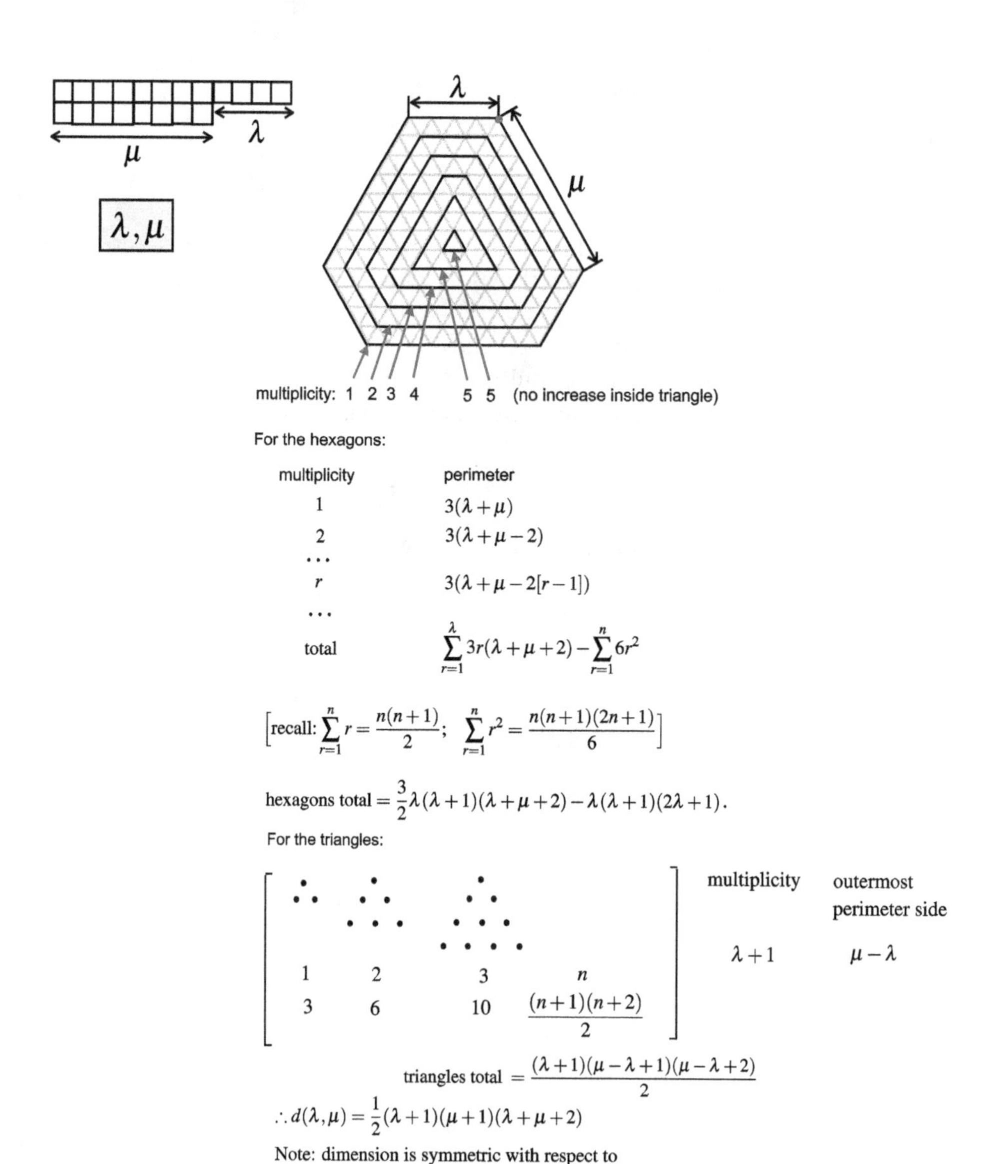

Figure 6.7. A generic depiction of $SU(3)$ weight diagrams identifying the multiplicity, 2-fold, 3-fold, 4-fold of the weight points. The outermost weight points have multiplicity 1. The 'highest-weight' state is indicated by a red dot. The lengths of the sides of the outermost hexagonal shape are directly related to the tableau as shown.

6.6.3 Dimension of an *SU*(3) irrep: Robinson 'hook-length' method (figure 6.8)

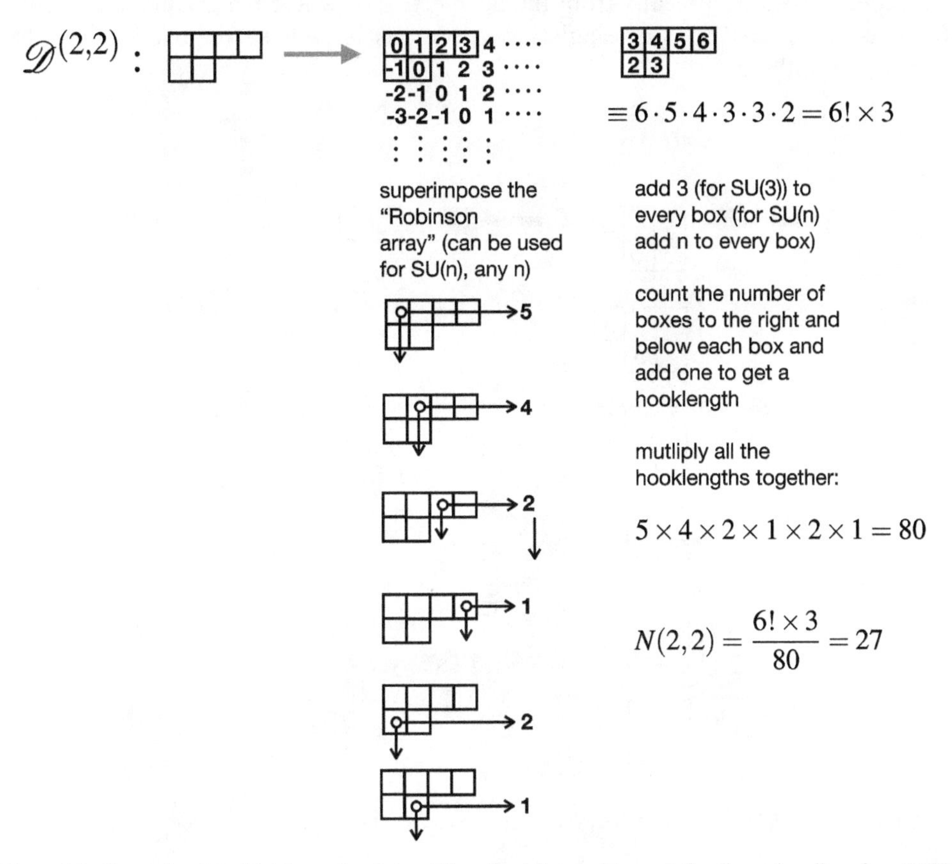

Figure 6.8. Example of the Robinson 'hook-length' method for working out the dimensionality of an *SU*(3) irrep.

6.6.4 *SU*(2) irreps contained in an *SU*(3) irrep

Example

Consider the *SU*(3) tableaux from the perspective of where the '3' quark labels can be entered (figure 6.9): (the singlet contributions have been removed where they contribute nothing).

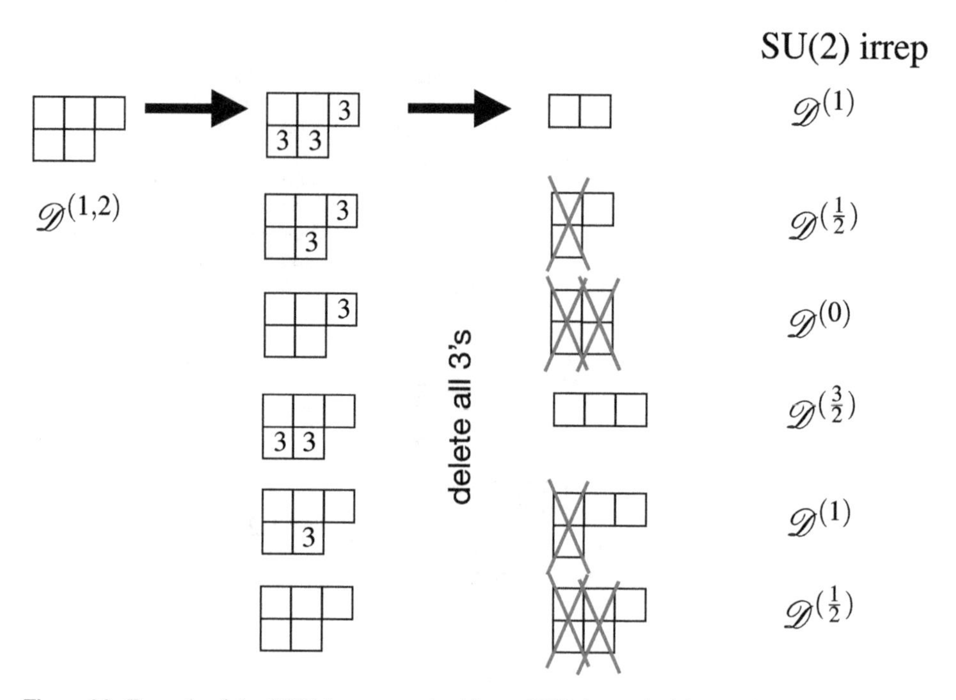

Figure 6.9. Example of the *SU*(2) irreps contained in an *SU*(3) irrep, elucidated using Young tableaux.

This process is easily identified in the $\mathcal{D}^{(1,2)}$ tableaux (figure 6.10):

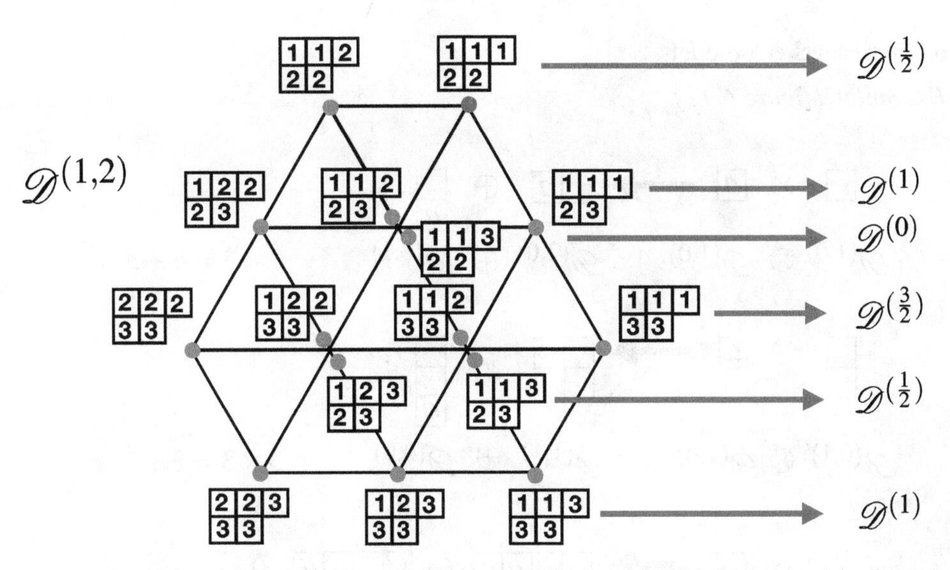

Figure 6.10. Example of the $SU(2)$ irreps contained in an $SU(3)$ irrep, elucidated using a weight diagram.

6.6.5 Kronecker products

Examples (figure 6.11)

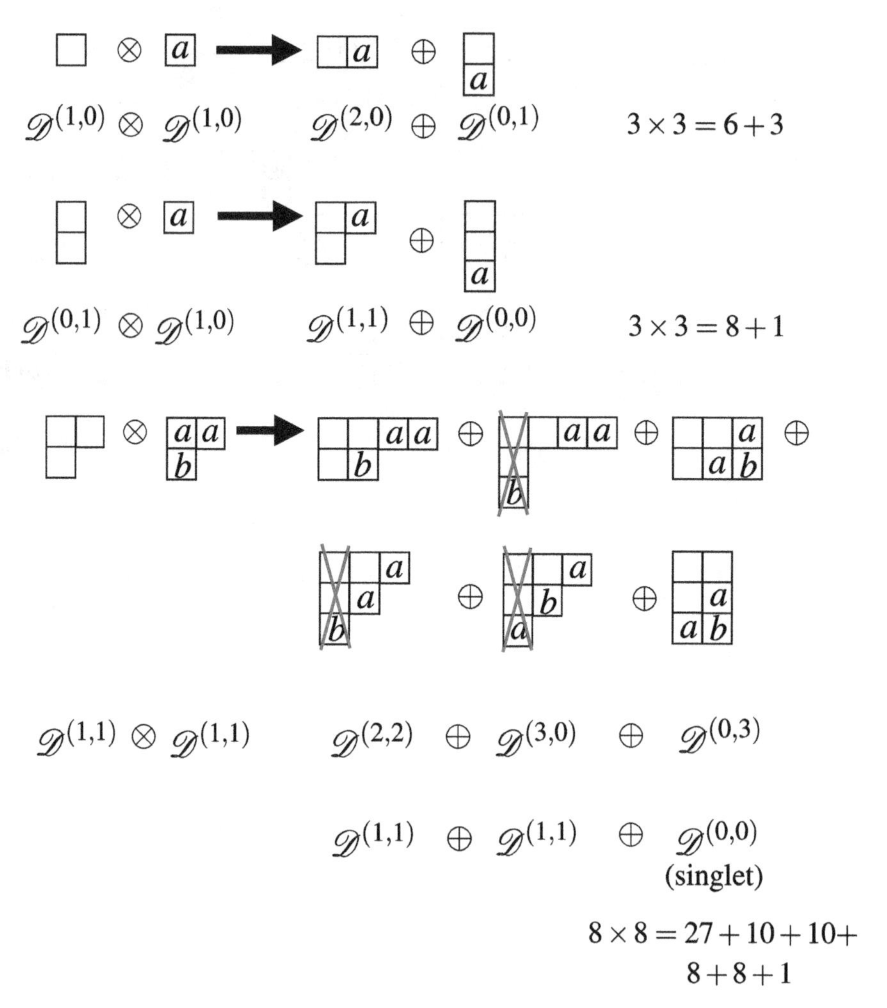

Figure 6.11. Example of the Kronecker product of two $SU(3)$ irreps.

*Rules (*SU(n)*)*

(a) Choose the simpler tableau (T') and enter a in each box of the first row, b in each box of the second row, c in each box of the third row, ….

(b) Add to tableau T one box marked with a (from T') in all possible allowed ways. Then add a second box marked with a to each of the obtained tableaux (from 1). with the restriction that two a's cannot appear in a single column. When all the a's have been added start with the b's, then the c's, …; again two b's are not allowed in a single column, …. There is an added restriction that, reading from *right to left*, first row → second row → etc., at any stage, cumulatively, $n(a) \geqslant n(b) \geqslant n(c) \geqslant \cdots$.

6.7 Introduction to Cartan theory of Lie algebras

The power of Lie algebraic theory when applied to quantum mechanics is that it breaks up the Hilbert space into irreps that have standardised characteristics: labelling quantum numbers, Clebsch–Gordan coefficients, tensor operators, etc. This reduces many quantum mechanical problems to just a few types. It reduces the Hilbert space for any finite many-body quantum system to subspaces defined by *su*(2) quantum numbers which label the total spin of each state of the system. Other reductions may be present: that is a leading challenge at the research frontier for all of nuclear structure study. It also provides a powerful modelling principle, i.e. solvable models can be constructed using a Lie algebra and tested against observation. With the recognition of this methodology, the question arises: 'can Lie algebraic theory be formulated in a standardised way?' The answer is yes, by using the formulation of Lie algebras introduced by Elié Cartan.

Cartan theory of Lie algebras has two essential conceptual features:

(a) For a set of generators $\{G_i\}$ defining a Lie algebra, viz.

$$[G_i, G_j] = \sum_k C_{ij}^k G_j, \tag{6.194}$$

the generators can be combined linearly to yield two types of operators —$\{H_i, H_j, \ldots\}$, where $[H_i, H_j] = 0$; and $\{E_{\pm\alpha}\}$ which act as raising ($E_{+\alpha}$) and lowering ($E_{-\alpha}$) operators or ladder operators.

(b) The raising and lowering operators can be treated as vectors with dimension equal to the rank of the Lie algebra. The rank of a Lie algebra equals the number of elements in $\{H_i\}$.

The identification of the subset $\{H_i, H_j, \ldots\}$ of the generators $\{G_i\}$ that obey

$$[H_i, H_j] = 0, \tag{6.195}$$

the so-called Cartan subalgebra, are found by inspection of the commutator brackets or Lie products of the $\{G_i\}$.

Removing the $\{H_i\}$ from the $\{G_i\}$, the remainder constitute the $\{E_{\pm\alpha}\}$. Generally the $E_{\pm\alpha}$ are formed from linear combinations of the G_i (H_i excluded). The $E_{\pm\alpha}$ obey

$$[H_i, E_{\pm\alpha}] = \pm r_i(\alpha) E_{\pm\alpha}, \quad \forall\, i, \alpha. \tag{6.196}$$

Examples of Cartan structure are manifest in the Lie algebras encountered already. The algebra *su*(2) is the simplest manifestation, which is almost trivial, but its role in Cartan theory is fundamental. Relevant details of *su*(2) for present purposes are depicted in figure 6.12. It is a rank-1 Lie algebra: the Cartan subalgebra contains one operator, viz. $\hat{J}_0$. The action of the ladder operators on the weight vectors is given by

$$\hat{J}_+|jm\rangle = \sqrt{(j-m)(j+m+1)}\,\hbar|jm+1\rangle, \tag{6.197}$$

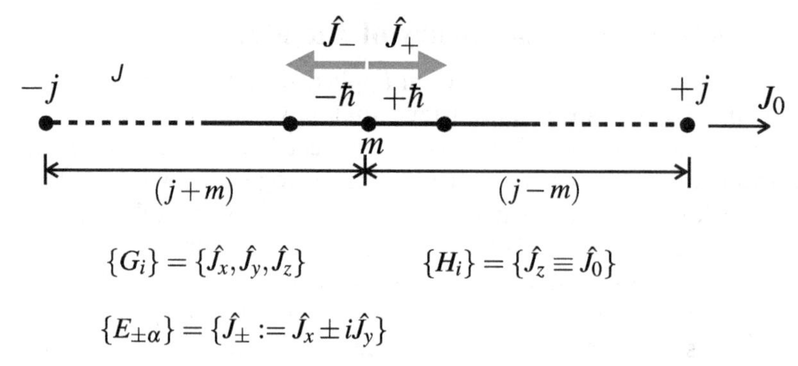

Figure 6.12. The Cartan structure of the Lie algebra *su*(2). The action of the 'root' vectors, $\hat{J}_\pm$ on the 'weight' points in the weight diagram are shown. The root vectors are usually termed ladder operators. The weight points define an irreducible *su*(2) representation via a specified value of j ($=0, \frac{1}{2}, 1, \frac{3}{2}, 2, \ldots$) and $m = +j, +j-1, +j-2, \ldots, -j+1, -j$ (the eigenvalues of $\hat{J}_0$ for $\hbar \equiv 1$). For a generic weight point m, it can be raised by $(j-m)$ steps and it can be lowered by $(j+m)$ steps.

$$\hat{J}_-|jm\rangle = \sqrt{(j+m)(j-m+1)}\,\hbar|jm-1\rangle. \tag{6.198}$$

This ensures that the representations are unitary, i.e. that $(\hat{J}_+|jm\rangle)^\dagger = \langle jm|\hat{J}_-$. Recall, this follows from $\langle jm|\hat{J}_-\hat{J}_+|jm\rangle = \langle jm|(\hat{J}^2 - \hat{J}_0^2 - \hat{J}_0\hbar)|jm\rangle$.

The algebras *so*(4) and *su*(3) have rank-2. As such they are already far from trivial. Relevant details for *so*(4) and *su*(3) follow.

6.7.1 Cartan structure of the *so*(4) Lie algebra

$$\{G_i\} = \{\hat{L}_x, \hat{L}_y, \hat{L}_z, \hat{A}_x, \hat{A}_y, \hat{A}_z\}, \tag{6.199}$$

$$\{H_i\} = \{\hat{L}_z, \hat{A}_z\}, \quad \{E_{\pm\alpha}\}\text{: not developed}, \tag{6.200}$$

or, via $\vec{M} := \frac{1}{2}(\vec{L} + \vec{K})$, $\vec{N} := \frac{1}{2}(\vec{L} - \vec{K})$, $\vec{K} := \sqrt{\frac{m}{2|E|}}\vec{A}$ (cf. equation (6.101) and (6.116)),

$$\{G_i\} = \{\hat{M}_x, \hat{M}_y, \hat{M}_z, \hat{N}_x, \hat{N}_y, \hat{N}_z\}, \tag{6.201}$$

$$\{H_i\} = \{\hat{M}_z, \hat{N}_z\}, \tag{6.202}$$

$$\{E_{\pm\alpha}\} = \{\hat{M}_\pm := \hat{M}_x \pm i\hat{M}_y; \hat{N}_\pm := \hat{N}_x \pm i\hat{N}_y\}. \tag{6.203}$$

Note: the 'M, N' form decomposes the *so*(4) algebra into two *su*(2) algebras, viz. $so(4) = su(2) \times su(2)$.

The *so*(4) roots are shown in figure 6.4. Manifestly, this decomposition reveals that the two *su*(2) subalgebras have root vectors that are orthogonal pairs. The simplest *so*(4) weight diagrams are shown in figure 6.3.

6.7.2 Cartan structure of the *su*(3) Lie algebra

$$\{G_i\} = \{\hat{L}_\nu, \hat{M}_\nu, \hat{N}_\nu; \nu = x, y, z, \text{ with } \hat{N}_x + \hat{N}_y + \hat{N}_z = \text{const.}\}, \tag{6.204}$$

cf. table 6.2 and equations (6.71)–(6.73) and (6.77);

$$\{H_i\} = \left\{\hat{L}_z \equiv \hat{L}_0,\ Q_0 := \frac{1}{\sqrt{6}}(2\hat{N}_z - \hat{N}_x - \hat{N}_y)\right\}; \tag{6.205}$$

$\{E_{\pm\alpha}\}$: via (cf. equation (6.9) and (6.74))

$$\hat{L}_\pm := \hat{L}_x \pm i\hat{L}_y, \tag{6.206}$$

$$\hat{Q}_{\pm 1} := -i\hat{M}_x \mp \hat{M}_y, \tag{6.207}$$

$$\hat{Q}_{\pm 2} := \frac{1}{2}(\hat{N}_x - \hat{N}_y \pm 2i\hat{M}_z), \tag{6.208}$$

and (cf. equation (6.85))

$$\hat{P}_\pm := \hat{L}_\pm \mp \frac{2}{\sqrt{mk}}\hat{Q}_{\pm 1}, \tag{6.209}$$

$$\hat{R}_\pm := \hat{L}_\pm \pm \frac{2}{\sqrt{mk}}\hat{Q}_{\pm 1}. \tag{6.210}$$

Thus,

$$\{E_{\pm\alpha}\} = \{\hat{P}_\pm, \hat{R}_\pm, \hat{Q}_\pm := \hat{Q}_{\pm 2}\}. \tag{6.211}$$

The *su*(3) roots are shown in figure 6.1. The simplest *su*(3) weight diagrams are shown in figure 6.2; many more *su*(3) weight diagrams are shown in figures 6.6(a)–(e).

The commutator bracket relations for the various forms of the *su*(3) generators can be assembled into a 'mosaic' (as shown for $\{\hat{L}_x, \hat{L}_y, \hat{L}_z, \hat{M}_x, \hat{M}_y, \hat{M}_z, \hat{N}_x, \hat{N}_y, \hat{N}_z\}$ in table 6.2) presented in tables 6.4, 6.5, and 6.7. We refer to the forms in table 6.2 as 'Cartesian', in table 6.4 as 'spherical tensor', and in table 6.5 as 'Cartan'. The Cartan form is directly related to equation (6.196) as shown in table 6.6. A further useful form, referred to as 'canonical', is presented in table 6.7. The term 'canonical' is in reference to $su(3) \supset su(2)$ subalgebra structure that is evident.

The Cartesian form provides a geometrical view of *su*(3), as depicted in figure 6.13. One observes the manifest vectorial character of the $\{E_{\pm\alpha}\}$ when this form is employed. The canonical example manifested in table 6.7 is one of three simple choices and its relationship to the Cartesian form is clarified by comparing figure 6.13 to figure 6.14. The algebraic relations are simply expressed using (cf. equation (6.172))

Table 6.4. Tabulation of the 28 commutator bracket relations for the operators defined in equations (6.205)–(6.208) where $\xi = \frac{mk}{4}$. Note: the operators manifest the spherical tensor form of equations (6.76) and (6.77); and cf. chapter 3.

	$\hat{L}_+$	$\hat{L}_-$	$\hat{Q}_0$	$\hat{Q}_{+1}$	$\hat{Q}_{-1}$	$\hat{Q}_{+2}$	$\hat{Q}_{-2}$
$\hat{L}_0$	$\hbar\hat{L}_+$	$-\hbar\hat{L}_-$	0	$\hbar\hat{Q}_{+1}$	$-\hbar\hat{Q}_{-1}$	$2\hbar\hat{Q}_{+2}$	$-2\hbar\hat{Q}_{-2}$
$\hat{L}_+$		$2\hbar\hat{L}_0$	$\sqrt{6}\hbar\hat{Q}_{+1}$	$2\hbar\hat{Q}_{+2}$	$\sqrt{6}\hbar\hat{Q}_0$	0	$2\hbar\hat{Q}_{-1}$
$\hat{L}_-$			$\sqrt{6}\hbar\hat{Q}_{-1}$	$\sqrt{6}\hbar\hat{Q}_0$	$2\hbar\hat{Q}_{-2}$	$2\hbar\hat{Q}_{+1}$	0
$\hat{Q}_0$				$-\sqrt{6}\xi\hbar\hat{L}_+$	$-\sqrt{6}\xi\hbar\hat{L}_-$	0	0
$\hat{Q}_{+1}$					$-2\xi\hbar\hat{L}_0$	0	$2\xi\hbar\hat{L}_-$
$\hat{Q}_{-1}$						$2\xi\hbar\hat{L}_+$	0
$\hat{Q}_{+2}$							$4\xi\hbar\hat{L}_0$

Table 6.5. Tabulation of the 28 commutator bracket relations for the operators defined in equations (6.205), (6.208)–(6.210) where $\kappa = \sqrt{\frac{3mk}{2}}$. This is the 'Cartan' form of *su*(3), and the manifestation of equation (6.196) is clear, cf. figure 6.12 and table 6.6. See equations (6.86)–(6.88), figure 6.1, and table 6.5.

	$\hat{Q}_0$	$\hat{P}_+$	$\hat{P}_-$	$\hat{Q}_+$	$\hat{Q}_-$	$\hat{R}_+$	$\hat{R}_-$
$\hat{L}_0$	0	$\hbar\hat{P}_+$	$-\hbar\hat{P}_-$	$2\hbar\hat{Q}_+$	$-2\hbar\hat{Q}_-$	$\hbar\hat{R}_+$	$-\hbar\hat{R}_-$
$\hat{Q}_0$		$\kappa\hbar\hat{P}_+$	$-\kappa\hbar\hat{P}_-$	0	0	$-\kappa\hbar\hat{R}_+$	$\kappa\hbar\hat{R}_-$
$\hat{P}_+$			$\frac{12\hbar}{\kappa}\hat{Q}_0 + 4\hbar\hat{L}_0$	0	$\frac{\kappa\hbar}{\sqrt{3}}\hat{R}_-$	$\frac{4\sqrt{6}\hbar}{\kappa}\hat{Q}_+$	0
$\hat{P}_-$				$\frac{\kappa\hbar}{\sqrt{3}}\hat{R}_+$	0	0	$-\frac{4\sqrt{6}\hbar}{\kappa}\hat{Q}_-$
$\hat{Q}_+$					$4\xi\hbar\hat{L}_0$	0	$-\frac{\kappa\hbar}{\sqrt{3}}\hat{P}_+$
$\hat{Q}_-$						$-\frac{\kappa\hbar}{\sqrt{3}}\hat{P}_-$	0
$\hat{R}_+$							$-\frac{12\hbar}{\kappa}\hat{Q}_0 + 4\hbar\hat{L}_0$

$$\hat{C}_{ij} := \frac{1}{2}\left(a_i^\dagger a_j + a_j^\dagger a_i\right) = a_i^\dagger a_j + \frac{1}{2}\delta_{ij}, \quad i, j = 1, 2, 3 = x, y, z; \tag{6.212}$$

whence

$$\hat{\omega}_0 := \frac{1}{2}(\hat{C}_{11} - \hat{C}_{22}), \tag{6.213}$$

Table 6.6. The $r_i(\alpha)$, cf. equation (6.196), for the Cartan subalgebra $\{H_i\} = \{\hat{L}_0, \hat{Q}_0\}$ and $\{E_{\pm\alpha}\} = \{\hat{P}_\pm, \hat{R}_\pm, \hat{Q}_\pm\}$. $\hbar = 1$, $\kappa = \sqrt{3}mk$. This Cartan structure is manifest in figure 6.1.

H_i		$r_i(\alpha)$	
	$\hat{P}_\pm$	$\hat{R}_\pm$	$\hat{Q}_{\pm 2}$
$\hat{L}_0$	± 1	± 1	± 2
$\hat{Q}_0$	$\pm\kappa$	$\mp\kappa$	0

Table 6.7. Tabulation of the 28 commutator bracket relations for the operators defined in equations (6.16)–(6.19). This is the canonical form of $su(3)$: $su(3) \supset su(2)$, where the $su(2)$ subalgebra $\{\omega_0, \omega_\pm\}$ is manifest.

	$\hat{\omega}_0$	$\hat{\omega}_+$	$\hat{\omega}_-$	$\hat{C}_{13}$	$\hat{C}_{31}$	$\hat{C}_{23}$	$\hat{C}_{32}$
$\hat{Q}_0$	0	0	0	$-3\hat{C}_{13}$	$3\hat{C}_{31}$	$-3\hat{C}_{23}$	$3\hat{C}_{32}$
$\hat{\omega}_0$		$\hat{\omega}_+$	$-\hat{\omega}_-$	$\frac{1}{2}\hat{C}_{13}$	$-\frac{1}{2}\hat{C}_{31}$	$-\frac{1}{2}\hat{C}_{23}$	$\frac{1}{2}\hat{C}_{32}$
$\hat{\omega}_+$			$2\hat{\omega}_0$	0	$-\hat{C}_{32}$	$\hat{C}_{13}$	0
$\hat{\omega}_-$				$\hat{C}_{23}$	0	0	$-\hat{C}_{31}$
$\hat{C}_{13}$					$\hat{C}_{11} - \hat{C}_{33}$	0	$\hat{C}_{12}$
$\hat{C}_{31}$						$-\hat{C}_{21}$	0
$\hat{C}_{23}$							$\hat{C}_{22} - \hat{C}_{33}$

$$\hat{\omega}_+ := \hat{C}_{12}, \tag{6.214}$$

$$\hat{\omega}_- := \hat{C}_{21}, \tag{6.215}$$

$$\hat{Q}_0 := \hat{C}_{11} + \hat{C}_{22} - 2\hat{C}_{33}. \tag{6.216}$$

The commutator brackets are evaluated straightforwardly using

$$[\hat{C}_{ij}, \hat{C}_{kl}] = \hat{C}_{il}\delta_{jk} - \hat{C}_{jk}\delta_{il}, \quad i, j, k, l = 1, 2, 3. \tag{6.217}$$

Figure 6.13 depicts a representation of $u(3)$. Its relationship to figure 6.14 is dictated by

$$\hat{N} = \hat{C}_{11} + \hat{C}_{22} + \hat{C}_{33} = \text{const.} \tag{6.218}$$

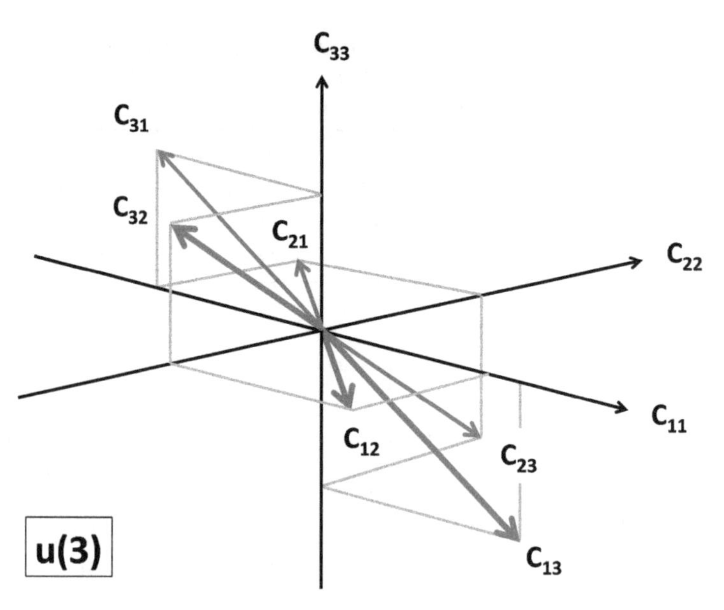

Figure 6.13. The nine operators, C_{ij}, $i, j = 1, 2, 3$ defining the algebra $u(3)$ with perspective on their vectorial character with respect to Cartesian axes, 1, 2 and 3.

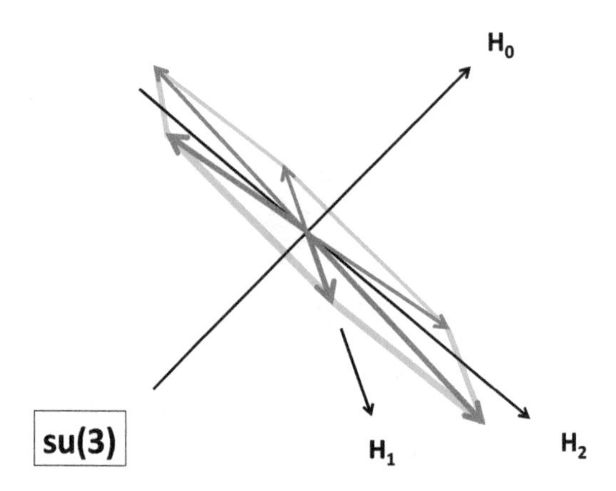

Figure 6.14. The six operators that define the root space of the $su(3)$ subalgebra of $u(3)$. This is specified by $H_0 := C_{11} + C_{22} + C_{33} = \text{const}$. It can be expressed as $u(3) = su(3) \times u(1)$.

The algebra $u(3)$ possesses three commuting operators viz. $\{\hat{C}_{11}, \hat{C}_{22}, \hat{C}_{33}\}$ and has rank-3. The constraint, equation (6.216), results in two commuting operators: these can be chosen to be $\{\hat{C}_{11}, \hat{C}_{22}\}$, or $\{\hat{\omega}_0, \hat{Q}_0\}$, or other linear combinations of $\hat{C}_{11}$, $\hat{C}_{22}$ and $\hat{C}_{33}$. Indeed, there are many other choices, cf. equation (6.195) where an $so(3)$ subalgebra is obtained using $\hat{L}_k := \hat{L}_{ij} := -i\hbar(\hat{C}_{ij} - \hat{C}_{ji})$, where $i, j, k = x, y, z$ yields the angular momentum algebra.

6.7.3 The generic Lie algebra

Starting from the set of operators $\{H_i, E_{\pm\alpha}\}$, formed from suitable linear combinations of the generators $\{G_i\}$ of a Lie algebra, the Cartan structure of the Lie algebra can be elucidated. The set of operators $\{H_i\}$ are mutually commuting and so they define simultaneous eigenvectors, viz.

$$H_i|\lambda_1\lambda_2\ldots\lambda_i\ldots\lambda_r\rangle = \lambda_i|\lambda_1\lambda_2\ldots\lambda_i\ldots\lambda_r\rangle, \tag{6.219}$$

where r is the rank of the Lie algebra and $i = 1, \ldots, r$. Then, from $[H_i, E_{+\alpha}] = +r_i(\alpha)E_{+\alpha}$, cf. equation (6.196),

$$\begin{aligned} H_iE_{+\alpha}|\lambda_1\lambda_2\ldots\lambda_i\ldots\lambda_r\rangle &= (E_{+\alpha}H_i + r_i(\alpha)E_{+\alpha})|\lambda_1\lambda_2\ldots\lambda_i\ldots\lambda_r\rangle \\ &= (E_{+\alpha}\lambda_i + r_i(\alpha)E_{+\alpha})|\lambda_1\lambda_2\ldots\lambda_i\ldots\lambda_r\rangle \\ &= (\lambda_i + r_i(\alpha))E_{+\alpha}|\lambda_1\lambda_2\cdots\lambda_i\ldots\lambda_r\rangle, \end{aligned} \tag{6.220}$$

i.e.

$$E_{+\alpha}|\lambda_1\lambda_2\ldots\lambda_i\ldots\lambda_r\rangle \approx |\lambda_1 + r_1(\alpha), \lambda_2 + r_2(\alpha), \ldots, \lambda_i + r_i(\alpha), \ldots, \lambda_r + r_r(\alpha)\rangle. \tag{6.221}$$

The λ_i and the $r_i(\alpha)$ can be regarded as vectors

$$\vec{\lambda} = \begin{pmatrix} \lambda_1 \\ \lambda_2 \\ \cdots \\ \lambda_i \\ \cdots \\ \lambda_r \end{pmatrix}, \quad \vec{r}(\alpha) = \begin{pmatrix} r_1(\alpha) \\ r_2(\alpha) \\ \cdots \\ r_i(\alpha) \\ \cdots \\ r_r(\alpha) \end{pmatrix}, \tag{6.222}$$

respectively. Equation (6.221) can be depicted as shown in figure 6.15. The weight space has dimension r, i.e. the rank of the Lie algebra, cf. figure 6.12. Thus,

$$E_{+\alpha}|\vec{\lambda}\rangle \approx |\vec{\lambda} + \vec{r}(\alpha)\rangle. \tag{6.223}$$

It follows directly that

$$r_i(\alpha) + r_i(\beta) = r_i(\alpha + \beta), \tag{6.224}$$

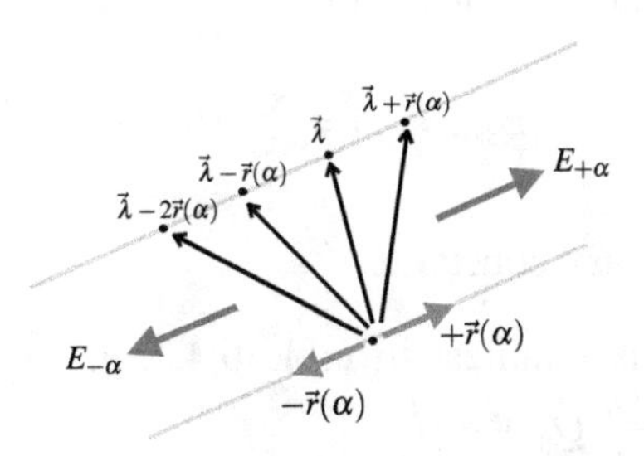

Figure 6.15. A depiction of the Cartan structure of weight vectors, $\vec{\lambda}$ and root vectors, $\vec{r}(\alpha)$ in weight space. Note the $su(2)$-like relationship for this 'α-component' of $\vec{\lambda}$.

$$r_i(N\alpha) = Nr_i(\alpha), \tag{6.225}$$

$$r_i(-\alpha) = -r_i(\alpha), \tag{6.226}$$

provided the laddering action of the $E_{\pm\alpha}$ is within bounds, i.e. $r_i(\alpha + \beta)$, $Nr_i(\alpha)$ exist (see below).

A class of Lie algebras of particular interest to physics is the *compact* Lie algebras, i.e. $\exists|\vec{\lambda}^{\max}\rangle$ and $|\vec{\lambda}^{\min}\rangle$ such that

$$E_{+\alpha}|\vec{\lambda}^{\max}\rangle = 0, \tag{6.227}$$

$$E_{-\alpha}|\vec{\lambda}^{\min}\rangle = 0, \quad \forall\alpha. \tag{6.228}$$

These are called 'highest-weight' and 'lowest-weight' states, respectively. The extraordinary feature of Cartan theory applied to compact Lie algebras is that it provides a simple demonstration that *there are only a few classes of such algebras.* This is achieved through a few elementary theorems. (Note that equations (6.227) and (6.228) dictate boundaries for the weight space and so conditions are imposed on the existence of solutions to equations (6.224) and (6.225).)

Theorem 6.7.1.

$$[E_{+\alpha}, E_{-\alpha}] = \sum_{i=1}^{r} \rho_i(\alpha) H_i, \tag{6.229}$$

where the $\rho_i(\alpha)$ are arbitrary constants.

Proof. Consider

$$\begin{aligned}
[H_i, [E_{+\alpha}, E_{-\alpha}]] &= [H_i, E_{+\alpha}E_{-\alpha}] - [H_i, E_{-\alpha}E_{+\alpha},] \\
&= E_{+\alpha}[H_i, E_{-\alpha}] + [H_i, E_{+\alpha}]E_{-\alpha} - E_{-\alpha}[H_i, E_{+\alpha}] - [H_i, E_{-\alpha}]E_{+\alpha} \\
&= E_{+\alpha}(-r_i(\alpha)E_{-\alpha}) + r_i(\alpha)E_{+\alpha}E_{-\alpha} - E_{-\alpha}r_i(\alpha)E_{+\alpha} - (-r_i(\alpha)E_{-\alpha})E_{+\alpha} \\
&= 0,
\end{aligned} \tag{6.230}$$

where equation (6.196) has been used; thus

$$[E_{+\alpha}, E_{-\alpha}] = \sum_{i=1}^{r} \rho_i(\alpha) H_i,$$

where the $\rho_i(\alpha)$ are arbitrary constants. □

Note, such structure is manifest in table 6.4, e.g. $[\hat{Q}_{+1}, \hat{Q}_{-1}] = -2\xi\hbar\hat{L}_0$ and in table 6.5, e.g. $[\hat{P}_+, \hat{P}_-] = \frac{12\hbar}{\kappa}\hat{Q}_0 + 4\hbar\hat{L}_0$.

Theorem 6.7.2.

$$[E_\alpha, E_\beta] = N_{\alpha\beta} E_{\alpha+\beta}, \tag{6.231}$$

provided $\vec{r}(\alpha + \beta)$ exists, where (cf. equation (6.224))

$$\vec{r}(\alpha + \beta) = \vec{r}(\alpha) + \vec{r}(\beta), \tag{6.232}$$

and $N_{\alpha\beta}$ is a constant.

Proof. Consider

$$\begin{aligned}[H_i, [E_\alpha, E_\beta]] &= [H_i, E_\alpha E_\beta] - [H_i, E_\beta E_\alpha] \\ &= E_\alpha[H_i, E_\beta] + [H_i, E_\alpha]E_\beta - E_\beta[H_i, E_\alpha] \\ &\quad - [H_i, E_\beta]E_\alpha,\end{aligned} \tag{6.233}$$

and, using equation (6.196)

$$\begin{aligned}\therefore [H_i, [E_\alpha, E_\beta]] &= E_\alpha r_i(\beta)E_\beta + r_i(\alpha)E_\alpha E_\beta - E_\beta r_i(\alpha)E_\alpha - r_i(\beta)E_\beta E_\alpha \\ &= (r_i(\alpha) + r_i(\beta))[E_\alpha, E_\beta],\end{aligned} \tag{6.234}$$

whence from equations (6.196) and (6.232)

$$[H_i, E_{\alpha+\beta}] = r_i(\alpha + \beta)E_{\alpha+\beta}, \tag{6.235}$$

which has equation (6.231) as a general solution. □

Note, such structure is manifest in table 6.4, e.g. $[\hat{L}_+, \hat{Q}_{+1}] = 2\hbar\hat{Q}_{+2}$ and in table 6.5, e.g. $[\hat{P}_+, \hat{R}_+] = 4\sqrt{6}\frac{\hbar}{k}\hat{Q}_+$. It can further be noted that the relationship, e.g. $[\hat{L}_+, \hat{Q}_{+1}] = 2\hbar\hat{Q}_{+2}$ conforms to the spherical tensor character of $\hat{Q}_{+2}$ cf. equation (6.76).

Theorem 6.7.3.

$$\frac{2\vec{\lambda} \cdot \vec{r}(\alpha)}{\vec{r}(\alpha) \cdot \vec{r}(\alpha)} = L - R, \tag{6.236}$$

where L is the number of times that $\vec{\lambda}$ can be lowered in increments of $\vec{r}(\alpha)$ and R is the number of times that $\vec{\lambda}$ can be raised in increments of $\vec{r}(\alpha)$.

Proof. Consider

$$\begin{aligned}\langle\vec{\lambda}|[E_{+\alpha}, E_{-\alpha}]|\vec{\lambda}\rangle &= \langle\vec{\lambda}|\left(\sum_i \rho_i(\alpha)H_i\right)|\vec{\lambda}\rangle \\ &= \vec{\rho}(\alpha) \cdot \vec{\lambda},\end{aligned} \tag{6.237}$$

where theorem 6.7.1 has been used with the choice $\langle\vec{\lambda}|\vec{\lambda}\rangle = 1$. From figure 6.15, noting the basic '$su(2)$' action of $E_{\pm\alpha}$, cf. figure 6.12, then for

$$H_i \leftrightarrow \hat{J}_0, \quad E_{\pm\alpha} \leftrightarrow \hat{J}_\pm, \quad |\vec{r}(\alpha)| \leftrightarrow \hbar, \quad |\vec{\lambda}\rangle \leftrightarrow |jm\rangle,$$

equation (6.237) is seen to be fulfilled by

$$[\hat{J}_+, \hat{J}_-] = 2\hat{J}_0\hbar, \tag{6.238}$$

i.e.

$$\langle jm|[\hat{J}_+, \hat{J}_-]|jm\rangle = \langle jm|2\hat{J}_0\hbar|jm\rangle = 2\, m\hbar\hbar. \tag{6.239}$$

Thus, for the $su(2)$ action of $E_{\pm\alpha}$, equation (6.237) holds for

$$\vec{\rho}(\alpha) = 2\vec{r}(\alpha). \tag{6.240}$$

Further, noting that $|jm\rangle$ can be raised $(j - m)$ times, i.e. $j - m \leftrightarrow R$, and it can be lowered $(j + m)$ times, i.e. $j + m \leftrightarrow L$, then from equation (6.239), from the right-hand side, $[L - R = j + m - (j - m) = 2\, m, \quad |\vec{r}(\alpha)| \leftrightarrow \hbar, \quad |\vec{\lambda}| \leftrightarrow m\hbar]$

$$(L - R)|\vec{r}(\alpha)||\vec{r}(\alpha)| = 2|\vec{r}(\alpha)||\vec{\lambda}|. \tag{6.241}$$

The result, equation (6.237) follows. □

Two further relationships follow from this result and equations (6.197) and (6.198), viz.

$$E_{+\alpha}|\vec{\lambda}\rangle = \sqrt{\frac{1}{2}R(L+1)}\,|\vec{r}(\alpha)||\vec{\lambda} + \vec{r}(\alpha)\rangle \tag{6.242}$$

and

$$E_{-\alpha}|\vec{\lambda}\rangle = \sqrt{\frac{1}{2}L(R+1)}\,|\vec{r}(\alpha)||\vec{\lambda} - \vec{r}(\alpha)\rangle. \tag{6.243}$$

Theorem 6.7.4.

$$\frac{2\vec{r}(\alpha)\cdot\vec{r}(\beta)}{\vec{r}(\alpha)\cdot\vec{r}(\alpha)} = \frac{2\vec{r}(\alpha)\cdot\vec{r}(\beta)}{\vec{r}(\beta)\cdot\vec{r}(\beta)} = n, \tag{6.244}$$

where n is an integer.

Proof. This result follows directly from the recognition that there is a formal identity between cf. equation (6.219)

$$H_i|\vec{\lambda}\rangle = \lambda_i|\vec{\lambda}\rangle,$$

and cf. equation (6.196),

$$[H_i, E_\alpha] = r_i(\alpha)E_\alpha,$$

which leads to

$$H_i|\vec{r}(\alpha)\rangle = r_i(\alpha)|\vec{r}(\alpha)\rangle, \tag{6.245}$$

i.e. if the representations

$$[H_i, E_\alpha] \leftrightarrow H_i|E_\alpha\rangle, \quad E_\alpha \leftrightarrow |E_\alpha\rangle \tag{6.246}$$

with

$$E_\alpha \leftrightarrow \vec{r}(\alpha) \tag{6.247}$$

are made. This is called the *adjoint* or *regular representation* for the Lie algebra. Here, the formal identity leads to the recognition that the root vectors can be weight vectors. (This is always realised: e.g. for $su(2)$, $\{\hat{J}_0, \hat{J}_{\pm 1}\} \leftrightarrow \{|jm\rangle, j = 1, m = 0, \pm 1\}$; and for $su(3)$, $\{L_0, Q_0, P_\pm, R_\pm, Q_{\pm 2}\} \leftrightarrow \{\lambda = 1, \mu = 1\}$, cf. figures 6.1 and 6.6(a) (where L_0, Q_0 are located at the centre of the Young tableau).) Thus, adopting the result of theorem 6.7.3 with $\vec{\lambda} \leftrightarrow \vec{r}(\beta)$, equation (6.244) follows, with the added recognition that α and β are interchangeable. □

Theorem 6.7.4 leads to very stringent limitations on the number of compact Lie algebraic structures that can exist. Consider

$$\vec{r}(\alpha)\cdot\vec{r}(\beta) = |\vec{r}(\alpha)||\vec{r}(\beta)|\cos\theta_{\alpha\beta}, \tag{6.248}$$

whence

$$\cos^2\theta_{\alpha\beta} = \frac{\vec{r}(\alpha)\cdot\vec{r}(\beta)\,\vec{r}(\alpha)\cdot\vec{r}(\beta)}{\vec{r}(\alpha)\cdot\vec{r}(\alpha)\,\vec{r}(\beta)\cdot\vec{r}(\beta)} = \frac{nn'}{4}, \tag{6.249}$$

where n and n' are integers. But $0 \leqslant \cos^2\theta_{\alpha\beta} \leqslant 1$,

$$\therefore\ \cos^2\theta_{\alpha\beta} = \frac{0}{4}, \frac{1}{4}, \frac{2}{4}, \frac{3}{4}, \text{ or } \frac{4}{4}. \tag{6.250}$$

Further, the ratios of the lengths of the root vectors are determined by

$$\vec{r}(\alpha)\cdot\vec{r}(\beta) = \frac{n}{2}\vec{r}(\alpha)\cdot\vec{r}(\alpha) = \frac{n'}{2}\vec{r}(\beta)\cdot\vec{r}(\beta), \tag{6.251}$$

$$\therefore\ \frac{|\vec{r}(\alpha)|^2}{|\vec{r}(\beta)|^2} = \frac{n'}{n}. \tag{6.252}$$

All possible pairwise contributions of root vectors can be tabulated as shown in table 6.8.

Table 6.8. Allowed root pair configurations for rank-2 Lie algebra.

$\cos^2 \theta_{\alpha\beta}$	$\cos \theta_{\alpha\beta}$	$\theta_{\alpha\beta}$	$\frac{\mid \vec{r}(\alpha) \mid}{\mid \vec{r}(\beta) \mid}$	Principal root vector diagram
$\frac{0}{4}$	0	90°, 270°	indeterminate	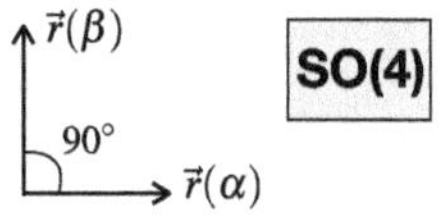 $\vec{r}(\beta)$, 90°, $\vec{r}(\alpha)$ SO(4)
$\frac{1}{4}$	$\pm\frac{1}{2}$	60°, 120° 240°, 300°	$\frac{1}{1}$	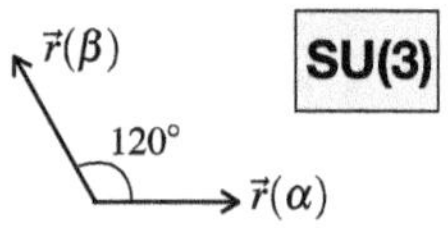 $\vec{r}(\beta)$, 120°, $\vec{r}(\alpha)$ SU(3)
$\frac{2}{4}$	$\pm\frac{1}{\sqrt{2}}$	45°, 135° 225°, 315°	$\frac{1}{\sqrt{2}}$ or $\frac{\sqrt{2}}{1}$	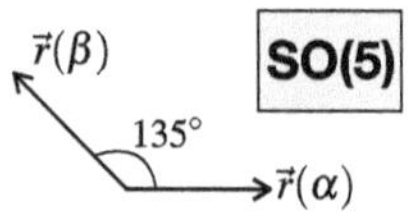 $\vec{r}(\beta)$, 135°, $\vec{r}(\alpha)$ SO(5)
$\frac{3}{4}$	$\pm\frac{\sqrt{3}}{2}$	30°, 150° 210°, 330°	$\frac{1}{\sqrt{3}}$ or $\frac{\sqrt{3}}{1}$	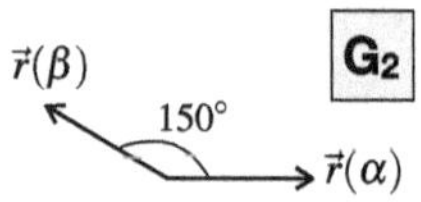 $\vec{r}(\beta)$, 150°, $\vec{r}(\alpha)$ G_2
$\frac{4}{4}$	± 1	0°, 180°	$\frac{1}{1}$[a]	 $\vec{r}(\alpha)$

[a] $\frac{2}{1}$ and $\frac{1}{2}$ are excluded because $\vec{r}(\beta) = \pm\, \vec{r}(\alpha)$. The case $\frac{4}{4}$ is rank-1.

The allowed root pair configurations are encoded in *principal root vector diagrams.* From these diagrams, the entire root space can be constructed for any configuration and any rank. This is done using the *Weyl reflection theorem.*

Theorem 6.7.5 (The Weyl reflection theorem). *If* $\vec{r}(\alpha)$ and $\vec{r}(\beta)$ *are two roots; then so is the root produced by reflecting* $\vec{r}(\beta)$ *in a plane perpendicular to* $\vec{r}(\alpha)$ *(and so is the root produced by reflecting* $\vec{r}(\alpha)$ *in a plane perpendicular to* $\vec{r}(\beta)$*).*

Proof. This is simply depicted as in figures 6.16–6.19: □

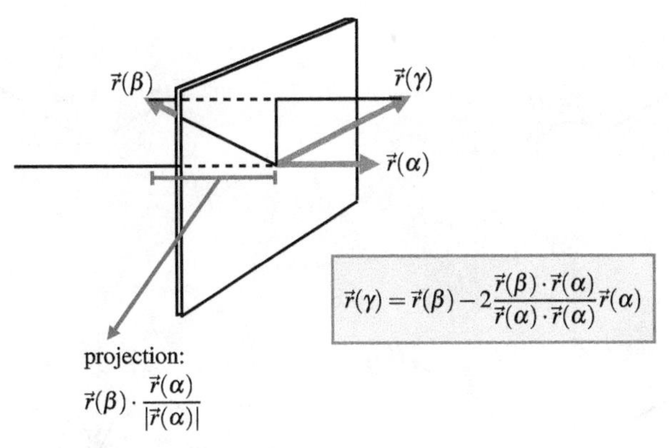

Figure 6.16. Depiction of the Weyl reflection theorem.

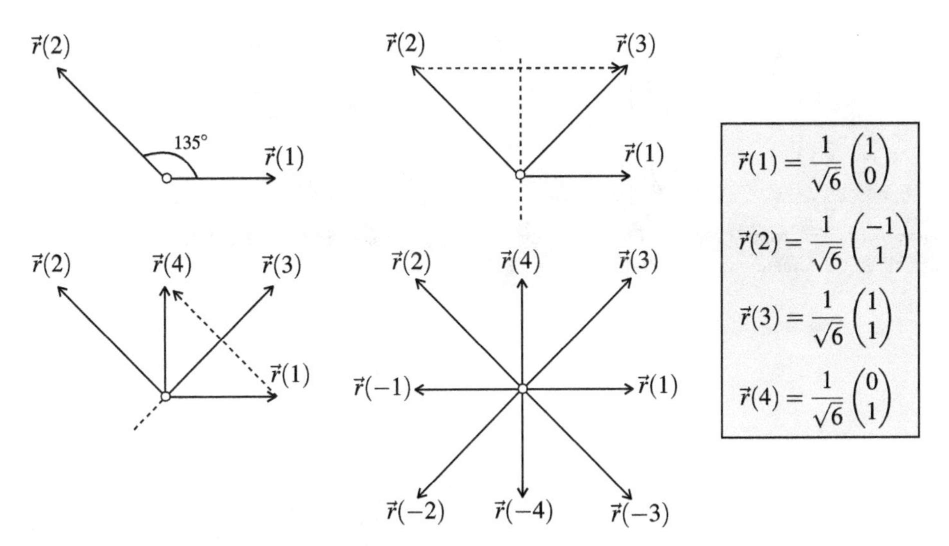

Figure 6.17. The root space for O(5) showing the construction using Weyl reflections. The box gives normalized coordinate values for $\vec{r}(i)$, $i = 1, \ldots, 4$; $\sum_i \vec{r}(i)^2 = 1$.

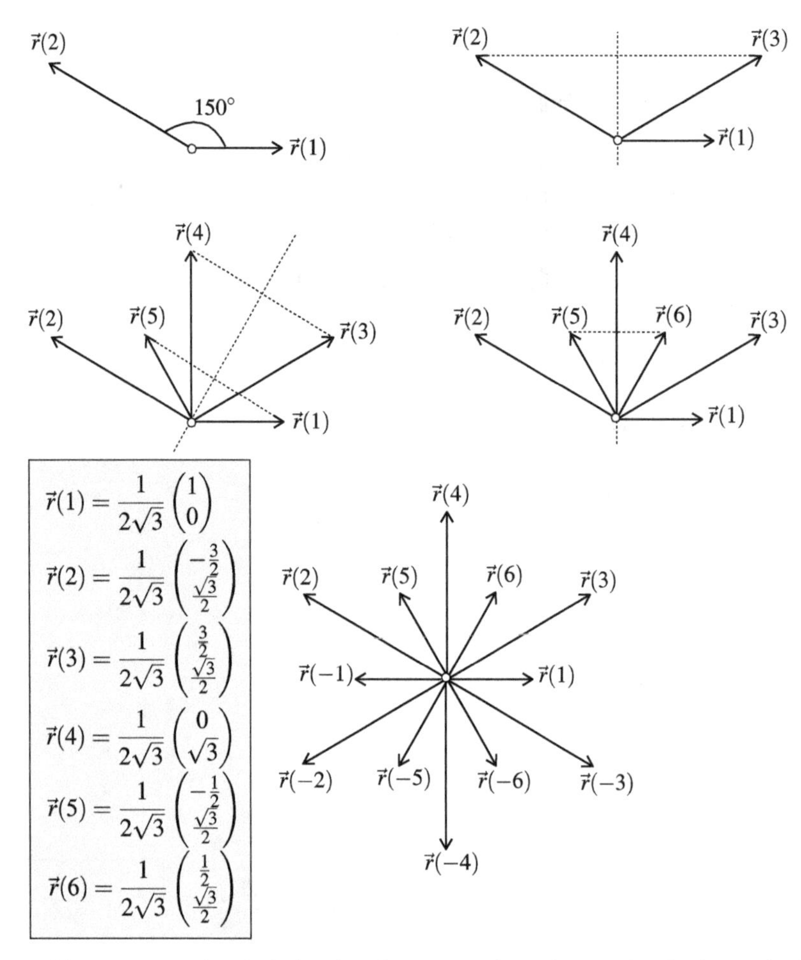

Figure 6.18. The root space for G(2) showing the construction using Weyl reflections. The box gives normalized coordinate values for $\vec{r}(i)$, $i = 1, \ldots, 6$; $\sum_i \vec{r}(i)^2 = 1$.

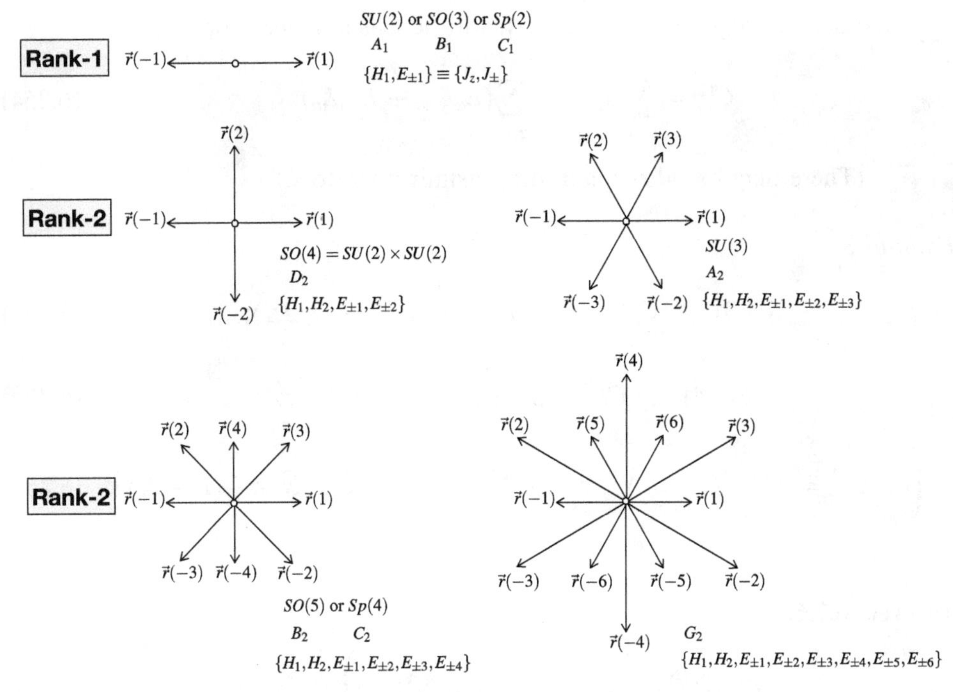

Figure 6.19. The root space for all of the compact Lie groups of rank-1 and rank-2. The Cartan structure is indicated along with the commonly used names and the classic names, viz. A_r, B_r, C_r, D_r, $r = 1, 2$.

6.7.4 Irrep quantum numbers: Cartan subalgebras and Casimir operators

Our interest in Lie algebraic structures in quantum mechanics is as a means of breaking-up a Hilbert space into irreps of a Lie group. The irrep quantum numbers provide labelling quantum numbers and at least some of these quantum numbers will have physical significance. The irrep dimensions provide the key to degeneracies for a given physical system.

We must now turn to the quantum number labels and dimensions of the irreps. The Cartan subalgebra, $\{H_1, H_2, \dots, H_r\}$ of a Lie algebra of rank r provides a set of commuting operators and hence a set of compatible quantum number labels. The Casimir operators associated with a Lie algebra are operators that commute with all the generators of the algebra. Two types of Casimir operators are commonly encountered in physics:

(a) For the unitary group $U(n)$, the 'oscillator' representation, $\{G_{ij} = a_i^\dagger a_j,\ i, j = 1, \dots n\}$ provides the all-commuting operator

$$C^{(1)}_{U(n)} = \sum_{i=1}^{n} G_{ii}. \tag{6.253}$$

(It is a commonly used $SU(n)$ irrep label.)

(b) Each Lie algebra possesses a quadratic Casimir operator

$$C^{(2)} = \sum_{i=1}^{r} H_i^2 + \frac{1}{2}\sum_{\alpha>0}(E_\alpha E_{-\alpha} + E_{-\alpha}E_\alpha). \tag{6.254}$$

(There may be other quadratic Casimir operators.)

Examples

$$SO(3) \qquad C^{(2)} = L^2 = L_z^2 + \frac{1}{2}(L_+L_- + L_-L_+); \tag{6.255}$$

$$SO(4) \qquad C_1^{(2)} = K^2 + L^2, \quad C_2^{(2)} = \vec{K}\cdot\vec{L} \tag{6.256}$$

$$\left(\text{or } C_1^{(2)} = M^2, \quad C_2^{(2)} = N^2, \text{ where } \vec{M} \equiv \frac{1}{2}(\vec{L}+\vec{K}), \quad \vec{N} \equiv \frac{1}{2}(\vec{L}-\vec{K})\right). \tag{6.257}$$

Theorem 6.7.6.

$$C^{(2)}|\vec{\lambda}^{\max}\rangle = \vec{\lambda}^{\max}\cdot\left(\vec{\lambda}^{\max} + \frac{1}{2}\sum_{\alpha>0}\vec{r}(\alpha)\right)|\vec{\lambda}^{\max}\rangle. \tag{6.258}$$

Proof From $[E_\alpha, E_{-\alpha}] = \sum_{i=1}^{r} r_i(\alpha)H_i$ (theorem 6.7.1)

$$\begin{aligned}
\therefore C^{(2)}\left|\vec{\lambda}^{\max}\right\rangle &= \sum_{i=1}^{r} H_i^2\left|\vec{\lambda}^{\max}\right\rangle + \frac{1}{2}\sum_{\alpha>0}\left(2E_{-\alpha}\cancelto{0}{E_\alpha} + \sum_{i=1}^{r} r_i(\alpha)H_i\right)\left|\vec{\lambda}^{\max}\right\rangle \\
&= \left(\sum_{i=1}^{r}(\lambda_i^{\max})^2 + \frac{1}{2}\sum_{\alpha>0}\sum_{i=1}^{r} r_i(\alpha)\lambda_i^{\max}\right)\left|\vec{\lambda}^{\max}\right\rangle \\
&= \left(\vec{\lambda}^{\max}\cdot\vec{\lambda}^{\max} + \frac{1}{2}\sum_{\alpha>0}\vec{r}(\alpha)\cdot\vec{\lambda}^{\max}\right)\left|\vec{\lambda}^{\max}\right\rangle \\
&= \vec{\lambda}^{\max}\cdot\left(\vec{\lambda}^{\max} + \frac{1}{2}\sum_{\alpha>0}\vec{r}(\alpha)\right)\left|\vec{\lambda}^{\max}\right\rangle.
\end{aligned} \tag{6.259}$$

□

Irrep quantum numbers

- $\{\lambda_i\}$, the eigenvalues of the $\{H_i\}$.
- $\vec{\lambda}^{\max}$, related to the eigenvalue of $C^{(2)}$ and fixed for an irrep.

For a compact Lie algebra there is a highest-weight state, $|\vec{\lambda}^{\max}\rangle$, and a lowest-weight state, $|\vec{\lambda}^{\min}\rangle$, in each irrep. A useful characterisation of $|\vec{\lambda}^{\max}\rangle$ is the number of times it can be lowered by each of the $\{E_{-\alpha}\}$. From equation (6.236),

$$\vec{\lambda}^{\max} \cdot \vec{r}(\alpha) = \frac{1}{2} L\vec{r}(\alpha) \cdot \vec{r}(\alpha). \tag{6.260}$$

Example: irreps for SU*(3)*

The Cartesian coordinates of the $SU(3)$ root diagram are shown in figure 6.20. The highest-weight state, $\vec{M}^{\lambda_1\lambda_2}$ $(\vec{M} \equiv \vec{\lambda}_{\max})$ is defined

$$\frac{\vec{M}^{\lambda_1\lambda_2} \cdot \vec{r}(\alpha)}{\vec{r}(\alpha) \cdot \vec{r}(\alpha)} = \frac{\lambda_1}{2}, \quad \frac{\vec{M}^{\lambda_1\lambda_2} \cdot \vec{r}(\beta)}{\vec{r}(\beta) \cdot \vec{r}(\beta)} = \frac{\lambda_2}{2}, \tag{6.261}$$

where

$$\vec{r}(\alpha) = \begin{pmatrix} \frac{1}{\sqrt{3}} \\ 0 \end{pmatrix}, \quad \vec{r}(\beta) = \begin{pmatrix} -\frac{1}{2\sqrt{3}} \\ \frac{1}{2} \end{pmatrix}. \tag{6.262}$$

Then, for

$$\vec{M}^{\lambda_1\lambda_2} = \begin{pmatrix} a \\ b \end{pmatrix}, \tag{6.263}$$

$$\frac{a\frac{1}{\sqrt{3}}}{\frac{1}{3}} = \frac{\lambda_1}{2}, \quad a = \frac{\lambda_1}{2\sqrt{3}}, \tag{6.264}$$

$$\frac{\left(a\left(-\frac{1}{2\sqrt{3}}\right) + b\left(\frac{1}{2}\right)\right)}{\frac{1}{3}} = \frac{\lambda_2}{2}, \quad b = \frac{1}{6}(\lambda_1 + 2\lambda_2), \tag{6.265}$$

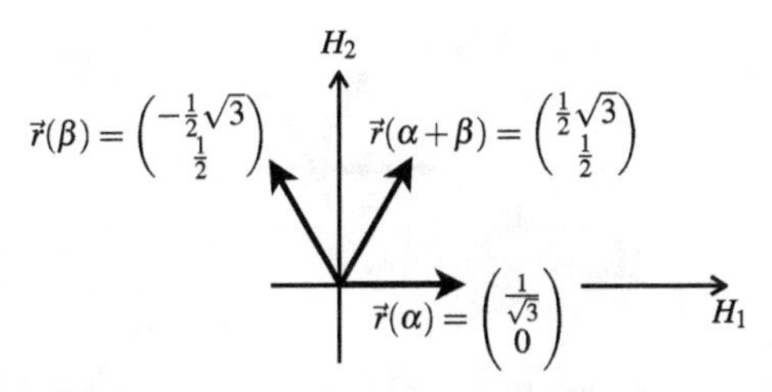

Figure 6.20. Cartesian coordinates of the $SU(3)$ root diagram.

i.e.

$$\vec{M}^{\textcircled{\lambda_1}\text{-}\textcircled{\lambda_2}} = \frac{1}{2\sqrt{3}} \begin{pmatrix} \lambda_1 \\ \frac{\lambda_1 + 2\lambda_2}{\sqrt{3}} \end{pmatrix}. \tag{6.266}$$

The maximum weight state in each irrep is given by (λ_1, λ_2)= (0,0), (1,0), (0,1), (2,0), (1,1), (0,2), ⋯.

$\mathcal{D}^{(1,0)}$: This is called the 'quark' representation (figure 6.21). The combination of $\vec{M}^{\textcircled{1}\text{-}\textcircled{0}}$ with $-\vec{r}(\beta)$ is excluded because $\lambda_2 = 0$ and, thus, from $\frac{\vec{M}^{\textcircled{\lambda_1}\text{-}\textcircled{\lambda_2}} \cdot \vec{r}(\beta)}{\vec{r}(\beta) \cdot \vec{r}(\beta)} = \frac{\lambda_2}{2}$, $\vec{M}^{\textcircled{1}\text{-}\textcircled{0}}$ cannot be lowered by $E_{-\beta}$. Figure 6.22 shows the ladder operators that cannot raise or lower the highest-weight states in the $SU(3)$ irreps $(\lambda_1, 0)$, $(0, \lambda_2)$ and (λ_1, λ_2).

For the eigenvalues of the Casimir operator $C^{(2)}$, equation (6.259), for the (λ_1, λ_2) irrep. from

$$\vec{M}^{\textcircled{\lambda_1}\text{-}\textcircled{\lambda_2}} = \frac{1}{2\sqrt{3}} \begin{pmatrix} \lambda_1 \\ \frac{\lambda_1 + 2\lambda_2}{\sqrt{3}} \end{pmatrix} = \begin{pmatrix} \frac{\frac{\lambda_1}{2}}{\sqrt{3}} \\ \frac{\frac{\lambda_1}{3} + \frac{2\lambda_2}{3}}{2} \end{pmatrix}, \tag{6.267}$$

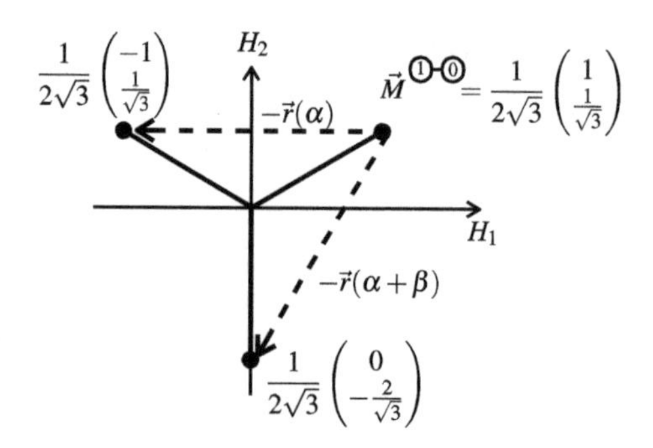

Figure 6.21. Cartesian coordinates of the fundamental $SU(3)$ weight diagram.

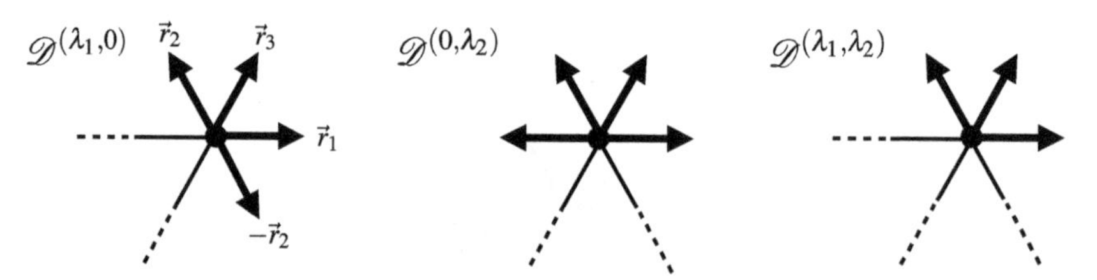

Figure 6.22. Depiction of ladder operators that cannot raise or lower the highest-weight states of the $SU(3)$ irreps $(\lambda_1, 0)$, $(0, \lambda_2)$ and (λ_1, λ_2).

$$\vec{r}_1 = \vec{r}(\alpha) = \begin{pmatrix} \dfrac{1}{\sqrt{3}} \\ 0 \end{pmatrix}, \quad \vec{r}_2 = \vec{r}(\beta) = \begin{pmatrix} -\dfrac{1}{2\sqrt{3}} \\ \dfrac{1}{2} \end{pmatrix}, \quad \vec{r}_3 = \vec{r}(\alpha+\beta) = \begin{pmatrix} \dfrac{1}{2\sqrt{3}} \\ \dfrac{1}{2} \end{pmatrix}, \tag{6.268}$$

$$|\vec{r}_1|^2 + |\vec{r}_2|^2 + |\vec{r}_3|^2 = 1, \tag{6.269}$$

$$\frac{2\vec{r}_1 \cdot \vec{r}_2}{|\vec{r}_1|^2} = -1, \quad \frac{2\vec{r}_1 \cdot \vec{r}_3}{|\vec{r}_1|^2} = +1, \quad \frac{2\vec{r}_2 \cdot \vec{r}_3}{|\vec{r}_2|^2} = +1, \tag{6.270}$$

$$\vec{M}^{(\lambda_1)\text{-}(\lambda_2)} - n\vec{r}_1 = \begin{pmatrix} \dfrac{\frac{\lambda_1}{2} - n}{\sqrt{3}} \\ \dfrac{\frac{\lambda_1}{3} + \frac{2\lambda_2}{3}}{2} \end{pmatrix} \quad : \quad n = 1, 2, \cdots, \lambda_1, \tag{6.271}$$

$$\vec{M}^{(\lambda_1)\text{-}(\lambda_2)} - m\vec{r}_2 = \begin{pmatrix} \dfrac{\frac{\lambda_1}{2} + \frac{m}{2}}{\sqrt{3}} \\ \dfrac{\frac{\lambda_1}{3} + \frac{2\lambda_2}{3} - m}{2} \end{pmatrix} \quad : \quad m = 1, 2, \cdots, \lambda_1. \tag{6.272}$$

Then, for

$$\vec{\lambda}^{\max} = \vec{M}^{(\lambda_1)\text{-}(\lambda_2)} = \frac{1}{2\sqrt{3}} \begin{pmatrix} \lambda_1 \\ \frac{\lambda_1 + 2\lambda_2}{\sqrt{3}} \end{pmatrix}, \tag{6.273}$$

and

$$\vec{r}(\alpha) = \begin{pmatrix} \dfrac{1}{\sqrt{3}} \\ 0 \end{pmatrix}, \quad \vec{r}(\beta) = \begin{pmatrix} -\dfrac{1}{2\sqrt{3}} \\ \dfrac{1}{2} \end{pmatrix}, \quad \vec{r}(\alpha+\beta) = \begin{pmatrix} \dfrac{1}{2\sqrt{3}} \\ \dfrac{1}{2} \end{pmatrix}, \tag{6.274}$$

$$\therefore \vec{R} = \vec{r}(\alpha) + \vec{r}(\beta) + \vec{r}(\alpha+\beta) = \begin{pmatrix} \dfrac{1}{\sqrt{3}} \\ 1 \end{pmatrix}, \tag{6.275}$$

and

$$\vec{\lambda}^{\max} \cdot (\vec{\lambda}^{\max} + \vec{R}) = \frac{1}{2\sqrt{3}} \begin{pmatrix} \lambda_1 & \dfrac{\lambda_1 + 2\lambda_2}{\sqrt{3}} \end{pmatrix} \begin{pmatrix} \dfrac{\lambda_1}{2\sqrt{3}} + \dfrac{1}{\sqrt{3}} \\ \dfrac{\lambda_1 + 2\lambda_2}{6} + 1 \end{pmatrix}, \tag{6.276}$$

$$\therefore \vec{\lambda}^{\max} \cdot (\vec{\lambda}^{\max} + \vec{R}) = \frac{1}{2\sqrt{3}} \left\{ \frac{\lambda_1^{\,2}}{2\sqrt{3}} + \frac{\lambda_1}{\sqrt{3}} + \frac{(\lambda_1 + 2\lambda_2)^2}{6\sqrt{3}} + \frac{\lambda_1 + 2\lambda_2}{\sqrt{3}} \right\}, \tag{6.277}$$

$$\therefore \vec{\lambda}^{\max} \cdot (\vec{\lambda}^{\max} + \vec{R}) = \frac{1}{9}(\lambda_1^2 + \lambda_2^2 + \lambda_1\lambda_2 + 3\lambda_1 + 3\lambda_2). \tag{6.278}$$

Multiplicities of weight points occur for some *su*(3) irreps (the 'hexagonal' ones). We do not use the formalism herein to elucidate these, but they emerge in a straightforward manner using Young tableaux, cf. figure 6.6(e).

Reference

[1] Rowe D J and Wood J L 2010 *Fundamentals of Nuclear Models: Foundational Models* (Singapore: World Scientific)

IOP Publishing

Quantum Mechanics for Nuclear Structure, Volume 2
An intermediate level view
Kris Heyde and John L Wood

Chapter 7

Perturbation theory and the variational method

A standard introduction to time-independent perturbation theory is given. This is sufficient for needs typically encountered in the study of nuclear structure. The procedure for handling the occurrence of degeneracies is described. Procedures for incorporating symmetry into calculations is sketched. The variational method is covered in a standard treatment.

Concepts: choice of basis; perturbation expansion of energies and state vectors; first-order perturbation theory; second-order perturbation theory; degeneracy; rotational symmetry; inversion symmetry; variational method.

7.1 Time-independent perturbation theory

Time-independent perturbation theory is a means of determining the energy eigenvalues and eigenkets of a system with a Hamiltonian

$$\hat{H} = \hat{H}_0 + \hat{V}, \qquad \hat{V} \neq \hat{V}(t), \tag{7.1}$$

where $|\hat{V}| \ll |\hat{H}_0|$ and the exact solutions for H_0 are known:

$$\hat{H}_0|n^{(0)}\rangle = E_n^{(0)}|n^{(0)}\rangle, \tag{7.2}$$

i.e. we seek solutions to

$$(\hat{H}_0 + \hat{V})|n\rangle = E_n|n\rangle. \tag{7.3}$$

The perturbation is $\hat{V}$. It is not necessary to include all perturbations in $\hat{V}$. For example in calculating the spin–spin interaction in hydrogen, it can be done while ignoring the spin–orbit interaction. Of course, to obtain the full fine and hyperfine splitting in hydrogen all terms must be calculated, but they can be calculated separately when they are all small.

doi:10.1088/978-0-7503-2171-6ch7

It is standard procedure to solve the problem in the form

$$(\hat{H}_0 + \lambda\hat{V})|n\rangle = E_n|n\rangle, \tag{7.4}$$

where λ is a continuous real parameter. This provides a means of keeping track of the number of times the perturbation enters into the calculation. (The parameter λ can be considered to vary from 0 to 1 and can be set equal to 1 at the end of the calculation.) Strictly speaking, we should index the energy eigenkets and energy eigenvalues by λ, viz.

$$(\hat{H}_0 + \lambda\hat{V})|n\rangle_\lambda = E_n^{(\lambda)}|n\rangle_\lambda, \tag{7.5}$$

but this will be understood to be so, in writing equation (7.4), to avoid cumbersome notation.

Perturbation theory presumes that solutions to

$$\hat{H}_0|n^{(0)}\rangle = E_n^{(0)}|n^{(0)}\rangle, \tag{7.6}$$

have been obtained. Then the set $\{|n^{(0)}\rangle\}$ is complete in the sense that

$$\sum_n |n^{(0)}\rangle\langle n^{(0)}| = \hat{I}. \tag{7.7}$$

We define

$$\Delta_n := E_n - E_n^{(0)}. \tag{7.8}$$

Then, the equation to be solved (approximately) is

$$(\hat{H}_0 + \lambda\hat{V})|n\rangle = (E_n^{(0)} + \Delta_n)|n\rangle \tag{7.9}$$

or

$$(E_n^{(0)} - \hat{H}_0)|n\rangle = (\lambda\hat{V} - \Delta_n)|n\rangle. \tag{7.10}$$

The basic strategy is to expand $|n\rangle$ and Δ_n in powers of λ:

$$|n\rangle = |n^{(0)}\rangle + \lambda|n^{(1)}\rangle + \lambda^2|n^{(2)}\rangle + \cdots, \tag{7.11}$$

$$\Delta_n = \lambda\Delta_n^{(1)} + \lambda^2\Delta_n^{(2)} + \cdots. \tag{7.12}$$

The first energy approximation is directly obtained by substituting equations (7.11) and (7.12) into equation (7.10):

$$(E_n^{(0)} - \hat{H}_0)(|n^{(0)}\rangle + \lambda|n^{(1)}\rangle + \cdots) = (\lambda\hat{V} - \lambda\Delta_n^{(1)} - \cdots)(|n^{(0)}\rangle + \lambda|n^{(1)}\rangle + \cdots), \tag{7.13}$$

and expanding,

$$(E_n^{(0)} - \hat{H}_0)|n^{(0)}\rangle + \lambda(E_n^{(0)} - \hat{H}_0)|n^{(1)}\rangle + \cdots = \lambda\hat{V}|n^{(0)}\rangle - \lambda\Delta_n^{(1)}|n^{(0)}\rangle + \cdots; \tag{7.14}$$

then, since $\hat{H}_0|n^{(0)}\rangle = E_n^{(0)}|n^{(0)}\rangle$, equating terms up to first order in λ,

$$(E_n^{(0)} - \hat{H}_0)|n^{(1)}\rangle = \hat{V}|n^{(0)}\rangle - \Delta_n^{(1)}|n^{(0)}\rangle. \tag{7.15}$$

Then, taking the inner product on both sides with $\langle n^{(0)}|$

$$\langle n^{(0)}|(E_n^{(0)} - \hat{H}_0)|n^{(1)}\rangle = \langle n^{(0)}|\hat{V}|n^{(0)}\rangle - \Delta_n^{(1)}\langle n^{(0)}|n^{(0)}\rangle, \tag{7.16}$$

and using

$$\langle n^{(0)}|\hat{H}_0 = E_n^{(0)}\langle n^{(0)}|, \quad \langle n^{(0)}|n^{(0)}\rangle = 1, \tag{7.17}$$

$$\therefore \Delta_n^{(1)} = \langle n^{(0)}|\hat{V}|n^{(0)}\rangle, \tag{7.18}$$

i.e. the first-order correction to the energy is just the expectation value of $\hat{V}$ for the state $|n^{(0)}\rangle$. This is the most elementary and fundamental result of perturbation theory.

For $|n^{(1)}\rangle$:

$$|n^{(1)}\rangle = \sum_m |m^{(0)}\rangle\langle m^{(0)}|n^{(1)}\rangle, \tag{7.19}$$

$$\therefore |n^{(1)}\rangle = |n^{(0)}\rangle\langle n^{(0)}|n^{(1)}\rangle + \sum_{m\neq n} |m^{(0)}\rangle\langle m^{(0)}|n^{(1)}\rangle. \tag{7.20}$$

Then, taking the inner product of both sides of equation (7.15) with $\langle m^{(0)}|$, where $m \neq n$,

$$\langle m^{(0)}|E_n^{(0)} - \hat{H}_0|n^{(1)}\rangle = \langle m^{(0)}|\hat{V}|n^{(0)}\rangle - \Delta_n^{(1)}\langle m^{(0)}|n^{(0)}\rangle, \tag{7.21}$$

and using

$$\langle m^{(0)}|\hat{H}_0 = E_m^{(0)}\langle m^{(0)}|, \quad \langle m^{(0)}|n^{(0)}\rangle = 0, \; m \neq n, \tag{7.22}$$

$$\therefore \langle m^{(0)}|n^{(1)}\rangle = \frac{\langle m^{(0)}|\hat{V}|n^{(0)}\rangle}{E_n^{(0)} - E_m^{(0)}}, \; m \neq n. \tag{7.23}$$

It is critical to note at this point that the present method is only valid in the *absence of degeneracies*, i.e. $E_n^{(0)} \neq E_m^{(0)}$. Perturbation theory for situations where degeneracy is involved will be considered shortly.

$$\therefore |n^{(1)}\rangle = |n^{(0)}\rangle\langle n^{(0)}|n^{(1)}\rangle + \sum_{m\neq n} |m^{(0)}\rangle\frac{\langle m^{(0)}|\hat{V}|n^{(0)}\rangle}{E_n^{(0)} - E_m^{(0)}}. \tag{7.24}$$

Thus, to order λ

$$\begin{aligned} |n\rangle &= |n^{(0)}\rangle + \lambda|n^{(1)}\rangle \\ &= (1 + \lambda a)|n^{(0)}\rangle + \lambda\sum_{m\neq n}\frac{\langle m^{(0)}|\hat{V}|n^{(0)}\rangle}{E_n^{(0)} - E_m^{(0)}}|m^{(0)}\rangle, \end{aligned} \tag{7.25}$$

where

$$a \equiv \langle n^{(0)}|n^{(1)}\rangle. \tag{7.26}$$

Normalization of $|n\rangle$ to order λ requires $a = 0$. From equation (7.23), $\langle n^{(0)}|n^{(1)}\rangle$ is undetermined and so we define $a \equiv 0$:

$$\langle n^{(0)}|n^{(1)}\rangle := 0, \tag{7.27}$$

and

$$|n^{(1)}\rangle = \sum_{m\neq n} |m^{(0)}\rangle \frac{\langle m^{(0)}|\hat{V}|n^{(0)}\rangle}{E_n^{(0)} - E_m^{(0)}}. \tag{7.28}$$

The second energy approximation is obtained from equations (7.10)–(7.12) expanded to order λ^2:

$$\begin{aligned}&(E_n^{(0)} - \hat{H}_0)(|n^{(0)}\rangle + \lambda|n^{(1)}\rangle + \lambda^2|n^{(2)}\rangle + \cdots)\\ &= (\lambda\hat{V} - \lambda\Delta_n^{(1)} - \lambda^2\Delta_n^{(2)} - \cdots)(|n^{(0)}\rangle + \lambda|n^{(1)}\rangle + \lambda^2|n^{(2)}\rangle + \cdots),\end{aligned} \tag{7.29}$$

$$\begin{aligned}&\therefore (E_n^{(0)} - \hat{H}_0)|n^{(0)}\rangle + \lambda(E_n^{(0)} - \hat{H}_0)|n^{(1)}\rangle + \lambda^2(E_n^{(0)} - \hat{H}_0)|n^{(2)}\rangle + \cdots\\ &= \lambda(\hat{V} - \Delta_n^{(1)})|n^{(0)}\rangle + \lambda^2\{(\hat{V} - \Delta_n^{(1)})|n^{(1)}\rangle - \Delta_n^{(2)}|n^{(0)}\rangle\} + \cdots,\end{aligned} \tag{7.30}$$

and using $\hat{H}_0|n^{(0)}\rangle = E_n^{(0)}|n^{(0)}\rangle$ and taking the inner product of both sides with $\langle n^{(0)}|$, to second order in λ,

$$\begin{aligned}&\lambda\langle n^{(0)}|(E_n^{(0)} - \hat{H}_0)|n^{(1)}\rangle + \lambda^2\langle n^{(0)}|(E_n^{(0)} - \hat{H}_0)|n^{(2)}\rangle\\ &= \lambda\langle n^{(0)}|\hat{V}|n^{(0)}\rangle - \lambda\Delta_n^{(1)}\langle n^{(0)}|n^{(0)}\rangle + \lambda^2\langle n^{(0)}|\hat{V}|n^{(1)}\rangle\\ &- \lambda^2\Delta_n^{(1)}\langle n^{(0)}|n^{(1)}\rangle - \lambda^2\Delta_n^{(2)}\langle n^{(0)}|n^{(0)}\rangle.\end{aligned} \tag{7.31}$$

Then, from $\langle n^{(0)}|\hat{H}_0 = E_n^{(0)}\langle n^{(0)}|$, $\langle n^{(0)}|n^{(0)}\rangle = 1$, $\langle n^{(0)}|n^{(1)}\rangle = 0$ (equation (7.27)), and $\Delta_n^{(1)} = \langle n^{(0)}|\hat{V}|n^{(0)}\rangle$ (equation (7.18)),

$$\therefore \Delta_n^{(2)} = \langle n^{(0)}|\hat{V}|n^{(1)}\rangle, \tag{7.32}$$

and for $|n^{(1)}\rangle$ from equation (7.28)

$$\therefore \Delta_n^{(2)} = \sum_{m\neq n} \frac{\langle n^{(0)}|\hat{V}|m^{(0)}\rangle\langle m^{(0)}|\hat{V}|n^{(0)}\rangle}{E_n^{(0)} - E_m^{(0)}}. \tag{7.33}$$

For $|n^{(2)}\rangle$:

$$|n^{(2)}\rangle = \sum_p |p^{(0)}\rangle\langle p^{(0)}|n^{(2)}\rangle, \tag{7.34}$$

$$\therefore |n^{(2)}\rangle = |n^{(0)}\rangle\langle n^{(0)}|n^{(2)}\rangle + \sum_{p\neq n} |p^{(0)}\rangle\langle p^{(0)}|n^{(2)}\rangle. \tag{7.35}$$

Then, taking the inner product of both sides of equation (7.30) with $\langle p^{(0)}|$, where $p \neq n$,

$$\begin{aligned}&\langle p^{(0)}|(E_n^{(0)} - \hat{H}_0)|n^{(0)}\rangle + \lambda\langle p^{(0)}|(E_n^{(0)} - \hat{H}_0)|n^{(1)}\rangle + \lambda^2\langle p^{(0)}|(E_n^{(0)} - \hat{H}_0)|n^{(2)}\rangle + \cdots \\ &= \lambda\langle p^{(0)}|\hat{V}|n^{(0)}\rangle - \lambda\Delta_n^{(1)}\langle p^{(0)}|n^{(0)}\rangle + \lambda^2\langle p^{(0)}|\hat{V}|n^{(1)}\rangle - \lambda^2\Delta_n^{(1)}\langle p^{(0)}|n^{(1)}\rangle - \lambda^2\Delta_n^{(2)}\langle p^{(0)}|n^{(0)}\rangle,\end{aligned} \tag{7.36}$$

and using $\hat{H}_0|n^{(0)}\rangle = E_n^{(0)}|n^{(0)}\rangle$, $\langle p^{(0)}|\hat{H}_0 = \langle p^{(0)}|E_p^{(0)}$, $\langle p^{(0)}|n^{(0)}\rangle = 0$, $p \neq n$,

$$\begin{aligned}&\therefore \lambda\left(E_n^{(0)} - E_p^{(0)}\right)\langle p^{(0)}|n^{(1)}\rangle + \lambda^2\left(E_n^{(0)} - E_p^{(0)}\right)\langle p^{(0)}|n^{(2)}\rangle \\ &= \lambda\langle p^{(0)}|\hat{V}|n^{(0)}\rangle + \lambda^2\langle p^{(0)}|\hat{V}|n^{(1)}\rangle - \lambda^2\Delta_n^{(1)}\langle p^{(0)}|n^{(1)}\rangle.\end{aligned} \tag{7.37}$$

But, from equation (7.23) with $m \to p$, the terms of order λ are equal. Thus, for the terms of order λ^2

$$\langle p^{(0)}|n^{(2)}\rangle = \frac{\langle p^{(0)}|\hat{V}|n^{(1)}\rangle}{\left(E_n^{(0)} - E_p^{(0)}\right)} - \Delta_n^{(1)}\frac{\langle p^{(0)}|n^{(1)}\rangle}{E_n^{(0)} - E_p^{(0)}}. \tag{7.38}$$

Then, using equation (7.28) for $|n^{(1)}\rangle$ and equation (7.23) with $m \to p$

$$\langle p^{(0)}|n^{(2)}\rangle = \frac{1}{E_n^{(0)} - E_p^{(0)}}\sum_{m\neq n}\langle p^{(0)}|\hat{V}|m^{(0)}\rangle\frac{\langle m^{(0)}|\hat{V}|n^{(0)}\rangle}{E_n^{(0)} - E_m^{(0)}} - \Delta_n^{(1)}\frac{\langle p^{(0)}|\hat{V}|n^{(0)}\rangle}{\left(E_n^{(0)} - E_p^{(0)}\right)^2}, \tag{7.39}$$

$$\begin{aligned}\therefore |n^{(2)}\rangle = |n^{(0)}\rangle\langle n^{(0)}|n^{(2)}\rangle &+ \sum_{p\neq n}\sum_{m\neq q}|p^{(0)}\rangle\frac{\langle p^{(0)}|\hat{V}|m^{(0)}\rangle\langle m^{(0)}|\hat{V}|n^{(0)}\rangle}{\left(E_n^{(0)} - E_p^{(0)}\right)\left(E_n^{(0)} - E_m^{(0)}\right)} \\ &- \sum_{p\neq n}|p^{(0)}\rangle\frac{\langle n^{(0)}|\hat{V}|n^{(0)}\rangle\langle p^{(0)}|\hat{V}|n^{(0)}\rangle}{\left(E_n^{(0)} - E_p^{(0)}\right)^2},\end{aligned} \tag{7.40}$$

where equation (7.18) has been used for $\Delta_n^{(1)}$. Thus, to order λ^2,

$$|n\rangle = |n^{(0)}\rangle + \lambda|n^{(1)}\rangle + \lambda^2|n^{(2)}\rangle, \tag{7.41}$$

$$\begin{aligned}\therefore |n\rangle = (1 + \lambda^2 b)|n^{(0)}\rangle &+ \lambda\sum_{m\neq n}|m^{(0)}\rangle\frac{V_{mn}}{\Delta_{mn}} \\ &+ \lambda^2\left\{\sum_{p\neq n}\sum_{m\neq n}|p^{(0)}\rangle\frac{V_{pm}V_{mn}}{\Delta_{np}\Delta_{nm}} - \sum_{p\neq n}|p^{(0)}\rangle\frac{V_{nn}V_{pn}}{(\Delta_{np})^2}\right\},\end{aligned} \tag{7.42}$$

where $V_{pm} = \langle p^{(0)}|\hat{V}|m^{(0)}\rangle$, etc., $\Delta_{np} = E_n^{(0)} - E_p^{(0)}$, etc., and

$$b = \langle n^{(0)}|n^{(2)}\rangle. \tag{7.43}$$

Normalization of $|n\rangle$ to order λ^2 requires

$$b = -\frac{1}{2}\sum_{m\neq n}\frac{|V_{nm}|^2}{(\Delta_{nm})^2}. \tag{7.44}$$

Hence,

$$|n^{(2)}\rangle = -\frac{1}{2}|n^{(0)}\rangle \sum_{m\neq n} \frac{|V_{nm}|^2}{(\Delta_{nm})^2} + \sum_{p\neq n}\sum_{m\neq n} |p^{(0)}\rangle \frac{V_{pm}V_{mn}}{\Delta_{np}\Delta_{nm}} - \sum_{p\neq n} |p^{(0)}\rangle \frac{V_{nn}V_{pn}}{(\Delta_{np})^2}. \quad (7.45)$$

This procedure is straightforwardly iterated to any order. For practical purposes, it is rare that orders higher than $|n^{(1)}\rangle$ and $\Delta_n^{(2)}$ are needed. Exact matrix diagonalization is always available as an alternative and, when degeneracies are present (see the next section), it must be used to solve at least part of the problem.

7.1.1 Exercises

7.1. Consider a two-level system with the Hamiltonian

$$\hat{H} = \hat{H}_0 + \hat{V},$$

where $\hat{H}_0|\Psi_1\rangle = E_1|\Psi_1\rangle$, $\hat{H}_0|\Psi_2\rangle = E_2|\Psi_2\rangle$, $\hat{V}|\Psi_1\rangle = |\Psi_2\rangle$, $\hat{V}|\Psi_2\rangle = |\Psi_1\rangle$, $E_1 \neq E_2$.

(a) Solve for the exact energy eigenvalues using matrix diagonalization.

(b) Solve for the energy eigenvalues using (non-degenerate) perturbation theory. How does your answer compare to the solution in (a)?

7.2. A one-dimensional harmonic oscillator with Hamiltonian

$$\hat{H} = \frac{\hat{p}^2}{2m} + \frac{1}{2}k\hat{x}^2$$

is subjected to a constant perturbation $\frac{1}{2}\alpha\hat{x}^2$.

(a) Solve for the energy eigenvalues using (non-degenerate) perturbation theory.

(b) Solve for the exact energy eigenvalues.

7.3. Solve for the energy eigenvalues and eigenvectors of

$$\hat{H} = \begin{pmatrix} 1 & \varepsilon & \varepsilon \\ \varepsilon & 2 & \varepsilon \\ \varepsilon & \varepsilon & 3 \end{pmatrix},$$

where ε is small, using (non-degenerate) perturbation theory.

7.2 Time-independent perturbation theory for systems with degeneracy

If degeneracies are present, e.g. $E_n^{(0)} = E_m^{(0)}$ for $m \neq n$ in equation (7.23), this is a catastrophe for the above treatment: zeros will appear in some of the denominators! This problem is avoided by first diagonalizing the matrix for the degenerate subspace. Usually, degenerate subspaces are not that large and so this task is simple. One only needs to diagonalize the degenerate subspace containing the state of interest! Other degenerate subspaces may be present, but their states are not degenerate with the state of interest.

Usually, diagonalization in the degenerate subspace containing the state of interest will remove the degeneracies associated with this state. Then one can proceed by using standard non-degenerate perturbation theory. However, it is possible that the degenerate subspace is already diagonal, e.g.

$$\hat{H} = \begin{pmatrix} E_1 & 0 & V_{13} \\ 0 & E_1 & V_{23} \\ V_{31} & V_{32} & E_2 \end{pmatrix}, \tag{7.46}$$

where

$$E_1 \equiv E_1^{(0)} + V_{11} = E_2^{(0)} + V_{22}, \tag{7.47}$$

$$E_2 \equiv E_3^{(0)} + V_{33}, \tag{7.48}$$

and

$$V_{ij} = \langle i|\hat{V}|\hat{j}\,\rangle. \tag{7.49}$$

In these circumstances, one first applies second-order perturbation theory, viz.

$$\underset{i,j\in D}{\Delta^{(2)}_{ij}} = \sum_{n\notin D} \frac{\langle i^{(0)}|\hat{V}|n^{(0)}\rangle\,\langle n^{(0)}|\hat{V}|j^{(0)}\rangle}{(E_D - E_n^{(0)})}, \tag{7.50}$$

where n is not an element of the subspace D and E_D is the energy of the degenerate subspace. Equation (7.50) is a generalisation of equation (7.33). It is discussed shortly. First an example is solved to illustrate the method.

7.3 An example of (second-order) degenerate perturbation theory

The Hamiltonian

$$\hat{H} = \begin{pmatrix} E_1 & 0 & b \\ 0 & E_1 & c \\ b^* & c^* & E_2 \end{pmatrix}, \tag{7.51}$$

is an example where a degeneracy cannot be removed in first order.

The exact solution is given by solutions to

$$\begin{vmatrix} E_1 - \lambda & 0 & b \\ 0 & E_1 - \lambda & c \\ b^* & c^* & E_2 - \lambda \end{vmatrix} = 0, \tag{7.52}$$

i.e.

$$(E_1 - \lambda)\{(E_1 - \lambda)(E_2 - \lambda) - |c|^2\} + b\{-(E_1 - \lambda)b^*\} = 0, \tag{7.53}$$

whence we obtain the first root, λ_1,

$$\lambda_1 = E_1, \tag{7.54}$$

and

$$E_1E_2 - |b|^2 - |c|^2 - (E_1 + E_2)\lambda + \lambda^2 = 0. \tag{7.55}$$

Equation (7.55) yields the roots

$$\lambda_2 = \frac{E_1 + E_2 + \sqrt{(E_1 + E_2)^2 - 4(E_1E_2 - |b|^2 - |c|^2)}}{2} \tag{7.56}$$

and

$$\lambda_3 = \frac{E_1 + E_2 - \sqrt{(E_1 + E_2)^2 - 4(E_1E_2 - |b|^2 - |c|^2)}}{2}. \tag{7.57}$$

Noting that

$$\sqrt{(E_1 + E_2)^2 - 4E_1E_2 + 4(|b|^2 + |c|^2)} = \sqrt{(E_1 - E_2)^2 + 4(|b|^2 + |c|^2)}, \tag{7.58}$$

for $|b|^2 + |c|^2 \ll E_1 - E_2$

$$\sqrt{(E_1 - E_2)^2 + 4(|b|^2 + |c|^2)} \approx (E_1 - E_2)\left\{1 + \frac{2(|b|^2 + |c|^2)}{(E_1 - E_2)^2}\right\}. \tag{7.59}$$

Thus

$$\lambda_2 \approx E_1 + \frac{|b|^2 + |c|^2}{(E_1 - E_2)} \tag{7.60}$$

and

$$\lambda_3 \approx E_2 - \frac{|b|^2 + |c|^2}{(E_1 - E_2)}. \tag{7.61}$$

If one attempts a solution using second-order perturbation theory in the form of equation (7.33), one obtains

$$\Delta_1^{(2)} = \frac{|V_{12}|^2}{E_1^{(0)} - E_2^{(0)}} + \frac{|V_{13}|^2}{E_1^{(0)} - E_3^{(0)}}, \tag{7.62}$$

and it is tempting to ignore the catastrophe of $E_1^{(0)} = E_2^{(0)} = E_1$ because $V_{12} = 0$. Whence, one obtains

$$\Delta_1^{(2)} = \frac{|b|^2}{(E_1 - E_2)}. \tag{7.63}$$

Similarly,

$$\Delta_2^{(2)} = \frac{|c|^2}{(E_1 - E_2)} \tag{7.64}$$

and

$$\Delta_3^{(2)} = \frac{|b|^2 + |c|^2}{(E_2 - E_1)} = \frac{-(|b|^2 + |c|^2)}{(E_1 - E_2)}. \tag{7.65}$$

Comparison of equations (7.54), (7.60) and (7.61) with (7.63), (7.64) and (7.65) (together with (7.51)) reveals failure! Evidently, dividing zero by zero is, as always, inadvisable!

However, if one attempts a solution using second-order perturbation theory in the form of equation (7.50) then the submatrix for the degenerate subspace can be written

$$\hat{H}' = \begin{pmatrix} E_1 + \Delta_{11}^{(2)} & \Delta_{12}^{(2)} \\ \Delta_{21}^{(2)} & E_1 + \Delta_{22}^{(2)} \end{pmatrix}, \tag{7.66}$$

where

$$\Delta_{11}^{(2)} = \frac{|V_{13}|^2}{E_1 - E_3^{(0)}} = \frac{|b|^2}{(E_1 - E_2)}, \tag{7.67}$$

$$\Delta_{12}^{(2)} = \frac{V_{13}V_{32}}{E_1 - E_3^{(0)}} = \frac{bc^*}{(E_1 - E_2)}, \tag{7.68}$$

$$\Delta_{21}^{(2)} = \Delta_{12}^{(2)*}, \tag{7.69}$$

and

$$\Delta_{22}^{(2)} = \frac{|V_{23}|^2}{E_1 - E_3^{(0)}} = \frac{|c|^2}{(E_1 - E_2)}. \tag{7.70}$$

Thus, the secular equation for $\hat{H}'$, viz.

$$(E_1 + \Delta_{11}^{(2)} - \lambda)(E_1 + \Delta_{22}^{(2)} - \lambda) - \Delta_{12}^{(2)}\Delta_{21}^{(2)} = 0, \tag{7.71}$$

becomes

$$\left\{E_1 - \lambda + \frac{|b|^2}{(E_1 - E_2)}\right\}\left\{E_1 - \lambda + \frac{|c|^2}{(E_1 - E_2)}\right\} - \frac{|b|^2|c|^2}{(E_1 - E_2)^2} = 0. \tag{7.72}$$

Hence,

$$(E_1 - \lambda)^2 + (E_1 - \lambda)\frac{(|b|^2 + |c|^2)}{(E_1 - E_2)} = 0, \tag{7.73}$$

which yields

$$\lambda_1 = E_1 \tag{7.74}$$

and

$$\lambda_2 = E_1 + \frac{|b|^2 + |c|^2}{(E_1 - E_2)}, \tag{7.75}$$

cf. equations (7.74) and (7.75) with equations (7.54) and (7.60) (also cf. equation (7.61) with equation (7.65)).

The derivation of equation (7.50) follows from equation (7.30) for $n = j$, $j \in D$,

$$\begin{aligned}&\left(E_j^{(0)} - \hat{H}_0\right)|j^{(0)}\rangle + \lambda\left(E_j^{(0)} - \hat{H}_0\right)|j^{(1)}\rangle + \lambda^2\left(E_j^{(0)} - \hat{H}_0\right)|j^{(2)}\rangle + \cdots \\ &= \lambda\left(\hat{V} - \Delta_j^{(1)}\right)|j^{(0)}\rangle + \lambda^2\left\{\left(\hat{V} - \Delta_j^{(1)}\right)|j^{(1)}\rangle - \Delta_j^{(2)}|j^{(0)}\rangle\right\} + \cdots.\end{aligned} \tag{7.76}$$

Then, taking the inner product on both sides with $\langle i^{(0)}|$, $i \in D$, and noting that $\hat{H}_0|j^{(0)}\rangle = E_j^{(0)}|j^{(0)}\rangle$, to second order in λ

$$\begin{aligned}&\lambda\langle i^{(0)}|\left(E_j^{(0)} - \hat{H}_0\right)|j^{(1)}\rangle + \lambda^2\langle i^{(0)}|\left(E_j^{(0)} - \hat{H}_0\right)|j^{(2)}\rangle \\ &= \lambda\langle i^{(0)}|\hat{V}|j^{(0)}\rangle - \lambda\Delta_j^{(1)}\langle i^{(0)}|j^{(0)}\rangle + \lambda^2\langle i^{(0)}|\hat{V}|j^{(1)}\rangle \\ &\quad - \lambda^2\Delta_j^{(1)}\langle i^{(0)}|j^{(1)}\rangle - \lambda^2\Delta_j^{(2)}\langle i^{(0)}|j^{(0)}\rangle.\end{aligned} \tag{7.77}$$

Thus, from $\langle i^{(0)}|\hat{H}_0 = \langle i^{(0)}|E_i^{(0)}$, $E_i^{(0)} = E_j^{(0)}$, $\langle i^{(0)}|\hat{V}|j^{(0)}\rangle = 0$, $\langle i^{(0)}|j^{(0)}\rangle = \delta_{ij}$, and $\langle j^{(0)}|j^{(1)}\rangle = 0$ (cf. equations (7.26) and (7.27) with $n = j$),

$$\langle i^{(0)}|\hat{V}|j^{(1)}\rangle = \Delta_j^{(1)}\langle i^{(0)}|j^{(1)}\rangle + \Delta_j^{(2)}\delta_{ij}. \tag{7.78}$$

Now, from equation (7.20) for $n = j$

$$|j^{(1)}\rangle = \sum_{k\in D}|k^{(0)}\rangle\langle k^{(0)}|j^{(1)}\rangle + \sum_{m\notin D}|m^{(0)}\rangle\langle m^{(0)}|j^{(1)}\rangle. \tag{7.79}$$

Then, from equation (7.15) for $n = j$,

$$\left(E_j^{(0)} - \hat{H}_0\right)|j^{(1)}\rangle = \hat{V}|j^{(0)}\rangle - \Delta_j^{(1)}|j^{(0)}\rangle, \tag{7.80}$$

taking the inner product on both sides with $\langle m^{(0)}|$, where $m \notin D$,

$$\langle m^{(0)}|\left(E_j^{(0)} - \hat{H}_0\right)|j^{(1)}\rangle = \langle m^{(0)}|\hat{V}|j^{(0)}\rangle - \Delta_j^{(1)}\langle m^{(0)}|j^{(0)}\rangle. \tag{7.81}$$

Thus, from $\langle m^{(0)}|\hat{H}_0 = \langle m^{(0)}|E_m^{(0)}$, $\langle m^{(0)}|j^{(0)}\rangle = 0$,

$$\langle m^{(0)}|j^{(1)}\rangle = \frac{\langle m^{(0)}|\hat{V}|j^{(0)}\rangle}{E_j^{(0)} - E_m^{(0)}} \tag{7.82}$$

and from equation (7.79),

$$\therefore |j^{(1)}\rangle \sum_{k\in D} |k^{(0)}\rangle\langle k^{(0)}|j^{(1)}\rangle + \sum_{m\notin D} |m^{(0)}\rangle \frac{\langle m^{(0)}|\hat{V}|j^{(0)}\rangle}{E_j^{(0)} - E_m^{(0)}}. \tag{7.83}$$

Finally, from equations (7.78) and (7.83),

$$\Delta_j^{(1)}\langle i^{(0)}|j^{(1)}\rangle + \Delta_j^{(2)}\delta_{ij} = \sum_{k\in D} \langle i^{(0)}|\hat{V}|k^{(0)}\rangle\langle k^{(0)}|j^{(1)}\rangle + \sum_{m\notin D} \frac{\langle i^{(0)}|\hat{V}|m^{(0)}\rangle\langle m^{(0)}|\hat{V}|j^{(0)}\rangle}{E_j^{(0)} - E_m^{(0)}}. \tag{7.84}$$

But, $\langle i^{(0)}|\hat{V}|m^{(0)}\rangle = 0$ and $E_j^{(0)} = E_D^{(0)}$,

$$\therefore \Delta_j^{(1)}\langle i^{(0)}|j^{(1)}\rangle + \Delta_j^{(2)}\delta_{ij} = \sum_{m\notin D} \frac{\langle i^{(0)}|\hat{V}|m^{(0)}\rangle\langle m^{(0)}|\hat{V}|j^{(0)}\rangle}{E_D^{(0)} - E_m^{(0)}}. \tag{7.85}$$

Defining $\Delta_j^{(2)}\delta_{ij} \equiv \Delta_{jj}^{(2)}$ and $\Delta_j^{(1)}\langle i^{(0)}|j^{(1)}\rangle \equiv \Delta_{ij}^{(2)}$, the desired result is obtained.

7.4 Perturbation theory and symmetry

The identification of the symmetries of a particular unperturbed system, i.e. of $\hat{H}_0$, greatly facilitates the delineation of the computational effort required to find the eigenvalues and eigenvectors of $\hat{H}_0 + \hat{V}$. This is true whether the solutions are obtained by a complete matrix diagonalization or by perturbation theory. This comes about because symmetry can tell us which matrix elements of $\hat{V}$ are zero. Two commonly occurring symmetries are used to illustrate this point.

Rotational symmetry is characteristic of central force problems and is manifested in the energy eigenkets being simultaneous eigenkets of $\hat{L}^2$ and (e.g.) $\hat{L}_z$. If $\hat{V}$ is expanded in terms of spherical tensors, this leads immediately to the identification of the zero matrix elements of $\hat{V}$ by use of the Wigner–Eckart theorem.

7.4.1 Example

$\hat{V} = \alpha\hat{z}$; $\hat{H}_0$: hydrogen atom, α: a constant; $\{|n^{(0)}\rangle\} = \{|nlm_lm_s\rangle\}$.

$$z = r\cos\theta = r\sqrt{\frac{4\pi}{3}}\,Y_{10}: \qquad \hat{z} = \hat{r}\sqrt{\frac{4\pi}{3}}\,\hat{T}_0^{(1)}.$$

$$\begin{aligned}\langle n'l'm'_lm'_s|\hat{V}|nlm_lm_s\rangle &= \langle n'l'm'_lm'_s|\alpha\hat{z}|nlm_lm_s\rangle \\ &= \langle n'l'm'_lm'_s|\hat{r}\sqrt{\frac{4\pi}{3}}\,\alpha\hat{T}_0^{(1)}|nlm_lm_s\rangle.\end{aligned}$$

Then, using the Wigner–Eckart theorem

$$\langle n'l'm'_lm'_s|\hat{V}|nlm_lm_s\rangle = \langle lm_l10|l'm'_l\rangle \times C,$$

where C is a constant. Evidently, we must have

$$m_l = m'_l, \quad l' = l, \; |l \pm 1|$$

to ensure a non-zero Clebsch–Gordan coefficient and therefore for a non-zero $\langle n'l'm'lm'_s|\hat{V}|nlm_lm_s\rangle$.

7.4.2 Inversion symmetry

Any system[1] in any state is either even or odd under space inversion. This property is called *parity*, π, where

$$\pi = +1 \text{ or } + \text{ (even)}, \tag{7.86}$$

$$\pi = -1 \text{ or } - \text{ (odd)}. \tag{7.87}$$

A unitary transformation $\hat{U}_p$ can be associated with space inversion. For a system in a state $|k\rangle$,

$$\hat{U}_p|k\rangle = \pm 1|k\rangle. \tag{7.88}$$

Then consider $\langle k|\hat{V}|k\rangle$: under space inversion

$$\langle k|\hat{U}_p^\dagger\hat{V}\hat{U}_p|k\rangle = \langle k|\hat{V}|k\rangle \tag{7.89}$$

because either $\hat{U}_p|k\rangle = +|k\rangle$, $\langle k|\hat{U}_p = +\langle k|$ or $\hat{U}_p|k\rangle = -|k\rangle$, $\langle k|\hat{U}_p^\dagger = -\langle k|$. But we can consider equation (7.89) as

$$\langle k|\hat{V}_p|k\rangle = \langle k|\hat{V}|k\rangle, \tag{7.90}$$

where

$$\hat{V}_p \equiv \hat{U}_p^\dagger\hat{V}\hat{U}_p. \tag{7.91}$$

If $\hat{V}_p = -\hat{V}$ then, from equation (7.90), the matrix elements of $\hat{V}$ are zero.

7.4.3 Example

$\hat{V} = \alpha\vec{r}_{\text{op}}$; α: a constant.

Evidently:

$$\hat{U}_p^\dagger\alpha\vec{r}_{\text{op}}\hat{U}_p = -\alpha\vec{r}_{op}.$$

Thus, from equation (7.89)

$$-\alpha\langle k|\vec{r}_{\text{op}}|k\rangle = \alpha\langle k|\vec{r}_{\text{op}}|k\rangle,$$
$$\therefore \langle k|\vec{r}_{\text{op}}|k\rangle = 0.$$

[1] Well, this is almost true of most systems! However, the *weak interaction* (which, for example, controls beta decay) produces minute amounts of parity admixing in most systems.

It follows that in the example given for rotational symmetry, since z is just $\vec{r}$ in a specified direction, we must have $\langle nlm_l m_s|\hat{z}|nlm_l m_s\rangle = 0$ (note this is the case $n' = n$, $l' = l$, $m'_l = m_l$, $m'_s = m_s$).

7.4.4 Exercises

7.4. Show that under space inversion

$$Y_{lm}(\theta, \phi) \rightarrow (-1)^l Y_{lm}(\theta, \phi). \tag{7.92}$$

(Hint: consider $x \rightarrow -x$, etc., and the transformation between Cartesian and spherical polar coordinates.)

7.5. Show that $\langle n'l'm'_l m'_s|\alpha\vec{r}|nlm_l m_s\rangle = 0$ if $l' + l + 1 =$ odd using the position representation and
   (a) the property of spherical harmonics obtained in exercise 7-4 (equation (7.92)).
   (b) equation (2.114).

7.5 The variational method

Perturbation theory can be used only when a major portion of the Hamiltonian can be isolated and solved exactly. Failing this, matrix diagonalization is the only general method that can give a solution. However, if only an estimate of the ground-state energy is needed, a very simple method is provided by the *variational method.* This provides a means of estimating an upper bound for the ground-state energy of the system. It is dependent on the following theorem:

Theorem 7.5.1. *For any ket $|\tilde{0}\rangle$, for a system with Hamiltonian $\hat{H}$ and ground-state energy E_0,*

$$\langle \hat{H}\rangle \equiv \frac{\langle\tilde{0}|\hat{H}|\tilde{0}\rangle}{\langle\tilde{0}|\tilde{0}\rangle} \geqslant E_0, \tag{7.93}$$

where the denominator is unity if $|\tilde{0}\rangle$ is normalized.

Proof. Expand $|\tilde{0}\rangle$ in terms of eigenkets of $\hat{H}$,

$$|\tilde{0}\rangle = \sum_{k=0}^{\infty} |k\rangle\langle k|\tilde{0}\rangle, \tag{7.94}$$

where

$$\hat{H}|k\rangle = E_k|k\rangle \tag{7.95}$$

(the proof does not depend on knowing what the eigenkets of $\hat{H}$ are, it only depends on their existence). Substituting equation (7.94) into (7.93)

$$\therefore \langle \hat{H} \rangle = \frac{\sum_{l=0}^{\infty}\sum_{k=0}^{\infty}\langle \tilde{0}|k\rangle \langle k|\hat{H}|l\rangle \langle l|\tilde{0}\rangle}{\sum_{l=0}^{\infty}\sum_{k=0}^{\infty}\langle \tilde{0}|k\rangle \langle k|l\rangle \langle l|\tilde{0}\rangle}$$
$$= \frac{\sum_{l=0}^{\infty}\sum_{k=0}^{\infty}\langle \tilde{0}|k\rangle E_l \overbrace{\langle k|l\rangle}^{\delta_{kl}} \langle l|\tilde{0}\rangle}{\sum_{l=0}^{\infty}\sum_{k=0}^{\infty}\langle \tilde{0}|k\rangle \overbrace{\langle k|l\rangle}^{\delta_{kl}} \langle l|\tilde{0}\rangle}$$
$$= \frac{\sum_{k=0}^{\infty} E_k |\langle k|\tilde{0}\rangle|^2}{\sum_{k=0}^{\infty} |\langle k|\tilde{0}\rangle|^2}. \tag{7.96}$$

Clearly,

$$\frac{\sum_{k=0}^{\infty} E_k |\langle k|\tilde{0}\rangle|^2}{\sum_{k=0}^{\infty} |\langle k|\tilde{0}\rangle|^2} \geqslant \frac{\sum_{k=0}^{\infty} E_0 |\langle k|\tilde{0}\rangle|^2}{\sum_{k=0}^{\infty} |\langle k|\tilde{0}\rangle|^2}, \tag{7.97}$$

where the equality sign holds if $|\tilde{0}\rangle = |0\rangle$. But the right-hand side of equation (7.97) is equal to E_0. Thus, the theorem follows. □

The practical consequence of the variational theorem is that if $\langle \hat{H} \rangle$ depends on some parameter, which can be introduced into the trial ground-state ket $|\tilde{0}\rangle$, then minimising $\langle \hat{H} \rangle$ with respect to the parameter will still fulfil $\langle \hat{H} \rangle \geqslant E_0$.

IOP Publishing

Quantum Mechanics for Nuclear Structure, Volume 2
An intermediate level view
Kris Heyde and John L Wood

Chapter 8

Time-dependent perturbation theory

A standard introduction to time-dependent perturbation theory is given. This is sufficient to introduce Fermi's golden rule as used for the coupling of nuclei to electromagnetic fields. A few details of the interaction picture and the Dyson series are given.

Concepts: interaction picture; Dyson series; Fermi's golden rule.

Time-dependent processes are the means by which things are made to happen in quantum mechanics[1]. Generally, such processes are 'weak', i.e. if the process is described by $V(t)$, $V(t) \ll H$, where H is the Hamiltonian describing the system. Thus, perturbation theory provides a procedure by which the time-dependence of the process can be described.

Time-dependent processes are discussed in Volume 1, chapter 9; here we follow on from the end of section 9.10. We introduce a picture, similar to the Heisenberg and Schrödinger pictures of time dependence, called the 'interaction picture'.

8.1 The interaction picture

Time-dependent phenomena can be incorporated directly into the Heisenberg and Schrödinger pictures (cf. Volume 1, section 9.5). A particularly useful formulation along the lines of these two descriptions is the so-called *interaction picture*, which is defined for the Hamiltonian

[1] We emphasize at the outset that with respect to the measurement process in quantum mechanics, while it has a dependence on time, it lies entirely outside any framework of description. All we possess in the measurement process is 'more uncertainty before the measurement, less uncertainty after the measurement'. The probabilistic content and change in state of knowledge is not a physical process. Such pictorial language as 'quantum jumps' and 'collapse of the wave function', while amusing as a way to dramatise the difference between the quantum world and our everyday world, are dangerously misleading: they do not describe physical processes. We leave it to the reader to decide what happens when a change occurs in their state of knowledge regarding the world around them. We recommend 'On Quantum Theory' by Berthold-Georg Englert, [1] for further, indeed essential, reading in these matters.

doi:10.1088/978-0-7503-2171-6ch8

$$\hat{H} = \hat{H}_0 + \hat{V}_s(t) \tag{8.1}$$

by

$$|\alpha, t = 0; t\rangle_I \equiv e^{\frac{i\hat{H}_0 t}{\hbar}} |\alpha, t = 0; t\rangle_s \tag{8.2}$$

and

$$\hat{A}_I \equiv e^{\frac{i\hat{H}_0 t}{\hbar}} \hat{A}_s e^{\frac{-i\hat{H}_0 t}{\hbar}}; \tag{8.3}$$

cf.

$$|\alpha\rangle_H \equiv e^{\frac{i\hat{H}t}{\hbar}} |\alpha, t = 0; t\rangle_s \tag{8.4}$$

and

$$\hat{A}_H \equiv e^{\frac{i\hat{H}t}{\hbar}} \hat{A}_s e^{\frac{-i\hat{H}t}{\hbar}}. \tag{8.5}$$

The content of the interaction picture can be realised by considering the action of $i\hbar\frac{\partial}{\partial t}$ on equation (8.2):

$$i\hbar\frac{\partial}{\partial t}|\alpha, t = 0; t\rangle_I = i\hbar\frac{\partial}{\partial t}\left\{e^{\frac{i\hat{H}_0 t}{\hbar}} |\alpha, t = 0; t\rangle_s\right\}, \tag{8.6}$$

$$\therefore i\hbar\frac{\partial}{\partial t}|\alpha, t = 0; t\rangle_I = -\hat{H}_0 e^{\frac{i\hat{H}_0 t}{\hbar}} |\alpha, t = 0; t\rangle_s + e^{\frac{i\hat{H}_0 t}{\hbar}} (\hat{H}_0 + \hat{V}_s(t))|\alpha, t = 0; t\rangle, \tag{8.7}$$

where

$$i\hbar\frac{\partial}{\partial t}|\alpha, t = 0; t\rangle_s = \hat{H}|\alpha, t = 0; t\rangle_s \tag{8.8}$$

has been used. Thus,

$$i\hbar\frac{\partial}{\partial t}|\alpha, t = 0; t\rangle_I = e^{\frac{i\hat{H}_0 t}{\hbar}} \hat{V}_s(t)|\alpha, t = 0; t\rangle_s, \tag{8.9}$$

$$\therefore i\hbar\frac{\partial}{\partial t}|\alpha, t = 0; t\rangle_I = e^{\frac{i\hat{H}_0 t}{\hbar}} \hat{V}_s(t) e^{\frac{-i\hat{H}_0 t}{\hbar}} e^{\frac{i\hat{H}_0 t}{\hbar}} |\alpha, t = 0, t\rangle_s, \tag{8.10}$$

$$\therefore i\hbar\frac{\partial}{\partial t}|\alpha, t = 0; t\rangle_I = \hat{V}_I(t)|\alpha, t = 0; t\rangle_I, \tag{8.11}$$

where equation (8.3) for $\hat{A}_I = \hat{V}_I(t)$ and equation (8.2) have been used. Equation (8.11) closely resembles the Schrödinger equation for the time evolution of a state ket, but with $\hat{H}$ replaced by $\hat{V}_I(t)$, i.e. $|\alpha, t = 0; t\rangle_I$ would be time independent for $\hat{V}_I(t) = 0$.

Further, from equation (8.3), differentiating with respect to time:

$$\frac{d\hat{A}_I}{dt} = \frac{i\hat{H}_0}{\hbar}e^{\frac{i\hat{H}_0 t}{\hbar}}\hat{A}_s e^{\frac{-i\hat{H}_0 t}{\hbar}} + e^{\frac{i\hat{H}_0 t}{\hbar}}\frac{dA_s}{dt}e^{\frac{-i\hat{H}_0 t}{\hbar}} + e^{\frac{i\hat{H}_0 t}{\hbar}}A_s\left(\frac{-i\hat{H}_0}{\hbar}\right)e^{\frac{-i\hat{H}_0 t}{\hbar}}; \tag{8.12}$$

whence using equation (8.3),

$$\frac{d\hat{A}_I}{dt} = \frac{i}{\hbar}\{\hat{H}_0\hat{A}_I - \hat{A}_I\hat{H}_0\} + e^{\frac{i\hat{H}_0 t}{\hbar}}\frac{d\hat{A}_s}{dt}e^{\frac{-i\hat{H}_0 t}{\hbar}}, \tag{8.13}$$

and recalling that $\hat{A}_s$ is time independent,

$$\therefore \frac{d\hat{A}_I}{dt} = \frac{1}{i\hbar}[\hat{A}_I, \hat{H}_0]. \tag{8.14}$$

Equation (8.14) closely resembles the Heisenberg equation for the time evolution of an operator, but with $\hat{H}$ replaced by $\hat{H}_0$. If $\hat{A}_I$ commutes with $\hat{H}_0$, then $\hat{A}_I$ is time independent.

The relationship between the Heisenberg, Schrödinger and interaction pictures is summarised in table 8.1. The interaction picture can be regarded as a hybrid of the other two.

The base kets of the interaction picture can be expanded in the basis $\{|n\rangle\}$ $(H_0|n\rangle = E_0|n\rangle)$ where,

$$|\alpha, t = 0; t\rangle_s = \sum_n c_n(t)e^{-\frac{iE_n t}{\hbar}}|n\rangle, \tag{8.15}$$

as

$$|\alpha, t = 0, t\rangle_I = \sum_n c_n(t)|n\rangle, \tag{8.16}$$

where the $c_n(t)$ are the *same* as the $c_n(t)$ in equation (8.15). This is seen by applying $e^{\frac{-i\hat{H}_0 t}{\hbar}}$ to both sides of equation (8.16) from the left:

$$e^{\frac{-i\hat{H}_0 t}{\hbar}}|\alpha, t = 0; t\rangle_I = \sum_n c_n(t)e^{\frac{-i\hat{H}_0 t}{\hbar}}|n\rangle, \tag{8.17}$$

then from equation (8.2) and (8.15)

Table 8.1. The relationship between the Heisenberg, Schrödinger and interaction pictures.

	Heisenberg picture	Interaction picture	Schrödinger picture
State ket	No change	Evolution determined by $\hat{V}_I(t)$	Evolution determined by $\hat{H}$
Observable	Evolution determined by $\hat{H}$	Evolution determined by $\hat{H}_0$	No change

$$|\alpha, t = 0; t\rangle_s = \sum_n c_n(t) e^{\frac{-i\hat{H}_0 t}{\hbar}} |n\rangle. \tag{8.18}$$

The equation for the $\dot{c}_m(t)$,

$$\dot{c}_m(t) = \frac{1}{i\hbar} \sum_n V_{mn}(t) e^{i\omega_{mn} t} c_n(t), \tag{8.19}$$

can be deduced from the interaction picture by taking the inner product of both sides of equation (8.11) with $\langle m|$ and using the completeness relation:

$$i\hbar \frac{\partial}{\partial t} \langle m|\alpha, t = 0; t\rangle_I = \sum_n \langle m|\hat{V}_I|n\rangle\langle n|\alpha, t = 0; t\rangle_I; \tag{8.20}$$

but from equation (8.2),

$$\begin{aligned} \langle m|V_I(t)|n\rangle &= \langle m|e^{\frac{i\hat{H}_0 t}{\hbar}} \hat{V}_s(t) e^{\frac{-i\hat{H}_0 t}{\hbar}} |n\rangle \\ &= V_{mn}(t) e^{\frac{i(E_m - E_n)t}{\hbar}}, \end{aligned} \tag{8.21}$$

and from equation (8.16)

$$c_m(t) = \langle m|\alpha, t = 0; t\rangle_I, \tag{8.22}$$

whence

$$i\hbar \frac{\partial}{\partial t} c_m(t) = \sum_n V_{mn}(t) e^{i\omega_{mn}} c_n(t), \tag{8.23}$$

cf. equation (8.19), where $\omega_{mn} = \frac{(E_m - E_n)}{\hbar}$.

Finally, we can define a time evolution operator in the interaction picture by:

$$|\alpha, t = 0; t\rangle_I \equiv \hat{U}_I(t, 0)|\alpha, t = 0\rangle_I. \tag{8.24}$$

Then, from equations (8.11) and (8.24):

$$i\hbar \frac{\partial}{\partial t} \hat{U}_I(t, 0) = \hat{V}_I(t)\hat{U}_I(t, 0). \tag{8.25}$$

Equation (8.25) has the initial condition

$$\hat{U}_I(t = 0, 0) = \hat{I}, \tag{8.26}$$

and is equivalent to the integral equation

$$\hat{U}_I(t, 0) = \hat{I} - \frac{i}{\hbar} \int_0^t \hat{V}_I(t')\hat{U}_I(t', 0) \mathrm{d}t'. \tag{8.27}$$

This leads, by iteration of equation (8.27), to the *Dyson series* for the time evolution operator in the interaction picture:

$$\hat{U}_I(t, 0) = \hat{I} - \frac{i}{\hbar}\int_0^t \hat{V}_I(t')\left\{\hat{I} - \frac{i}{\hbar}\int_0^{t'} \hat{V}_I(t'')\hat{U}_I(t'', 0)\mathrm{d}t''\right\}\mathrm{d}t', \tag{8.28}$$

$$\begin{aligned}\therefore\ \hat{U}_I(t, 0) = \hat{I} - \frac{i}{\hbar}\int_0^t \mathrm{d}t'\hat{V}_I(t') + \left(\frac{-i}{\hbar}\right)^2\int_0^t \mathrm{d}t'\int_o^{t'} \mathrm{d}t''\hat{V}_I(t')\hat{V}_I(t'') \\ + \cdots + \left(\frac{-i}{\hbar}\right)^n\int_0^t \mathrm{d}t'\int_0^{t'} \mathrm{d}t''\cdots\int_0^{t(n-1)} \mathrm{d}t^{(n-1)}\hat{V}_I(t')\hat{V}_I(t'')\cdots\hat{V}_I(t^n).\end{aligned} \tag{8.29}$$

Approximate solutions are obtained for equation (8.29) by terminating the series at any desired point. (The question of the convergence of the Dyson series is not discussed here beyond specifying that $\hat{V}_I(t)$ must be ‘sufficiently small’.)

Once $\hat{U}_I(t, 0)$ is found, the time evolution of any state ket can be predicted. For example, if the initial state ket at $t = 0$ is an energy eigenket of $\hat{H}_0$, then

$$|m, t = 0; t\rangle_I = \hat{U}_I(t, 0)|m, t = 0\rangle \tag{8.30}$$

$$= \sum_n |n\rangle\langle n|\hat{U}_I(t, 0)|m, t = 0\rangle \tag{8.31}$$

and comparing with equation (8.16),

$$\langle n|\hat{U}_I(t, 0)|m, t = 0\rangle = c_n(t), \tag{8.32}$$

i.e. the matrix elements of $\hat{U}_I(t, 0)$ in the $\{|n\rangle\}$ basis of the energy eigenkets of $\hat{H}_0$ are just the transition amplitudes for going from $t = 0$ to $t = t$.

8.2 Time-dependent perturbation theory

Time-dependent perturbation theory is a means of determining probabilities of change and rates of change in systems governed by Hamiltonians of the form

$$\hat{H}(t) = \hat{H}_0 + \hat{V}(t), \tag{8.33}$$

when $|\hat{V}(t)| \ll |\hat{H}_0|$. This is very useful because many time-dependent problems in quantum mechanics are of this type. For example, a system described by $\hat{H}_0$ could interact with another system such as a colliding projectile or an electromagnetic field, where $\hat{V}(t)$ describes the interaction. It is presumed that the solutions to

$$\hat{H}_0|n\rangle = E_n|n\rangle \tag{8.34}$$

can be found.

We consider the time-dependent Hamiltonian in the form

$$\hat{H}(t) = \hat{H}_0 + \lambda\hat{V}(t), \tag{8.35}$$

where λ is a continuous real parameter, $0 \leqslant \lambda \leqslant 1$, that is introduced to keep track of the number of times the perturbation enters the calculation. We define the $c_n(t)$ in the set of coupled differential equations,

$$\dot{c}_j(t) = \frac{1}{i\hbar}\sum_k \lambda V_{jk}(t)e^{i\omega_{jk}t}c_k(t), \tag{8.36}$$

cf. equations (8.19) and

$$i\hbar\begin{pmatrix}\dot{c}_1\\ \dot{c}_2\\ \dot{c}_3\\ \vdots\end{pmatrix} = \begin{pmatrix} V_{11} & V_{12}e^{i\omega_{12}t} & \cdots & \cdot \\ V_{21}e^{-i\omega_{12}t} & V_{22} & & \\ \vdots & \vdots & V_{33} & \cdot \\ \vdots & \vdots & \vdots & \end{pmatrix}\begin{pmatrix}c_1\\ c_2\\ c_3\\ \vdots\end{pmatrix}, \tag{8.37}$$

in terms of the following power series in λ:

$$c_k(t) = c_k^{(0)}(t) + \lambda c_k^{(1)}(t) + \lambda^2 c_k^{(2)}(t) + \cdots. \tag{8.38}$$

Substituting equation (8.38) into equation (8.36) and equating the coefficients of equal powers of λ:

$$\dot{c}_j^{(0)}(t) = 0, \tag{8.39}$$

$$\dot{c}_j^{(1)}(t) = \frac{1}{i\hbar}\sum_k V_{jk}(t)e^{i\omega_{jk}t}c_k^{(0)}(t), \tag{8.40}$$

$$\cdots$$

$$\dot{c}_j^{(r)}(t) = \frac{1}{i\hbar}\sum_k V_{jk}(t)e^{i\omega_{jk}t}c_k^{(r-1)}(t), \tag{8.41}$$

$$\cdots$$

Thus, the set of coupled equations, equation (8.36) have been decoupled and equations (8.39)–(8.41) can be successively integrated, in principle, to any desired order.

For an initial state $|a\rangle$ with energy E_a, integrating equation (8.39):

$$c_j^{(0)} = \text{constant} = c_j^{(0)}(t=0) = \delta_{ja}; \tag{8.42}$$

then equation (8.40) can be written

$$\dot{c}_j^{(1)}(t) = \frac{1}{i\hbar}V_{ja}(t)e^{i\omega_{ja}t}, \tag{8.43}$$

and integrating,

$$c_j^{(1)}(t) = \frac{1}{i\hbar}\int_0^t V_{ja}(t')e^{i\omega_{ja}t'}\mathrm{d}t'. \tag{8.44}$$

Equation (8.44) is the fundamental equation of time-dependent perturbation theory. Note that $c_j^{(1)}(t)$ is the Fourier component of $V_{ja}(t)$ with frequency ω_{ja}. The probability that at time t the system will be found in the state $|j\rangle$, i.e. the probability for the transition $|a\rangle \to |j\rangle$, is then given by

$$P_{ja}(t) = \left| c_j^{(1)}(t) \right|^2. \tag{8.45}$$

This is the lowest order approximation to $P_{ja}(t)$. To second order,

$$P_{ja}(t) = \left| c_j^{(1)}(t) + c_j^{(2)}(t) \right|^2: \tag{8.46}$$

$c_j^{(2)}(t)$ is obtained by integrating (cf. equation (8.41))

$$\dot{c}_j^{(2)}(t) = \frac{1}{i\hbar} \sum_k V_{jk}(t) e^{i\omega_{jk}t} c_k^{(1)}(t), \tag{8.47}$$

where $c_k^{(1)}(t)$ is given by equation (8.44) (with $j = k$); whence

$$c_j^{(2)}(t) = \frac{1}{(i\hbar)^2} \sum_k \int_0^t \mathrm{d}t' \int_0^{t'} \mathrm{d}t'' e^{i\omega_{jk}t'} V_{jk}(t') e^{i\omega_{ka}t''} V_{ka}(t''). \tag{8.48}$$

Thus, we have an iterative procedure for decomposing the transition amplitude: the zeroth-order amplitude is $c_j^{(0)}(t)$ which, from equation (8.42), is the amplitude for the system to remain unchanged; the first-order amplitude is $c_j^{(1)}(t)$ which describes transitions direct from the initial to the final state; the second-order amplitude is $c_j^{(2)}(t)$ which describes two-step transitions. The two-step transitions occur via any intermediate state from the set $\{|n\rangle\}$ of eigenstates of $\hat{H}_0$. These intermediate states should be regarded as virtual, i.e. they are not observed. In fact, energy conservation is not even a condition on their involvement.

The various orders of the transition amplitude can be obtained directly from the Dyson series, equation (8.29), by using equation (8.31),

$$c_n(t) = \langle n | \hat{U}_I(t, 0) | m \rangle, \tag{8.49}$$

equation (8.36) with $\lambda = 1$ and $k = n$,

$$c_n(t) = c_n^{(0)}(t) + c_n^{(1)}(t) + c_n^{(2)}(t) + \cdots, \tag{8.50}$$

and equation (8.4) with $\hat{A} = \hat{V}(t)$,

$$\hat{V}_I(t) = e^{\frac{i\hat{H}_0 t}{\hbar}} \hat{V}_s(t) e^{\frac{-i\hat{H}_0 t}{\hbar}}. \tag{8.51}$$

For example, the first-order term in the Dyson series gives

$$\begin{aligned} c_n(t) &= \frac{-i}{\hbar}\int_0^t \mathrm{d}t'\langle n|e^{\frac{i\hat{H}_0 t'}{\hbar}}\hat{V}_s(t')e^{\frac{-i\hat{H}_0 t'}{\hbar}}|m\rangle \\ &= \frac{-i}{\hbar}\int_0^t \mathrm{d}t'\langle n|e^{\frac{iE_n t'}{\hbar}}\hat{V}_s(t')e^{\frac{-iE_m t'}{\hbar}}|m\rangle \\ &= \frac{-i}{\hbar}\int_0^t \mathrm{d}t'\langle n|\hat{V}_s(t')|m\rangle e^{\frac{i(E_n-E_m)t'}{\hbar}}, \end{aligned} \tag{8.52}$$

from which equation (8.44) follows directly for $n = j$, $m = a$.

8.3 Constant perturbations and Fermi's golden rule

As an application of time-dependent perturbation theory, consider a constant perturbation turned on at $t = 0$:

$$V(t) = 0, \;\; t \leqslant 0, \tag{8.53}$$

$$V(t) = V(\text{a constant}), \;\; t > 0, \tag{8.54}$$

for some system. Further, the system is in the state $|s\rangle$ with probability unity for $t \leqslant 0$. Then

$$c_n^{(0)}(t) = c_n^{(0)}(0) = \delta_{sn}, \tag{8.55}$$

$$c_n^{(1)}(t) = \frac{-i}{\hbar}V_{ns}\int_0^t e^{i\omega_{ns}t'}\mathrm{d}t', \tag{8.56}$$

$$\therefore c_n^{(1)}(t) = \frac{V_{ns}}{E_n - E_s}(1 - e^{i\omega_{ns}t}), \tag{8.57}$$

and

$$|\, c_n^{(1)}(t)|^2 = \frac{|V_{ns}|^2}{(E_n - E_s)^2}(2 - 2\cos\omega_{ns}t), \tag{8.58}$$

$$\therefore |\, c_n^{(1)}(t)|^2 = \frac{4|V_{ns}|^2}{(E_n - E_s)^2}\sin^2\left\{\frac{(E_n - E_s)t}{2\hbar}\right\}. \tag{8.59}$$

Evidently, the transition probability for $s \to n$ depends on $|V_{ns}|^2$ and $\frac{1}{(E_n - E_s)^2}$.

To gain further insight into $|c_n^{(1)}(t)|^2$, defining

$$\frac{|E_n - E_s|}{2\hbar} := \omega, \tag{8.60}$$

$$\therefore |\, c_n^{(1)}(t)|^2 = \frac{|V_n s|^2 t^2}{\hbar^2}\frac{\sin^2\omega t}{(\omega t)^2}. \tag{8.61}$$

The expression on the right-hand side of equation (8.61) is plotted as a function of ω in figure 8.1. The transition probability is clearly very small unless $\omega < \frac{\pi}{t}$, i.e. $|E_n - E_s| < \frac{h}{t}$. Hence, as time increases, it is more and more probable that the final state will have the same energy as the initial state. In considering the meaning of this result, it seems to imply that for large t nothing will happen unless $E_n = E_s$ and hence only transitions between degenerate states in the system could occur! However, this is not a problem when one considers the 'other' system to which the system is coupled. The other system is the sink or source of energy and we have in mind a system with a smoothly-varying continuous energy spectrum, e.g. a radiation field or a projectile. Strictly speaking, both systems (e.g. an atom plus a radiation field) should have their Hamiltonians included in $\hat{H}_0$. This is not commonly done because the system serving as the energy source–sink can be viewed as not being of primary interest. (Also, leaving it out of the description avoids the difficulties of working with continuous basis states and problems of orthogonality and completeness.) To make sense of the result that for large t, $E_n = E_s$, we now append the source–sink energy ε to the energy E of the system under discussion and specify

$$|(E_n + \varepsilon_n) - (E_s + \varepsilon_s)| < \frac{h}{t}, \tag{8.62}$$

whence for large t, the condition for a transition is

$$E_n + \varepsilon_n = E_s + \varepsilon_s. \tag{8.63}$$

Equation (8.63) is just the statement of the conservation of energy! It is convenient to reinterpret E_n and E_s in our development as containing ε_n and ε_s, respectively.

We can argue with complete generality that the possible final states of any time-dependent process form a continuum. This is manifestly true for interactions between a system (with discrete levels) and a radiation field or a moving projectile. Thus, we define the density of final states in the interval $(E, E + \mathrm{d}E)$ to be $\rho(E)\mathrm{d}E$,

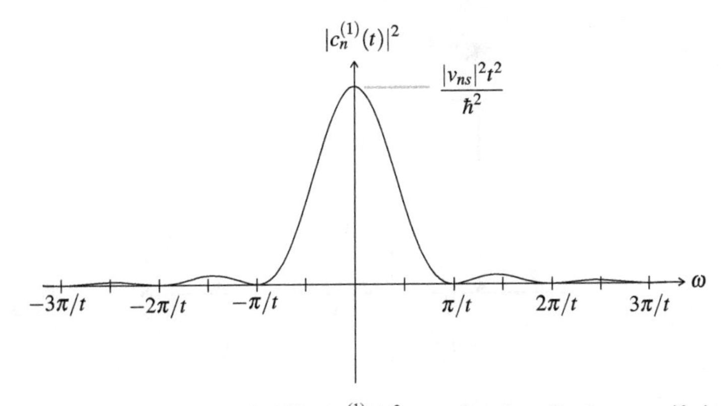

Figure 8.1. The first-order transition probability, $|c_n^{(1)}(t)|^2$, as a function of ω for a specified value of t. The maxima at $\frac{3\pi}{2t}$ and $\frac{5\pi}{2t}$ have amplitudes $0.045|V_n s|^2\frac{t^2}{\hbar^2}$ and $0.016|V_n s|^2\frac{t^2}{\hbar^2}$, respectively.

where $\rho(E)$ is assumed to be a smoothly-varying continuous function. Then for the total transition probability, P

$$P = \sum_{\substack{n \\ E_n \approx E_s}} | c_n^{(1)}(t)|^2 \Rightarrow \int_{E_s - \frac{h}{t}}^{E_s + \frac{h}{t}} \mathrm{d}E_n \rho(E_n) | c_n^{(1)}(t)|^2, \tag{8.64}$$

i.e. we have replaced a sum over n in the vicinity of n where $E_n \approx E_s$ by an integral over E_n in the interval $E_s - \frac{h}{t} \lesssim E_n \lesssim E_s + \frac{h}{t}$. Therefore, from equations (8.61) and (8.60),

$$P = \int_{E_s - \frac{h}{t}}^{E_s + \frac{h}{t}} \mathrm{d}E_n \rho(E_n) 4 | V_{ns}|^2 t^2 \frac{\sin^2(E_n - E_s)\frac{t}{2\hbar}}{[(E_n - E_s)t]^2}, \tag{8.65}$$

$$\therefore P = \int_{E_s - \frac{h}{t}}^{E_s + \frac{h}{t}} \mathrm{d}E_n \rho(E_n) 2 | V_{ns}|^2 \frac{[1 - \cos(E_n - E_s)\frac{t}{\hbar}]}{[E_n - E_s]^2}. \tag{8.66}$$

We are interested in the rate of change of P with time

$$\frac{\mathrm{d}P}{\mathrm{d}t} = \int_{E_s - \frac{h}{t}}^{E_s + \frac{h}{t}} \mathrm{d}E_n \rho(E_n) \frac{2 | V_{ns}|^2}{\hbar} \frac{\sin(E_n - E_s)\frac{t}{\hbar}}{[E_n - E_s]}. \tag{8.67}$$

The integral in equation (8.67) can be computed using

$$\lim_{g \to \infty} \frac{\sin(gx)}{x} = \pi \delta(x). \tag{8.68}$$

The justification of equation (8.68) can be argued using figure 8.2.

Thus, noting that $\int \delta(ax)\mathrm{d}x = \frac{1}{a} \int \delta(y)\mathrm{d}y$,

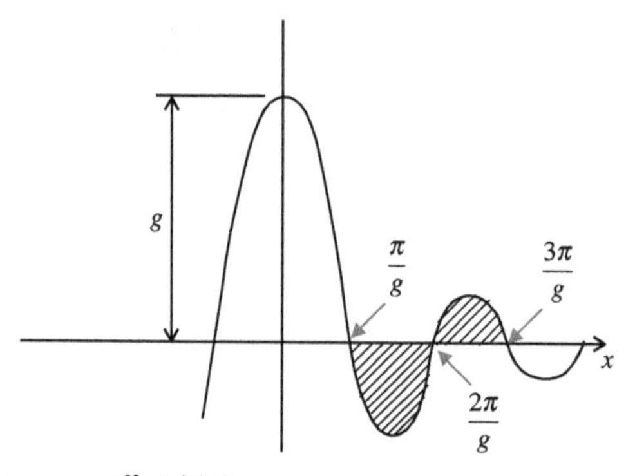

Figure 8.2. A depiction of $\lim_{g\to\infty} \int_{-\infty}^{\infty} \frac{g \sin(gx)}{gx} \mathrm{d}x$. Pairs of shaded areas cancel for $g \to \infty$, and the central peak at $x = 0$ with height g and width $\frac{\pi}{g}$ remains. Its area is $\approx \pi$ and equals π for $g \to \infty$.

$$\therefore \lim_{t\to\infty} \frac{\sin(E_n - E_s)\frac{t}{\hbar}}{\frac{[E_n - E_s]}{\hbar}} = \pi\delta(E_n - E_s), \tag{8.69}$$

and

$$\frac{\mathrm{d}P}{\mathrm{d}t} = \frac{2\pi}{\hbar}|V_{ns}|^2\rho(E_n)_{E_n\approx E_s}. \tag{8.70}$$

Note that for very sharp $\frac{\sin(E_n - E_s)\frac{t}{\hbar}}{[E_n - E_s]}$, and $\rho(E_n)$, V_{ns} ~ const., and can be taken out from under the integral. Equation (8.70) is usually called *Fermi's Golden Rule*.

Exercise

8.1. Derive the Golden Rule for

$$\hat{V}(t) = \hat{V}e^{i\omega t} + \hat{V}^{\dagger}e^{-i\omega t},$$

where ω is a real constant.

Reference

[1] Englert B-G 2013 On quantum theory *Eur. Phys. J. D* **67** 238

IOP Publishing

Quantum Mechanics for Nuclear Structure, Volume 2
An intermediate level view

Kris Heyde and John L Wood

Chapter 9

Electromagnetic fields in quantum mechanics

A standard introduction to the electromagnetic (EM) field in quantum mechanics is given. This employs second quantization formalism, cf. chapter 4. The interaction of the EM field with matter is outlined. An application to the interaction of the hydrogen atom with the EM field and calculation of an excited state lifetime is illustrated in full detail. The calculation of the density of modes in the EM field is given in appendix B.

Concepts: Maxwell's equations; scalar and vector potentials; gauge transformation; Coulomb gauge; harmonic oscillator description; EM vacuum; EM interaction Hamiltonian; induced and spontaneous emission; absorption; dipole approximation.

9.1 The quantization of the electromagnetic field

The starting point of any discussion of electromagnetic fields is Maxwell's equations. For a discussion of the quantization of the electromagnetic field it is sufficient to consider a region of space where there are no charges or currents, i.e. the source(s) of the field are implied to be outside of the region. For such an empty region of space, Maxwell's equations are

$$\vec{\nabla} \times \vec{E} = -\frac{\partial \vec{B}}{\partial t}, \tag{9.1}$$

$$\frac{1}{\mu_0}\vec{\nabla} \times \vec{B} = \varepsilon_0 \frac{\partial \vec{E}}{\partial t}, \tag{9.2}$$

$$\vec{\nabla} \cdot \vec{E} = 0, \tag{9.3}$$

$$\vec{\nabla} \cdot \vec{B} = 0; \tag{9.4}$$

doi:10.1088/978-0-7503-2171-6ch9

where μ_0 and ε_0 are the permeability and permittivity of free space, respectively, $\vec{E}$ and $\vec{B}$ are the electric and magnetic fields, respectively; and the units are MKS or SI.

The electric and magnetic fields are commonly re-expressed in terms of the scalar and vector potentials, ϕ and $\vec{A}$, respectively:

$$\vec{E} = -\vec{\nabla}\phi - \frac{\partial \vec{A}}{\partial t}, \tag{9.5}$$

$$\vec{B} = \vec{\nabla} \times \vec{A}. \tag{9.6}$$

The potentials ϕ, $\vec{A}$ are not unique. The relations between $\vec{E}$, $\vec{B}$ and ϕ, $\vec{A}$ in equations (9.5) and (9.6), are unaffected by the changes:

$$\vec{A} \to \vec{A}' = \vec{a} + \vec{\nabla}\chi, \tag{9.7}$$

$$\phi \to \phi' = \phi - \frac{\partial \chi}{\partial t}, \tag{9.8}$$

where $\chi = \chi(\vec{r}, t)$ is an arbitrary scalar function. This change in ϕ and $\vec{A}$ is called a *gauge transformation.*

For the quantization of the electromagnetic field, we work in the *Coulomb gauge* defined by:

$$\vec{\nabla} \cdot \vec{A} = 0, \tag{9.9}$$

$$\phi = 0. \tag{9.10}$$

This choice leads to

$$\nabla^2 \vec{A} - \frac{1}{c^2}\frac{\partial^2 \vec{A}}{\partial t^2} = 0. \tag{9.11}$$

Equation (9.11) looks suspiciously like a wave equation! The wave properties of $\vec{A}$ are conveniently elucidated by making an expansion in a Fourier series:

$$\vec{A}(\vec{r}, t) = \frac{1}{\sqrt{V}}\sum_{\vec{k}}\left\{\vec{A}_{\vec{k}}(t)e^{i\vec{k}\cdot\vec{r}} + \vec{A}^*_{\vec{k}}(t)e^{-i\vec{k}\cdot\vec{r}}\right\}, \tag{9.12}$$

where we specify that the plane waves in the expansion are real and satisfy periodic boundary conditions within a large cube of side $L = V^{\frac{1}{3}}$. Thus, the components of the wavevector $\vec{k}$ take the values:

$$k_x = \frac{2\pi n_x}{L}, \quad k_y = \frac{2\pi n_y}{L}, \quad k_z = \frac{2\pi n_z}{L}, \tag{9.13}$$

with

$$n_x, n_y, n_z = 0, 1, 2, \ldots. \tag{9.14}$$

The choice of the Coulomb gauge, equation (9.10), gives

$$\vec{k} \cdot \vec{A}_{\vec{k}}(t) = \vec{k} \cdot \vec{A}^*_{\vec{k}}(t) = 0, \tag{9.15}$$

i.e. the Fourier coefficients are perpendicular to the propagation vector $\vec{k}$. Consequently, $\vec{A}_{\vec{k}}(t)$ has only two components. A vector with this property is described as *transverse*. The absence of a third (or longitudinal) component to the electromagnetic field can be attributed to its gauge symmetry.

We can write

$$\vec{A}_{\vec{k}}(t) \equiv \sum_{\alpha} \hat{\varepsilon}_{\vec{k}\alpha} c_{\vec{k}\alpha}(t), \quad \alpha = 1, 2, \tag{9.16}$$

where α indexes the two independent transverse components or states of polarization for each Fourier component $\vec{k}$. Substituting equation (9.12) into (9.11),

$$k^2 \vec{A}_{\vec{k}}(t) + \frac{1}{c^2} \frac{\partial^2 \vec{A}_{\vec{k}}(t)}{\partial t^2} = 0, \tag{9.17}$$

where the Fourier components with different $\vec{k}$ are independent. A similar equation is obtained for $\vec{A}^*_{\vec{k}}(t)$. Hence, the Fourier components oscillate harmonically with frequencies

$$\omega_{\vec{k}} \equiv c|\vec{k}| \equiv \omega_k, \tag{9.18}$$

and they can be expressed as

$$\vec{A}_{\vec{k}}(t) = \vec{A}_{\vec{k}}(0) e^{-i\omega_k t}; \tag{9.19}$$

whence, from equation (9.16)

$$c_{\vec{k}\alpha}(t) = c_{\vec{k}\alpha}(0) e^{-i\omega_k t}. \tag{9.20}$$

Thus,

$$\vec{E}(\vec{r}, t) = \frac{i}{\sqrt{V}} \sum_{\vec{k},\alpha} \omega_k \left\{ c_{\vec{k}\alpha}(0) \hat{\varepsilon}_{\vec{k}\alpha} e^{i\vec{k}\cdot\vec{r} - i\omega_k t} - c^*_{\vec{k}\alpha}(0) \hat{\varepsilon}_{\vec{k}\alpha} e^{-i\vec{k}\cdot\vec{r} + i\omega_k t} \right\}, \tag{9.21}$$

and

$$\vec{B}(\vec{r}, t) = \frac{i}{\sqrt{V}} \sum_{\vec{k},\alpha} \vec{k} x \left\{ c_{\vec{k}\alpha}(0) \hat{\varepsilon}_{\vec{k}\alpha} e^{i\vec{k}\cdot\vec{r} - i\omega_k t} + c^*_{\vec{k}\alpha}(0) \hat{\varepsilon}_{\vec{k}\alpha} e^{-i\vec{k}\cdot\vec{r} + i\omega_k t} \right\}, \tag{9.22}$$

where the $\hat{\varepsilon}_{\vec{k}\alpha}$ are real. This is the famous prediction due to Maxwell that equations (9.1)–(9.4) give rise to electromagnetic waves.

To carry out the quantization of the electromagnetic field, we then turn to its Hamiltonian:

$$H = \frac{1}{2} \int_V \left(\varepsilon_0 E^2 + \frac{1}{\mu_0} B^2 \right) \mathrm{d}V. \tag{9.23}$$

Substituting equations (9.21) and (9.22) into equation (9.23) and integrating,

$$\therefore\ H = 2\varepsilon_0 \sum_{\vec{k}\alpha} \omega_k^2 |c_{\vec{k}\alpha}(t)|^2. \tag{9.24}$$

Equation (9.24) can be transformed to standard Hamiltonian form by defining:

$$Q_{\vec{k}\alpha}(t) \equiv \sqrt{\varepsilon_0}\left(c_{\vec{k}\alpha}(t) + c^*_{\vec{k}\alpha}(t)\right), \tag{9.25}$$

$$P_{\vec{k}\alpha}(t) \equiv -i\omega_k\sqrt{\varepsilon_0}\left(c_{\vec{k}\alpha}(t) - c^*_{\vec{k}\alpha}(t)\right). \tag{9.26}$$

Then

$$c_{\vec{k}\alpha}(t) = \frac{1}{2\omega_k\sqrt{\varepsilon_0}}(\omega_k Q_{\vec{k}\alpha}(t) + iP_{\vec{k}\alpha}(t)), \tag{9.27}$$

and

$$H = \frac{1}{2}\sum_{\vec{k}\alpha}\left(P^2_{\vec{k}\alpha} + \omega_k^2 Q^2_{\vec{k}\alpha}\right), \tag{9.28}$$

where the time dependence is omitted because H is independent of time.

Equation (9.28) looks like a sum over independent harmonic oscillators. This is borne out by the identification of $P_{\vec{k}\alpha}$ and $Q_{\vec{k}\alpha}$ as canonically conjugate 'momenta' and 'positions' by noting that:

$$P_{\vec{k}\alpha} = \frac{\partial Q_{\vec{k}\alpha}}{\partial t}, \tag{9.29}$$

which follows from equations (9.25), (9.26) and (9.20); and

$$\dot{Q}_{\vec{k}\alpha} = \frac{\partial H}{\partial P_{\vec{k}\alpha}}, \tag{9.30}$$

$$\dot{P}_{\vec{k}\alpha} = -\frac{\partial H}{\partial Q_{\vec{k}\alpha}}, \tag{9.31}$$

i.e. $P_{\vec{k}\alpha}$ and $Q_{\vec{k}\alpha}$ satisfy Hamilton's equations.

We now adopt the standard recipe for quantization by assuming that the commutator bracket relations for mechanical position and momentum apply also to the electromagnetic position and momentum, viz.

$$[\hat{P}_{\vec{k}\alpha}, \hat{P}_{\vec{k}'\alpha'}] = [\hat{Q}_{\vec{k}\alpha}, \hat{Q}_{\vec{k}'\alpha'}] = 0, \tag{9.32}$$

$$[\hat{Q}_{\vec{k}\alpha}, \hat{P}_{\vec{k}'\alpha'}] = i\hbar\delta_{\vec{k}\vec{k}'}\delta_{\alpha\alpha'}. \tag{9.33}$$

Then, all of the results for the one-dimensional harmonic oscillator can be adopted (in the form of a sum of independent oscillators). Thus, defining raising and lowering operators:

$$b_{\vec{k}\alpha} \equiv \frac{1}{\sqrt{2\hbar\omega_k}}(\omega_k \hat{Q}_{\vec{k}\alpha} + i\hat{P}_{\vec{k}\alpha}), \tag{9.34}$$

$$b^\dagger_{\vec{k}\alpha} \equiv \frac{1}{\sqrt{2\hbar\omega_k}}(\omega_k \hat{Q}_{\vec{k}\alpha} - i\hat{P}_{\vec{k}\alpha}), \tag{9.35}$$

$$\left[b_{\vec{k}\alpha}, b^\dagger_{\vec{k}'\alpha'}\right] = \delta_{\vec{k}\vec{k}'}\delta_{\alpha\alpha'}, \tag{9.36}$$

$$\hat{H} = \sum_{\vec{k},\alpha}\left(b^\dagger_{\vec{k}\alpha} b_{\vec{k}\alpha} + \frac{1}{2}\right)\hbar\omega_k, \tag{9.37}$$

and

$$\hat{N}_{\vec{k}\alpha} = b^\dagger_{\vec{k}\alpha} b_{\vec{k}\alpha}. \tag{9.38}$$

The operators $\{\hat{N}_{\vec{k}\alpha}\}$ form a complete set of commuting operators. The eigenkets are written in the form

$$|n_{\vec{k}_1\alpha_1}, n_{\vec{k}_1\beta_1}, n_{\vec{k}_2\alpha_2} \ldots n_{\vec{k}_i\alpha_i} \ldots\rangle,$$

where the ordering must be specified, but can be freely chosen; and the independent pairs of states of polarization are labelled α and β. Then

$$b^\dagger_{\vec{k}_i\alpha_i}|n_{\vec{k}_1\alpha_1} n_{\vec{k}_1\beta_1} \ldots n_{\vec{k}_i\alpha_i} \ldots\rangle = \sqrt{n_{\vec{k}_i\alpha_i} + 1}\,|n_{\vec{k}_1\alpha_1} n_{\vec{k}_1\beta_1} \ldots (n_{\vec{k}_i\alpha_i} + 1) \ldots\rangle, \tag{9.39}$$

$$b_{\vec{k}_i\alpha_i}|n_{\vec{k}_1\alpha_1} n_{\vec{k}_1\beta_1} \ldots n_{\vec{k}_i\alpha_i} \ldots\rangle = \sqrt{n_{\vec{k}_i\alpha_i}}\,|n_{\vec{k}_1\alpha_1} n_{\vec{k}_1\beta_1} \ldots (n_{\vec{k}_i\alpha_i} - 1) \ldots\rangle, \tag{9.40}$$

$$\hat{H}|n_{\vec{k}_1\alpha_1} n_{\vec{k}_1\beta_1} \ldots n_{\vec{k}_i\alpha_i} \ldots\rangle = \sum_{i=1}^{\infty}(n_{\vec{k}_i\alpha_i} + n_{\vec{k}_i\beta_i} + 1)\hbar\omega_{k_i}|n_{\vec{k}_1\alpha_1} n_{\vec{k}_1\beta_1} \ldots n_{\vec{k}_i\alpha_i} \ldots\rangle, \tag{9.41}$$

$$|n_{\vec{k}_1\alpha_1} n_{\vec{k}_1\beta_1} \ldots n_{\vec{k}_i\alpha_i} \ldots\rangle = \prod_{i=1}^{\infty} \frac{\left(b^\dagger_{\vec{k}_i\alpha_i}\right)^{n_{\vec{k}_i\alpha_i}}}{\sqrt{n_{\vec{k}_i\alpha_i}!}} \frac{\left(b^\dagger_{\vec{k}_i\beta_i}\right)^{n_{\vec{k}_i\beta_i}}}{\sqrt{n_{\vec{k}_i\beta_i}!}} |00\ldots\rangle, \tag{9.42}$$

and

$$\langle n_{\vec{k}_1\alpha_1} n_{\vec{k}_1\beta_1} \ldots n_{\vec{k}_i\alpha_i} \ldots | n_{\vec{k}_1\alpha_1} n_{\vec{k}_1\beta_1} \ldots n_{\vec{k}_i\alpha_i} \ldots\rangle = \prod_{i,j=1}^{\infty} \delta_{n_{\vec{k}_i\alpha_i} n_{\vec{k}_j\alpha_j}} \delta_{n_{\vec{k}_i\beta_i} n_{\vec{k}_j\beta_j}}. \tag{9.43}$$

The components of the electromagnetic field labelled by $\vec{k}_i$ (with their associated pairs of independent states of polarization α_i, β_i) are called *normal modes*. The quanta represented by $n_{\vec{k}_i\alpha_i}$, etc., are called *photons*, a term introduced by G N Lewis in 1926. The ket $|00\cdots\rangle$ appearing in equation (9.42) is called the *electromagnetic vacuum* or simply *the vacuum*. The vacuum is the state that is devoid of field quanta of any kind—electromagnetic (photons), strong (gluons), weak (intermediate vector bosons), gravitational (gravitons), and any (as yet) unknown fields. Although there are zero photons (or field quanta of any kind) in the vacuum, the vacuum is not empty! It contains the zero-point energy; however this cannot be considered in terms of photons.

The electromagnetic field is more than just photons. Photons are a convenient concept for what is added to or subtracted from the electromagnetic field, e.g. through the process of emission or absorption of energy by atoms, molecules, nuclei; but no photons does not mean no electromagnetic field. One might question whether or not the vacuum has a real existence: indeed, the vacuum leads to observable effects. Most notably, it leads to the spontaneous emission of electromagnetic radiation from excited states of quantum systems. At a more subtle (i.e. difficult to observe) level the vacuum leads to quantum electrodynamic effects such as the *Lamb shift*, the *anomalous magnetic moment* of the electron, and the *Casimir effect* (the existence of an attractive force between two closely-spaced conducting plates). Most dramatically of all, it lies at the heart of a major subarea of physics called *cavity quantum electrodynamics* [1, 2].

The creation and annihilation operators, $b^\dagger_{\vec{k}\alpha}$, $b_{\vec{k}\alpha}$, provide an explicit 'quantum language' in which all quantities can be expressed. Their time evolution is immediately obtained from the Heisenberg picture:

$$\frac{\mathrm{d}b_{\vec{k}\alpha}(t)}{\mathrm{d}t} = \frac{1}{i\hbar}[b_{\vec{k}\alpha}(t), \hat{H}] = -i\omega_k b_{\vec{k}\alpha}(t), \tag{9.44}$$

whence

$$b_{\vec{k}\alpha}(t) = b_{\vec{k}\alpha}(0)e^{-i\omega_k t}, \tag{9.45}$$

and similarly

$$b^\dagger_{\vec{k}\alpha}(t) = b^\dagger_{\vec{k}\alpha}(0)e^{i\omega_k t}. \tag{9.46}$$

Thus, from equations (9.34) and (9.35), equation (9.27) can be written

$$c_{\vec{k}\alpha}(t) = \sqrt{\frac{\hbar}{2\varepsilon_0\omega_k}}\, b_{\vec{k}\alpha}(0)e^{-i\omega_k t}. \tag{9.47}$$

From this, using equations (9.16) and (9.12), the vector field operator can be written:

$$\vec{A}(\vec{r}, t)_{\mathrm{op}} = \sum_{\vec{k},\alpha}\sqrt{\frac{\hbar}{2\varepsilon_0 V\omega_k}}\,\hat{\varepsilon}_{\vec{k}\alpha}\left\{b_{\vec{k}\alpha}(0)e^{i(\vec{k}\cdot\vec{r}-\omega_k t)} + b^\dagger_{\vec{k}\alpha}(0)e^{-i(\vec{k}\cdot\vec{r}-\omega_k t)}\right\}. \tag{9.48}$$

Further using equations (9.20), (9.21) and (9.47), $\vec{E}(\vec{r}, t)_{\mathrm{op}}$ can be written:

$$\vec{E}(\vec{r}, t)_{\mathrm{op}} = i\sum_{\vec{k},\alpha}\sqrt{\frac{\hbar\omega_k}{2\varepsilon_0 V}}\,\hat{\varepsilon}_{\vec{k}\alpha}\left\{b_{\vec{k}\alpha}(0)e^{i(\vec{k}\cdot\vec{r}-\omega_k t)} - b^\dagger_{\vec{k}\alpha}(0)e^{-i(\vec{k}\cdot\vec{r}-\omega_k t)}\right\}; \tag{9.49}$$

and, using equations (9.20), (9.22), and (9.47), $\vec{B}(\vec{r}, t)_{\text{op}}$ can be written:

$$\vec{B}(\vec{r}, t)_{\text{op}} = i\sum_{\vec{k},\alpha}\sqrt{\frac{\hbar}{2\varepsilon_0 V\omega_k}}(\vec{k} \times \hat{\varepsilon}_{\vec{k}\alpha})\left\{b_{\vec{k}\alpha}(0)e^{i(\vec{k}\cdot\vec{r}-\omega_k t)} - b^{\dagger}_{\vec{k}\alpha}(0)e^{-i(\vec{k}\cdot\vec{r}-\omega_k t)}\right\}. \tag{9.50}$$

9.2 The interaction of the electromagnetic field with matter

The Hamiltonian for a field interacting with matter can be written,

$$\hat{H} = \hat{H}_{\text{field}} + \hat{H}_{\text{matter}} + \hat{H}_{\text{int}}. \tag{9.51}$$

We have in mind an atom such as a hydrogen atom constituting the matter part and an electromagnetic field with which this atom is interacting. Thus,

$$\hat{H}_{\text{field}} = \hat{H}_{\text{em}} = \frac{1}{2}\int_V\left(\varepsilon_0\hat{E}^2 + \frac{1}{\mu_0}\hat{B}^2\right\}\mathrm{d}V, \tag{9.52}$$

and, e.g. for such an H atom,

$$\hat{H}_{\text{matter}} = \hat{H}_{\text{H atom}} = \frac{\hat{P}^2}{2m} + \frac{\hat{p}^2}{2\mu} - \frac{e^2}{4\pi\varepsilon_0\hat{r}}, \tag{9.53}$$

where equation (9.52) is identical to equation (9.23); and equation (9.53) contains the kinetic energy resulting from the centre-of-mass motion, the kinetic energy resulting from the relative motion, and the potential energy resulting from the electrostatic interaction between the proton and the electron, respectively.

Provided $\hat{H}_{\text{int.}}$ is small[1], we can regard it as a perturbation and, for

$$\hat{H} = \hat{H}_0 + \hat{H}_{\text{int.}}, \tag{9.54}$$

$$\hat{H}_0 = \hat{H}_{\text{em}} + \hat{H}^{\text{rel.}}_{\text{H atom}} + \hat{H}^{\text{C of M}}_{\text{H atom}}, \tag{9.55}$$

we have already obtained solutions to $\hat{H}_{\text{em}}$ and $\hat{H}^{\text{C of M}}_{\text{H atom}}$, viz.

$$E_{\text{em}} = \sum_{\vec{k},\alpha}\left(n_{\vec{k}\alpha} + \frac{1}{2}\right)\hbar\omega_k \tag{9.56}$$

(from equation (9.37)), and

$$E^{\text{rel.}}_{\text{H atom}} = -\frac{Ry}{n^2} = \frac{-13.6\text{ eV}}{n^2}, \quad n = 1, 2, 3, \ldots \tag{9.57}$$

(where equation (9.57) is identical to equations (6.131) and (6.132)).

[1] The Coulomb potential for hydrogen in its ground state is $\frac{E_1}{ea_1} = \frac{13.6}{e \times 5.29 \times 10^{-11}} = 2.6 \times 10^{11}\text{ V m}^{-1}$.

The expression for $\hat{H}_{\text{int.}}$ is obtained from the Hamiltonian in classical electrodynamics for a charged particle moving in an electromagnetic field,

$$H = \frac{1}{2m}(\vec{p} - q\vec{A})^2 + q\phi; \tag{9.58}$$

where the particle has mass m, electrical charge q, and kinematic momentum $\vec{p}$; the field is characterised by vector and scalar fields $\vec{A}$ and ϕ, respectively, at the location of the particle; and the units are MKS or SI. With the standard prescription for the quantum mechanical position representation, $\vec{p} \to -i\hbar\vec{\nabla}$, equation (9.58) can be expanded to give

$$\hat{H} = \frac{-\hbar^2}{2m}\nabla^2 + \frac{i\hbar q}{2m}(\vec{\nabla} \cdot \vec{A}_{\text{op}} + \vec{A}_{\text{op}} \cdot \vec{\nabla}) + \frac{q^2}{2m}A_{\text{op}}^2 + q\phi_{\text{op}}. \tag{9.59}$$

In the Coulomb gauge

$$\vec{\nabla} \cdot \vec{A}_{\text{op}} = 0, \tag{9.60}$$

whence

$$\begin{aligned} \vec{\nabla} \cdot (\vec{A}_{\text{op}}\psi) &= (\vec{\nabla} \cdot \vec{A}_{\text{op}})\psi + (\vec{A}_{\text{op}} \cdot \vec{\nabla})\psi \\ &= \vec{A}_{\text{op}} \cdot \vec{\nabla}\psi. \end{aligned} \tag{9.61}$$

Further, except for the most intense laser beams, the term containing A_{op}^2 is negligible. Thus,

$$\hat{H}_{\text{int.}} = \frac{i\hbar q}{m}\vec{A}_{\text{op}} \cdot \vec{\nabla} + q\phi_{\text{op}}. \tag{9.62}$$

Only the $\vec{A}_{\text{op}} \cdot \vec{\nabla}$ term in equation (9.62) will couple to electromagnetic radiation. The $q\phi_{\text{op}}$ term describes coupling to an electrostatic field. Thus, we will usually write

$$\hat{H}_{\text{int.}} = \frac{-q}{m}\vec{A}_{\text{op}} \cdot \vec{p}_{\text{op}}, \tag{9.63}$$

where $-i\hbar\vec{\nabla}$ has been replaced with $\vec{p}_{\text{op}}$. Equation (9.63) also describes the interaction between a charged particle and a magnetic field. For the hydrogen atom, there will be interactions between electromagnetic radiation and both the proton and the electron. However, the $\frac{1}{m}$ dependence of $\hat{H}_{\text{int.}}$ has the consequence that the interaction with the proton is negligible.

The above 'derivation' of $\hat{H}_{\text{int.}}$ does not allow for the particle having spin. Dirac has suggested an elegant recipe that incorporates spin in a way that gives results in agreement with experiment, viz.

$$\frac{p_{\text{op}}^2}{2m} \to \frac{(\vec{\sigma} \cdot \vec{p}_{\text{op}})^2}{2m}, \tag{9.64}$$

and for

$$\vec{p}_{\text{op}} \to \vec{p}_{\text{op}} - q\vec{A}_{\text{op}}, \tag{9.65}$$

$$\frac{p_{\text{op}}^2}{2m} \to \frac{1}{2m}\vec{\sigma}\cdot(\vec{p}_{\text{op}} - q\vec{A}_{\text{op}})\vec{\sigma}\cdot(\vec{p}_{\text{op}} - q\vec{A}_{\text{op}}), \tag{9.66}$$

where $\vec{\sigma}$ is the Pauli spin operator. Recall the identity (equation (1.35)):

$$(\vec{\sigma}\cdot\vec{a})(\vec{\sigma}\cdot\vec{b}) = \vec{a}\cdot\vec{b}\hat{I} + i\vec{\sigma}\cdot(\vec{a}\times\vec{b}). \tag{9.67}$$

Thus, in equation (9.64)

$$(\vec{\sigma}\cdot\vec{p}_{\text{op}})^2 = p_{\text{op}}^2, \tag{9.68}$$

i.e. one recovers the familiar form of the momentum operator from this recipe. However, for equation (9.66), using equation (9.67),

$$\begin{aligned}\frac{p_{\text{op}}^2}{2m} &\to \frac{1}{2m}(\vec{p}_{\text{op}} - q\vec{A}_{\text{op}})^2 + \frac{i\vec{\sigma}}{2m}\cdot\left\{(\vec{p}_{\text{op}} - q\vec{A}_{\text{op}})\times(\vec{p}_{\text{op}} - q\vec{A}_{\text{op}})\right\}\\ &\to \frac{1}{2m}(\vec{p}_{\text{op}} - q\vec{A}_{\text{op}})^2 - \frac{iq\vec{\sigma}}{2m}\cdot(\vec{p}_{\text{op}}\times\vec{A}_{\text{op}} + \vec{A}_{\text{op}}\times\vec{p}_{\text{op}}).\end{aligned} \tag{9.69}$$

Then, from $\vec{p}_{\text{op}} \to -i\hbar\vec{\nabla}$, $\vec{\nabla}\times\vec{A}_{\text{op}} = \vec{B}_{\text{op}}$ (equation (9.6)), $\vec{a}\times\vec{b} = -\vec{b}\times\vec{a}$, and

$$\begin{aligned}-i\hbar\vec{\nabla}\times\vec{A}_{\text{op}}\psi &= -i\hbar(\vec{\nabla}\times\vec{A}_{\text{op}})\psi + i\hbar\vec{A}_{\text{op}}\times\vec{\nabla}\psi\\ &= -i\hbar\vec{B}_{\text{op}}\psi - \vec{A}_{\text{op}}\times\vec{p}_{\text{op}}\psi,\end{aligned} \tag{9.70}$$

$$\therefore \frac{\hat{p}_{\text{op}}^2}{2m} \to \frac{1}{2m}(\vec{p}_{\text{op}} - q\vec{A}_{\text{op}})^2 - \frac{q\hbar}{2m}\vec{\sigma}\cdot\vec{B}_{\text{op}}, \tag{9.71}$$

For an interaction between the electromagnetic field and a charged particle with spin $\frac{1}{2}$ we have, therefore:

$$\hat{H}_{\text{int.}} = q\phi_{\text{op}} - \frac{q}{m}\vec{A}_{\text{op}}\cdot\vec{p}_{\text{op}} - \frac{q\hbar}{2m}\vec{\sigma}\cdot\vec{B}_{\text{op}}. \tag{9.72}$$

The operators $\vec{A}_{\text{op}}$ and $\vec{B}_{\text{op}}$ are given by equations (9.48) and (9.50). The operator $\vec{p}_{\text{op}}$ only acts on the spatial degrees of freedom of the particle; the operator $\vec{\sigma}$ only acts on the spin degrees of freedom of the particle; the operators $\vec{A}_{\text{op}}$ and $\vec{B}_{\text{op}}$ (cf. equations (9.48) and (9.49)) depend on the spatial degrees of freedom of the particle and operate on the degrees of freedom of the electromagnetic field.

9.3 The emission and absorption of photons by atoms

The emission and absorption of photons by atoms (or by molecules or nuclei) is a fundamental process of great importance to experimental quantum mechanics.

We will be concerned with state vectors that are the direct product of energy eigenstates of the atom and energy eigenstates of the electromagnetic field (cf. equation (9.42)). The interaction between the atom and the electromagnetic field resulting from the $\vec{A}_{\text{op}} \cdot \vec{p}_{\text{op}}$ term in equation (9.72) is described by

$$\begin{aligned}\langle t; n_{\vec{k}_i\alpha_i} - 1|\hat{H}_{int}|s; n_{\vec{k}_i\alpha_i}\rangle &= -\frac{q}{m}\langle t; n_{\vec{k}_i\alpha_i} - 1| \\ &\times \sqrt{\frac{\hbar}{2\varepsilon_0 V\omega_{k_i}}}\, b_{\vec{k}_i,\alpha_i}(0)e^{i(\vec{k}_i\cdot\vec{r}-\omega_{k_i}t)}\varepsilon_{\vec{k}_i\alpha_i} \cdot \vec{p}_{\text{op}}|s; n_{\vec{k}_i\alpha_i}\rangle \\ &= -\frac{q}{m}\sqrt{\frac{n_{\vec{k}_i\alpha_i}\hbar}{2\varepsilon_0 V\omega_{k_i}}}\langle t|e^{i\vec{k}_i\cdot\vec{r}}\hat{\varepsilon}_{\vec{k}_i\alpha_i} \cdot \vec{p}_{\text{op}}|s\rangle e^{-i\omega_{k_i}t},\end{aligned} \tag{9.73}$$

for the process where the atom undergoes the excitation $|s\rangle \rightarrow |t\rangle$ by absorbing a photon from the mode $\vec{k}_i$ with polarization α_i. We have in mind the response of a single electron, otherwise we must include a summation over all electrons and would replace $e^{i\vec{k}_i\cdot\vec{r}}\hat{\varepsilon}_{\vec{k}_i,\alpha_i} \cdot \vec{p}_{\text{op}}$ by $\sum_\nu e^{i\vec{k}_i\cdot\vec{r}_\nu}\hat{\varepsilon}_{\vec{k}_i,\alpha_i} \cdot \vec{p}_{\nu_{\text{op}}}$ where ν labels the electrons.

The result in equation (9.73) uses

$$b_{\vec{k}_i,\alpha_i}(0)|n_{\vec{k}_i\alpha_i}\rangle = \sqrt{n_{\vec{k}_i\alpha_i}}|n_{\vec{k}_i\alpha_i} - 1\rangle. \tag{9.74}$$

We have implicitly used the fact that only a photon with the right energy and polarization will be absorbed from the field, i.e.

$$E_t - E_s = \hbar\omega_{k_i}. \tag{9.75}$$

The second term in $\vec{A}(\vec{r}, t)_{\text{op}}$ (equation (9.48)), i.e. the term containing $b^\dagger_{\vec{k}_i,\alpha_i}(0)$, is responsible for emission. The matrix element for the photon part for emission is

$$\left\langle n_{\vec{k}_i\alpha_i} + 1 \,\middle|\, b^\dagger_{\vec{k}_i,\alpha_i} \,\middle|\, n_{\vec{k}_i\alpha_i}\right\rangle = \sqrt{n_{\vec{k}_i\alpha_i} + 1}. \tag{9.76}$$

Evidently, this is non-zero even when $n_{\vec{k}_i\alpha_i} = 0$: this is the case for *spontaneous emission*. The matrix element for the spatial part for emission is

$$\langle t; n_{\vec{k}_i\alpha_i} + 1|\hat{H}_{int}|s; n_{\vec{k}_i\alpha_i}\rangle = -\frac{q}{m}\sqrt{\frac{(n_{\vec{k}_i\alpha_i} + 1)\hbar}{2\varepsilon_0 V\omega_{k_i}}}\langle t|e^{-i\vec{k}_i\cdot\vec{r}}\hat{\varepsilon}_{\vec{k}_i,\alpha_i} \cdot \vec{p}_{\text{op}}|s\rangle e^{i\omega_k t}. \tag{9.77}$$

We can apply Fermi's golden rule directly, using equation (9.77), for spontaneous emission ($n_{\vec{k}_i\alpha_i} = 0$):

$$\frac{\mathrm{d}P}{\mathrm{d}t} = \frac{2\pi}{\hbar}\frac{e^2}{m^2}\frac{\hbar}{2\varepsilon_0 V\omega_{k_i}}|\langle t|e^{-i\vec{k}_i\cdot\vec{r}}\hat{\varepsilon}_{\vec{k}_i,\alpha_i} \cdot \vec{p}_{\text{op}}|s\rangle|^2\rho(E_t)_{E_t\approx E_s}. \tag{9.78}$$

It is necessary to obtain an expression for $\rho(E_t)_{E_t \approx E_s}$. Because the atomic states are discrete, this is determined just by the density of modes in the electromagnetic field. In an energy interval $\{E, E + \mathrm{d}E\} = \{\hbar\omega, \hbar(\omega + \mathrm{d}\omega)\}$ this is given, for emission into a solid angle $\mathrm{d}\Omega$, by (see appendix B)

$$\rho_{\hbar\omega,\mathrm{d}\Omega} = \frac{V\omega^2}{(2\pi)^3}\frac{\mathrm{d}\Omega}{\hbar c^3}. \tag{9.79}$$

Then, the rate of emission into a solid angle $\mathrm{d}\Omega$ is

$$\left.\frac{\mathrm{d}P}{\mathrm{d}t}\right)_{\mathrm{d}\Omega} = \frac{|\vec{k}_i|e^2\omega^2}{8\pi^2\varepsilon_0 m^2\hbar c^2}|\langle t|e^{-i\vec{k}_i\cdot r}\hat{\varepsilon}_{\vec{k}_i,\alpha_i}\cdot\vec{p}_{\mathrm{op}}|s\rangle|^2\mathrm{d}\Omega, \tag{9.80}$$

where equation (9.18) has been used; and note that V has cancelled.

Now, in typical atomic transitions

$$ƛ_{\mathrm{photon}} \equiv \frac{1}{|\vec{k}|} \gg r_{\mathrm{atom}}, \tag{9.81}$$

where r_{atom} is the atomic radius; e.g. for the $2p \to 1s$ transition in hydrogen (Lyman α transition), $\lambda = 121.5$ nm and $r_{\mathrm{atom}} \approx 0.2$ nm. Thus, we can make the approximation

$$e^{-i\vec{k}_i\cdot\vec{r}} = 1 - i\vec{k}_i\cdot\vec{r} - \frac{1}{2}(\vec{k}_i\cdot\vec{r})^2 + \cdots \approx 1. \tag{9.82}$$

This is called the *electric dipole approximation.*

$$\therefore \left.\frac{\mathrm{d}P}{\mathrm{d}t}\right|_{\mathrm{d}\Omega} = \frac{|\vec{k}_i|e^2\omega^2}{8\pi^2\varepsilon_0 m^2\hbar c^2}|\hat{\varepsilon}_{\vec{k}_i,\alpha_i}\cdot\langle t|\vec{p}_{\mathrm{op}}|s\rangle|^2\mathrm{d}\Omega. \tag{9.83}$$

Hence, we must compute the matrix element $\langle t|\vec{p}_{\mathrm{op}}|s\rangle$.

The matrix element $\langle t|\vec{p}_{\mathrm{op}}|s\rangle$ is derived by using the relationship:

$$[\vec{p}^2, \hat{r}_{\mathrm{op}}] = -2i\hbar\vec{p}_{\mathrm{op}}, \tag{9.84}$$

whence

$$\left[\frac{\hat{p}^2}{2m}, \vec{r}_{\mathrm{op}}\right] = \frac{-i\hbar}{m}\vec{p}_{\mathrm{op}}, \tag{9.85}$$

and

$$[\hat{H}_0, \vec{r}_{\mathrm{op}}] = \frac{-i\hbar}{m}\vec{p}_{\mathrm{op}}, \tag{9.86}$$

for

$$\hat{H}_0 = \frac{\hat{p}^2}{2m} + V(\hat{r}). \tag{9.87}$$

Thus,

$$\langle t|\vec{p}_{\mathrm{op}}|s\rangle = \frac{im}{\hbar}\langle t|[\hat{H}_0, \vec{r}_{\mathrm{op}}]|s\rangle, \tag{9.88}$$

$$\therefore \langle t|\vec{p}_{\mathrm{op}}|s\rangle = \frac{im}{\hbar}(E_t - E_s)\langle t|\vec{r}_{\mathrm{op}}|s\rangle. \tag{9.89}$$

We can immediately write down the selection rules for electric dipole radiation by recognising $\vec{r}$ to be a spherical tensor of rank one that is odd under parity, hence

$$|L_t - L_s| = 1, 0 \quad 0 \to 0 \text{ forbidden}, \tag{9.90}$$

$$\Delta\pi = \text{yes}. \tag{9.91}$$

Then, using (cf. equation (9.18))

$$|\vec{k}_i| = \frac{\omega_{k_i}}{c} = \frac{E_t - E_s}{\hbar c}, \tag{9.92}$$

$$\left.\frac{\mathrm{d}P}{\mathrm{d}t}\right|_{\mathrm{d}\Omega} = \frac{e^2}{8\pi^2\varepsilon_0\hbar^4c^3}(E_t - E_s)^3\left|\hat{\varepsilon}_{\vec{k}_i\alpha_i}\cdot\langle t|\vec{r}_{\mathrm{op}}|s\rangle\right|^2 \mathrm{d}\Omega. \tag{9.93}$$

To calculate the total transition rate, we must integrate over all solid angles and sum over the two states of polarization. Defining $\vec{r}_{ts} \equiv \langle t|\vec{r}_{\mathrm{op}}|s\rangle$, and with reference to figure 9.1,

$$\hat{\varepsilon}_{\vec{k}_i,\alpha_i}\cdot\vec{r}_{ts} = |\vec{r}_{ts}|\cos\Theta_{\alpha_i} = |\vec{r}_{ts}|\sin\theta\sin\phi, \tag{9.94}$$

and

$$\hat{\varepsilon}_{\vec{k}_i,\beta_i}\cdot\vec{r}_{ts} = |\vec{r}_{ts}|\cos\Theta_{\beta_i} = |\vec{r}_{ts}|\sin\theta\cos\phi. \tag{9.95}$$

Thus, the sum over the polarizations is

$$|\hat{\varepsilon}_{\vec{k}_i\alpha_i}\cdot\vec{r}_{ts}|^2 + |\hat{\varepsilon}_{\vec{k}_i\beta_i}\cdot\vec{r}_{ts}|^2 = |\vec{r}_{ts}|^2\sin^2\theta. \tag{9.96}$$

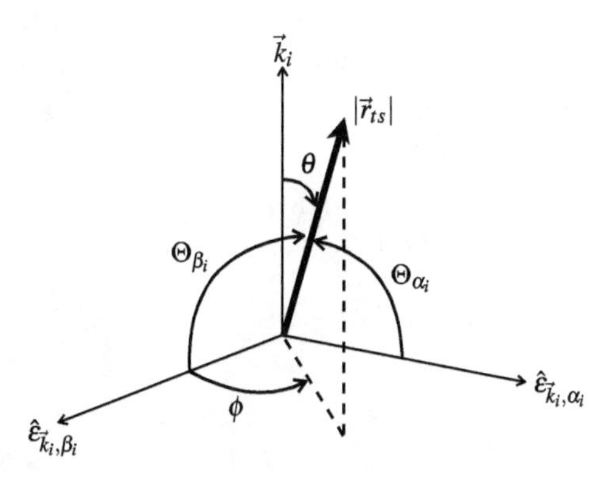

Figure 9.1. A depiction of the sum over the two states of polarization for equation (9.93).

Then, integrating over all solid angles:

$$\begin{aligned}\frac{\mathrm{d}P}{\mathrm{d}t} &= \frac{e^2}{8\pi^2\varepsilon_0\hbar^4c^3}(E_t - E_s)^3|r_{ts}|^2\int_0^{2\pi}\mathrm{d}\phi\int_0^{\pi}\sin^2\theta\sin\theta\,\mathrm{d}\theta \\ &= \frac{e^2}{8\pi^2\varepsilon_0\hbar^4c^3}(E_t - E_s)^3|r_{ts}|^2 2\pi\int_0^{\pi}(1-\cos^2\theta)\mathrm{d}(-\cos\theta)\end{aligned} \tag{9.97}$$

and the integral has the value $\frac{4}{3}$, whence

$$\frac{\mathrm{d}P}{\mathrm{d}t} = \frac{e^2}{3\pi\varepsilon_0\hbar^4c^3}(E_t - E_s)^3|\langle t|\vec{r}_{\mathrm{op}}|s\rangle|^2. \tag{9.98}$$

To proceed any further, the matrix element $\langle t|\vec{r}_{\mathrm{op}}|s\rangle$ must be computed. For a central force problem, this can be carried out in the position representation by making use of the results of section 1.9 for the angular coordinates. First, we can choose $r = z = r\cos\theta$ by defining the direction of emission of the photon to be the z-axis. Then using $\cos\theta = \sqrt{\frac{4\pi}{3}}\,Y_{10}(\theta, \phi)$, $|s\rangle = |n'l'm'\rangle$, $|t\rangle = |nlm\rangle$:

$$\begin{aligned}\langle t|\vec{r}_{\mathrm{op}}|s\rangle \leftrightarrow &\int_0^{2\pi}\mathrm{d}\phi\int_0^{\pi}\sin\theta\,\mathrm{d}\theta\,Y^*_{lm}(\theta, \phi)_t\sqrt{\frac{4\pi}{3}}Y_{10}(\theta, \phi)Y_{l'm'}(\theta, \phi)_s \\ &\times\int_0^{\infty}r^2\mathrm{d}r R^*_{nl}(r)_t r R_{n'l'}(r)_s,\end{aligned} \tag{9.99}$$

$$\begin{aligned}\therefore\ \langle t|\vec{r}_{\mathrm{op}}|s\rangle \leftrightarrow (-1)^m&\sqrt{(2l'+1)(2l+1)}\begin{pmatrix}1 & l' & l\\ 0 & 0 & 0\end{pmatrix}\begin{pmatrix}1 & l' & l\\ 0 & m' & -m\end{pmatrix} \\ &\times\int_0^{\infty}r^2\mathrm{d}r R^*_{nl}(r)_t r R_{n'l'}(r)_s,\end{aligned} \tag{9.100}$$

where equation (2.114) has been used to evaluate the angular dependent part. Evidently,

$$m' = m,\ \ l = |l' \pm 1|. \tag{9.101}$$

To proceed beyond this, we must specify the type of central force problem and determine $(E_t - E_s)^2$ (equation (9.98)) and the radial integral in equation (9.100). As an example, we will consider the $2p \to 1s$ transition in hydrogen, i.e. we will calculate the lifetime of the hydrogen $2p$ state. Then,

$$E_t - E_s = \frac{me^4}{8\varepsilon_0^2h^2}\left(\frac{1}{1^2} - \frac{1}{2^2}\right)\quad (t = 1, s = 2) \tag{9.102}$$

and (cf. Volume 1, appendix B)

$$\left.\frac{\mathrm{d}P}{\mathrm{d}t}\right|_{2p\to 1s} = \frac{3^2\pi^3 m^3 e^{14}}{2^{11}\varepsilon_0^7 h^{10} c^3}\left\{\sqrt{3}\begin{pmatrix}1 & 1 & 0\\ 0 & 0 & 0\end{pmatrix}\begin{pmatrix}1 & 1 & 0\\ 0 & 0 & 0\end{pmatrix}\right\}^2$$
$$\times\left\{\int_0^\infty \frac{1}{\sqrt{24a_0^3}}\frac{r}{a_0}e^{-\frac{r}{2a_0}}r\frac{2}{\sqrt{a_0^3}}e^{-\frac{r}{a_0}}r^2\,\mathrm{d}r\right\}^2 \tag{9.103}$$
$$\therefore\left.\frac{\mathrm{d}P}{\mathrm{d}t}\right|_{2p\to 1s} = \frac{3^2\pi^3 m^3 e^{14}}{2^{11}\varepsilon_0^7 h^{10} c^3}\frac{1}{3}\left\{\frac{1}{\sqrt{6}}\frac{1}{a_0^4}\int_0^\infty e^{-\frac{3r}{2a_0}}r^4\,\mathrm{d}r\right\}^2.$$

The integral in equation (9.103) is of the standard form $\int_0^\infty e^{-\alpha r}r^n\,\mathrm{d}r = \frac{n!}{\alpha^{n+1}}$.

$$\therefore\left.\frac{\mathrm{d}P}{\mathrm{d}t}\right|_{2p\to 1s} = \frac{3\pi^3 m^3 e^{14}}{2^{11}\varepsilon_0^7 h^{10} c^3}\left\{\frac{1}{\sqrt{6}}\frac{1}{a_0^4}4!\left(\frac{2a_0}{3}\right)^5\right\}^2; \tag{9.104}$$

and using

$$a_0 = \frac{\varepsilon_0 h^2}{\pi m e^2}, \tag{9.105}$$

$$\therefore\left.\frac{\mathrm{d}P}{\mathrm{d}t}\right|_{2p\to 1s} = \frac{2^4\pi m e^{10}}{3^8\varepsilon_0^5 h^6 c^3} = \frac{2^8}{3^8}\frac{mc^2\alpha^5}{\hbar}, \tag{9.106}$$

where

$$\alpha = \frac{e^2}{2\varepsilon_0 hc} = \frac{1}{137.035\,999\,084}, \tag{9.107}$$

this gives

$$\left.\frac{\mathrm{d}P}{\mathrm{d}t}\right|_{2p\to 1s} = 6.2683\times 10^8\ \mathrm{s}^{-1}, \tag{9.108}$$

whence the lifetime of the $2p$ state in hydrogen is

$$\tau_{2p}^{\text{theory}} = \frac{1}{6.2683\times 10^8} = 1.5953\ \text{ns}. \tag{9.109}$$

This can be compared with an experimental value of $\tau_{2p}^{\text{expt}} = 1.600 \pm 0.004$ ns [3] and 1.60 ± 0.01 ns [4].

References

[1] Haroche S and Raimond J-M 1993 *Sci. Am.* **268** 54
[2] Haroche S and Kleppner D 1989 *Phys. Today* **42** 24
[3] Bickel W S and Goodman A S 1966 *Phys. Rev.* **148** 1
[4] Chupp E L *et al* 1968 *Phys. Rev.* **175** 44

Chapter 10

Epilogue

In the first volume we have endeavoured to introduce the reader to the basic elements of quantum mechanics in a form that can eventually be put to use in the theory of nuclear structure. It is essentially a one-body formulation of quantum mechanics. In this second volume, basic steps into many-body formulations are made starting with the representation and coupling of spin-angular momentum states and associated operators. Herein, we have emphasized the algebraic structure of quantum theory. Algebraic structure is an essential aspect of the mathematical formulation of many-body quantum systems.

We have avoided introducing facets of nuclear structure in the formalism because this necessitates familiarity with nuclear data. There will be forthcoming books in the series that will introduce nuclear data, and separately that will introduce quantum mechanical modelling for nuclei. The present material provides the reader with much of the formalism that is needed to describe actual nuclei. Application to many-body quantum systems requires a substantial body of experimental data to guide model building. This is a matter of the historical record: model building for nuclear structure has advanced hand-in-hand with the accumulation of nuclear data.

Many-body quantum systems are in general too complex for *ab initio* approaches unless there is some idea of the manner in which such systems self organize. This has been true of condensed matter systems. Thus, high-temperature superconductors were not predicted to exist. The quantum Hall effects were not anticipated. Currently, graphene continues to surprise investigators.

A number of techniques in quantum mechanics, that play a key role in the study of nuclear structure, have been deferred. Most notably, the variational method will play a central role in some of the forthcoming books planned in this e-book series 'Nuclear Spectroscopy and Nuclear Structure'. In particular, when describing how mean-fields can be constructed in nuclei, one starts from a given effective nucleon–nucleon interaction and a suitable basis and seeks a linear combination of basis states that minimizes the energy. This goes under the name 'self-consistent many-body theory' with the Hartree–Fock method as the leading technique. This determines the optimal mean-field experienced by each nucleon treated as an independent particle.

Algebraic methods provide a powerful approach to modelling collective dynamics in many-body quantum systems. The leading feature of such models is that they are solvable. The details provided in the present two volumes take only the first rudimentary steps into this aspect of the quantum mechanical formalism. Indeed, developments of key techniques for handling algebraic structures have barely kept up with model building. Some examples are treated in detail, in the monograph by David Rowe and J L Wood [1]. Algebraic modelling of nuclear bound states is a planned feature in future books in the series.

A further major issue in nuclear structure physics is the need for a unified view of bound states and unbound states, as manifested in nuclear reaction theory. A quantum mechanical treatment of nuclear reaction theory requires input information for the structure of the target nucleus and the projectile nucleus. Further, key details of nuclear structure depend on reactions for their elucidation. As such, the present two-volume work does not cover key quantum mechanical techniques used in formulating a theory of reactions, even at the most elementary level. This limitation has been intentional and future books in the series are planned to handle this.

We leave the reader with the following closing thought. The theory of quantum systems introduces a language that ensures we do not specify more than can be known by observation: the mathematical structure enforces this limitation in knowledge.

Reference

[1] Rowe D J and Wood J L 2010 *Fundamentals of Nuclear Models: Foundational Models* (Singapore: World Scientific)

IOP Publishing

Quantum Mechanics for Nuclear Structure, Volume 2

An intermediate level view

Kris Heyde and John L Wood

Appendix A

Clebsch–Gordan coefficients and 3-*j* symbols

A.1 Clebsch–Gordan coefficients (tables A.1–A.4)

A.2 3-*j* symbols (table A.5)

A.3 Tables of 3-*j* symbol numerical values

There are two outstanding sets:

(a) 'The 3-*j* and 6-*j* symbols' [1].

(b) 'Numerical tables for angular correlation computations in α-, β-, and γ-spectroscopy: 3-*j*, 6-*j*, 9-*j* symbols, F- and Γ-coefficients' [2].

A.4 A worked example using 3-*j* symbols

To evaluate the expansion coefficients in (cf. figure 2.3)

$$|43\rangle = |2122\rangle\langle 2122|43\rangle + |2221\rangle\langle 2221|43\rangle.$$

From equation (2.75)

$$\langle 2122|43\rangle = (-1)^{-2+2-3}\sqrt{2\times 4+1}\begin{pmatrix} 2 & 2 & 4 \\ 1 & 2 & -3 \end{pmatrix} = -3\begin{pmatrix} 2 & 2 & 4 \\ 1 & 2 & -3 \end{pmatrix}$$

$$= -3(-1)^{4+2+2}\begin{pmatrix} 4 & 2 & 2 \\ -3 & 2 & 1 \end{pmatrix};$$

and

$$\langle 2221|43\rangle = (-1)^{-2+2-3}\sqrt{2\times 4+1}\begin{pmatrix} 2 & 2 & 4 \\ 2 & 1 & -3 \end{pmatrix}$$

$$= -3\begin{pmatrix} 4 & 2 & 2 \\ -3 & 2 & 1 \end{pmatrix}.$$

Then, from

$$\begin{pmatrix} J+2 & J & 2 \\ M & -M-1 & 1 \end{pmatrix}$$

$$= 2(-1)^{J-M}\left\{\frac{(J+M+2)(J-M+2)(J-M+1)(J-M)}{(2J+5)(2J+4)(2J+3)(2J+2)(2J+1)}\right\}^{\frac{1}{2}}$$

with $J = 2$, $M = -3$,

$$\begin{pmatrix} 4 & 2 & 2 \\ -3 & 2 & 1 \end{pmatrix} = 2(-1)^5\left\{\frac{(1)\cancel{(7)}\cancel{(6)}\cancel{(5)}}{(9)(8)\cancel{(7)}\cancel{(6)}\cancel{(5)}}\right\}^{\frac{1}{2}} = -\frac{1}{3\sqrt{2}},$$

$$\therefore \langle 2122|43\rangle = \frac{1}{\sqrt{2}}, \quad \langle 2221|43\rangle = \frac{1}{\sqrt{2}}.$$

Table A.1. $(j_1 m_1, \frac{1}{2} m_2 | jm)$.

$j =$	$m_2 = \frac{1}{2}$	$m_2 = -\frac{1}{2}$
$j_1 + \frac{1}{2}$	$\sqrt{\frac{j+m}{2j_1+1}}$	$\sqrt{\frac{j-m}{2j_1+1}}$
$j_1 - \frac{1}{2}$	$-\sqrt{\frac{j-m+1}{2j_1+1}}$	$\sqrt{\frac{j+m+1}{2j_1+1}}$

Table A.2. $(j_1 m_1, 1 m_2 | jm)$.

$j =$	$m_2 = 1$	$m_2 = 0$	$m_2 = -1$
$j_1 + 1$	$\sqrt{\frac{(j+m-1)(j+m)}{(2j_1+1)(2j_2+2)}}$	$\sqrt{\frac{(j-m)(j+m)}{(2j_1+1)(j_1+1)}}$	$\sqrt{\frac{(j-m-1)(j-m)}{(2j_1+1)(2j_1+2)}}$
j_1	$-\sqrt{\frac{(j+m)(j-m+1)}{2j_1(j_1+1)}}$	$\frac{m}{\sqrt{j_1(j_1+1)}}$	$\sqrt{\frac{(j-m)(j+m+1)}{2j_1(j_1+1)}}$
$j_1 - 1$	$\sqrt{\frac{(j-m+1)(j-m+2)}{2j_1(2j_1+1)}}$	$-\sqrt{\frac{(j-m+1)(j+m+1)}{j_1(2j_1+1}})$	$\sqrt{\frac{(j+m+2)(j+m+1)}{2j_1(2j_1+1)}}$

Table A.3. $(j_1 m_1, \frac{3}{2} m_2 | jm)$.

$j =$	$m_2 = \frac{3}{2}$	$m_2 = \frac{1}{2}$
$j_1 + \frac{3}{2}$	$\sqrt{\frac{(j+m-2)(j+m-1)(j+m)}{(2j_1+1)(2j_1+2)(2j_1+3)}}$	$\sqrt{\frac{3(j+m-1)(j+m)(j-m)}{(2j_1+1)(2j_1+2)(2j_1+3)}}$
$j_1 + \frac{1}{2}$	$-\sqrt{\frac{3(j+m-1)(j+m)(j-m+1)}{2j_1(2j_1+1)(2j_1+3)}}$	$-(j-3m+1)\sqrt{\frac{(j+m)}{2j_1(2j_1+1)(2j_1+3)}}$
$j_1 - \frac{1}{2}$	$\sqrt{\frac{3(j+m)(j-m+1)(j-m+2)}{(2j_1-1)(2j_1+1)(2j_1+2)}}$	$-(j+3m)\sqrt{\frac{j-m+1}{(2j_1-1)(2j_1+1)(2j_1+2)}}$
$j_1 - \frac{3}{2}$	$-\sqrt{\frac{(j-m+1)(j-m+2)(j-m+3)}{2j_1(2j_1-1)(2j_1+1)}}$	$\sqrt{\frac{3(j+m+1)(j-m+1)(j-m+2)}{(2j_1)(2j_1-1)(2j_1+1)}}$
$j =$	$m_2 = -\frac{1}{2}$	$m_2 = -\frac{3}{2}$
$j_1 + \frac{3}{2}$	$\sqrt{\frac{3(j+m)(j-m-1)(j-m)}{(2j_1+1)(2j_1+2)(2j_1+3)}}$	$\sqrt{\frac{(j-m-2)(j-m-1)(j-m)}{(2j_1+1)(2j_1+2)(2j_1+3)}}$
$j_1 + \frac{1}{2}$	$(j+3m+1)\sqrt{\frac{j-m}{2j_1(2j_1+1)(2j_1+3)}}$	$\sqrt{\frac{3(j+m+1)(j-m-1)(j-m)}{2j_1(2j_1+1)(2j_1+3)}}$
$j_1 - \frac{1}{2}$	$-(j-3m)\sqrt{\frac{(j+m+1)}{(2j_1-1)(2j_1+1)(2j_1+2)}}$	$\sqrt{3\frac{(j+m+1)(j+m+2)(j-m)}{(2j_1-1)(2j_1+1)(2j_1+2)}}$
$j_1 - \frac{3}{2}$	$-\sqrt{\frac{3(j+m+1)(j+m+2)(j-m+1)}{2j_1(2j_1-1)(2j_1+1)}}$	$\sqrt{\frac{(j+m+1)(j+m+2)(j+m+3)}{(2j_1)(2j_1-1)(2j_1+1)}}$

Table A.4. $(j_1 m_1,\ 2 m_2 | j m)$.

$j =$	$m_2 = 2$	$m_2 = 1$
$j_1 + 2$	$\sqrt{\frac{(j+m-3)(j+m-2)(j+m-1)(j+m)}{(2j_1+1)(2j_1+2)(2j_1+3)(2j_1+4)}}$	$\sqrt{\frac{(j-m)(j+m)(j+m-1)(j+m-2)}{(2j_1+1)(j_1+1)(2j_1+3)(j_1+2)}}$
$j_1 + 1$	$-\sqrt{\frac{(j+m-2)(j+m-1)(j+m)(j-m+1)}{2j_1(j_1+1)(j_1+2)(2j_1+1)}}$	$-(j-2m+1)\sqrt{\frac{(j+m)(j+m-1)}{2j_1(j_1+1)(j_1+2)(2j_1+1)}}$
j_1	$\sqrt{\frac{3(j+m-1)(j+m)(j-m+1)(j-m+2)}{(2j_1-1)2j_1(j_1+1)(2j_1+1)}}$	$(1-2m)\sqrt{\frac{3(j-m+1)(j+m)}{(2j_1-1)2j_1(j_1+1)(2j_1+3)}}$
$j_1 - 1$	$-\sqrt{\frac{(j+m)(j-m+1)(j-m+2)(j-m+3)}{2(j_1-1)j_1(j_1+1)(2j_1+1)}}$	$(j+2m)\sqrt{\frac{(j-m+2)(j-m+1)}{2(j_1-1)j_1(j_1+1)(2j_1+1)}}$
$j_1 - 2$	$\sqrt{\frac{(j-m+1)(j-m+2)(j-m+3)(j-m+4)}{(2j_1-2)(2j_1-1)2j_1(2j_1+1)}}$	$-\sqrt{\frac{(j-m+3)(j-m+2)(j-m+1)(j+m+1)}{(j_1-1)(2j_1-1)j_1(2j_1+1)}}$

$j =$	$m_2 = 0$
$j_1 + 2$	$\sqrt{\frac{3(j-m)(j-m-1)(j+m)(j+m-1)}{(2j_1+1)(2j_1+2)(2j_1+3)(j_1+2)}}$
$j_1 + 1$	$m\sqrt{\frac{3(j-m)(j+m)}{j_1(j_1+1)(j_1+2)(2j_1+1)}}$
j_1	$\frac{3m^2 - j(j+1)}{\sqrt{(2j_1-1)j_1(j_1+1)(2j_1+3)}}$
$j_1 - 1$	$-m\sqrt{\frac{3(j-m+1)(j+m-1)}{(j_1-1)j_1(j_1+1)(2j_1+1)}}$
$j_1 - 2$	$\sqrt{\frac{3(j-m+2)(j-m+1)(j+m+2)(j+m+1)}{(2j_1-2)(2j_1-1)j_1(2j_1+1)}}$

$j =$	$m_2 = -1$	$m_2 = -2$
$j_1 + 2$	$\sqrt{\frac{(j-m)(j-m-1)(j-m-2)(j+m)}{(j_1+1)(j_1+2)(2j_1+1)(2j_1+3)}}$	$\sqrt{\frac{(j-m-3)(j-m-2)(j-m-1)(j-m)}{(2j_1+1)(2j_1+2)(2j_1+3)(2j_1+4)}}$
$j_1 + 1$	$(j+2m+1)\sqrt{\frac{(j-m)(j-m-1)}{j_1(j_1+2)(2j_1+1)(2j_1+2)}}$	$\sqrt{\frac{(j-m-2)(j-m-1)(j-m)(j+m+1)}{j_1(j_1+1)(2j_1+1)(2j_1+4)}}$
j_1	$(2m+1)\sqrt{\frac{3(j-m)(j+m+1)}{j_1(2j_1-1)(2j_1+2)(2j_1+3)}}$	$\sqrt{\frac{3(j-m-1)(j-m)(j+m+1)(j+m+2)}{j_1(2j_1-1)(2j_1+2)(2j_1+3)}}$
$j_1 - 1$	$-(j-2m)\sqrt{\frac{(j+m+2)(j+m+1)}{j_1(j_1-1)(2j_1+1)(2j_1+2)}}$	$\sqrt{\frac{(j-m)(j+m+1)(j+m+2)(j+m+3)}{j_1(j_1-1)(2j_1+1)(2j_1+2)}}$
$j_1 - 2$	$-\sqrt{\frac{(j-m+1)(j+m+3)(j+m+2)(j+m+1)}{j_1(j_1-1)(2j_1-1)(2j_1+1)}}$	$\sqrt{\frac{(j+m+1)(j+m+2)(j+m+3)(j+m+4)}{2j_1(2j_1-1)(2j_1-2)(2j_1+1)}}$

Table A.5. Some 3-j coefficients. Their use requires permutations of columns so that the maximum J is on the left.

3-j coefficient	Value	Condition
$\begin{pmatrix} j_1 & j_2 & j_3 \\ 0 & 0 & 0 \end{pmatrix}$	$(-1)^{\frac{J}{2}}\sqrt{\frac{(j_1+j_2-j_3)!(j_1+j_3-j_2)!(j_2+j_3-j_1)!}{(j_1+j_2+j_3+1)!}}\frac{(\frac{1}{2}J)!}{(\frac{1}{2}J-j_1)!(\frac{1}{2}J-j_2)!(\frac{1}{2}J-j_3)!}$	J even
$\begin{pmatrix} j_1 & j_2 & j_3 \\ 0 & 0 & 0 \end{pmatrix}$	0	J odd
$\begin{pmatrix} J+\frac{1}{2} & J & \frac{1}{2} \\ M & -M-\frac{1}{2} & \frac{1}{2} \end{pmatrix}$	$(-1)^{(J-M-\frac{1}{2})}\sqrt{\frac{J-M+\frac{1}{2}}{(2J+2)(2J+1)}}$	$(J+\frac{1}{2}, J, \frac{1}{2})$
$\begin{pmatrix} J+1 & J & 1 \\ M & -M-1 & 1 \end{pmatrix}$	$(-1)^{(J-M-1)}\sqrt{\frac{(J-M)(J-M+1)}{(2J+3)(2J+2)(2J+1)}}$	$(J+1, J, 1)$
$\begin{pmatrix} J+1 & J & 1 \\ M & -M & 0 \end{pmatrix}$	$(-1)^{(J-M-1)}\sqrt{\frac{(J+M+1)(J-M+1)\cdot 2}{(2J+3)(2J+2)(2J+1)}}$	$(J+1, J, 1)$
$\begin{pmatrix} J & J & 1 \\ M & -M-1 & 1 \end{pmatrix}$	$(-1)^{(J-M)}\sqrt{\frac{(J-M)(J+M+1)\cdot 2}{(2J+2)(2J+1)(2J)}}$	$(J, J, 1)$
$\begin{pmatrix} J & J & 1 \\ M & -M & 0 \end{pmatrix}$	$(-1)^{(J-M)}\frac{M}{\sqrt{(2J+1)(J+1)(J)}}$	$(J, J, 1)$
$\begin{pmatrix} J+\frac{3}{2} & J & \frac{3}{2} \\ M & -M-\frac{3}{2} & \frac{3}{2} \end{pmatrix}$	$(-1)^{(J-M+\frac{1}{2})}\sqrt{\frac{(J-M-\frac{1}{2})(J-M+\frac{1}{2})(J-M+\frac{3}{2})}{(2J+4)(2J+3)(2J+2)(2J+1)}}$	$(J+\frac{3}{2}, J, \frac{3}{2})$
$\begin{pmatrix} J+\frac{3}{2} & J & \frac{3}{2} \\ M & -M-\frac{1}{2} & \frac{1}{2} \end{pmatrix}$	$(-1)^{(J-M+\frac{1}{2})}\sqrt{\frac{3(J-M+\frac{1}{2})(J-M+\frac{3}{2})(J+M+\frac{3}{2})}{(2J+4)(2J+3)(2J+2)(2J+1)}}$	$(J+\frac{3}{2}, J, \frac{3}{2})$
$\begin{pmatrix} J+\frac{1}{2} & J & \frac{3}{2} \\ M & -M-\frac{3}{2} & \frac{3}{2} \end{pmatrix}$	$(-1)^{(J-M-\frac{1}{2})}\sqrt{\frac{3(J-M-\frac{1}{2})(J-M+\frac{1}{2})(J-M+\frac{3}{2})}{(2J+3)(2J+2)(2J+1)2J}}$	$(J+\frac{1}{2}, J, \frac{3}{2})$
$\begin{pmatrix} J+\frac{1}{2} & J & \frac{3}{2} \\ M & -M-\frac{1}{2} & \frac{1}{2} \end{pmatrix}$	$(-1)^{(J-M-\frac{1}{2})}\sqrt{\frac{J-M+\frac{1}{2}}{(2J+3)(2J+2)(2J+1)2J}}$	$(J+\frac{1}{2}, J, \frac{3}{2})$
$\begin{pmatrix} J+2 & J & 2 \\ M & -M-2 & 2 \end{pmatrix}$	$(-1)^{(J-M)}\sqrt{\frac{(J-M-1)(J-M)(J-M+1)(J-M+2)}{(2J+5)(2J+4)(2J+3)(2J+2)(2J+1)}}$	$(J+2, J, 2)$
$\begin{pmatrix} J+2 & J & 2 \\ M & -M-1 & 1 \end{pmatrix}$	$2(-1)^{(J-M)}\sqrt{\frac{(J+M+2)(J-M+2)(J-M+1)(J-M)}{(2J+5)(2J+4)(2J+3)(2J+2)(2J+1)}}$	$(J+2, J, 2)$
$\begin{pmatrix} J+2 & J & 2 \\ M & -M & 0 \end{pmatrix}$	$(-1)^{(J-M)}\sqrt{\frac{6(J+M+2)(J+M+1)(J-M+2)(J-M+1)}{(2J+5)(2J+4)(2J+3)(2J+2)(2J+1)}}$	$(J+2, J, 2)$
$\begin{pmatrix} J+1 & J & 2 \\ M & -M-2 & 2 \end{pmatrix}$	$2(-1)^{(J-M+1)}\sqrt{\frac{(J-M-1)(J-M)(J-M+1)(J+M+2)}{(2J+4)(2J+3)(2J+2)(2J+1)2J}}$	$(J+1, J, 2)$

(*Continued*)

Table A.5. (*Continued*)

3-*j* coefficient	Value	Condition
$\begin{pmatrix} J+1 & J & 2 \\ M & -M-1 & 1 \end{pmatrix}$	$(-1)^{(J-M+1)}2(J+2M+2)\sqrt{\frac{(J-M+1)(J-M)}{(2J+4)(2J+3)(2J+2)(2J+1)2J}}$	$(J+1, J, 2)$
$\begin{pmatrix} J+1 & J & 2 \\ M & -M & 0 \end{pmatrix}$	$(-1)^{(J-M+1)}2M\sqrt{\frac{6(J+M+1)(J-M+1)}{(2J+4)(2J+3)(2J+2)(2J+1)2J}}$	$(J+1, J, 2)$
$\begin{pmatrix} J & J & 2 \\ M & -M-2 & 0 \end{pmatrix}$	$(-1)^{(J-M)}\sqrt{\frac{6(J-M-1)(J-M)(J+M+1)(J+M+2)}{(2J+3)(2J+2)(2J+1)(2J)(2J-1)}}$	$(J, J, 2)$
$\begin{pmatrix} J & J & 2 \\ M & -M-1 & 1 \end{pmatrix}$	$(-1)^{(J-M)}(1+2M)\sqrt{\frac{6(J+M+1)(J-M)}{(2J+3)(2J+2)(2J+1)(2J)(2J-1)}}$	$(J, J, 2)$
$\begin{pmatrix} J & J & 2 \\ M & -M & 0 \end{pmatrix}$	$(-1)^{(J-M)}\frac{2\left(3M^2-J(J+1)\right)}{\sqrt{(2J+3)(2J+2)(2J+1)(2J)(2J-1)}}$	$(J, J, 2)$

References

[1] Rotenberg M, Bivins R, Metropolis N and Wooten J K Jr 1959 *The 3-j and 6-j symbols* (Cambridge, MA: Technology Press, MIT)

[2] Hellwege K-H 1968 *Landolt-Börnstein Numerical Data and Functional Relationships in Science and Technology, Group I* vol 3 (Berlin: Springer)

IOP Publishing

Quantum Mechanics for Nuclear Structure, Volume 2
An intermediate level view
Kris Heyde and John L Wood

Appendix B

The mode density for the electromagnetic field

For a specified volume of space V, possible values of $\vec{k}$ form a lattice as shown in figure B.1. From

$$k_x = \frac{2\pi n_x}{L}, \quad k_y = \frac{2\pi n_y}{L}, \quad k_z = \frac{2\pi n_z}{L}, \tag{B.1}$$

$$L = V^{\frac{1}{3}}, \tag{B.2}$$

each point in the lattice is surrounded by an empty volume (in k-space!) of $(\frac{2\pi}{L})^3$. Then, the number of modes $\mathrm{d}N_{k,k+\mathrm{d}k}$ with wave vectors between k and $k + \mathrm{d}k$ equals the number of lattice points in a spherical shell of radius k and thickness $\mathrm{d}k$,

$$\therefore\ \mathrm{d}N_{k,k+\mathrm{d}k} = \frac{4\pi k^2 \mathrm{d}k}{\left(\frac{2\pi}{L}\right)^3}. \tag{B.3}$$

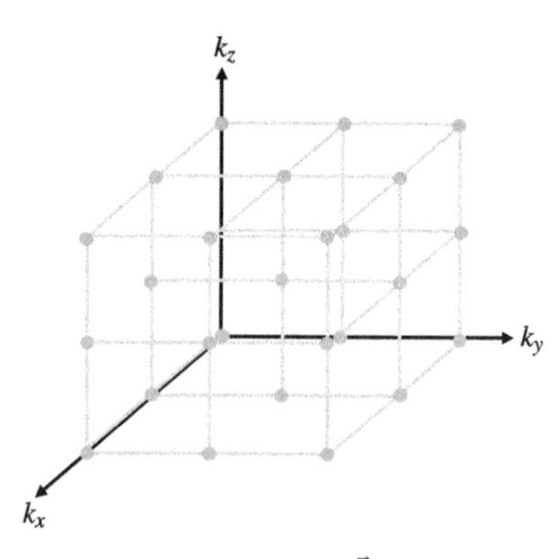

Figure B.1. A depiction of the distribution of values of $\vec{k} = (k_x, k_y, k_z)$ for the electromagnetic field.

doi:10.1088/978-0-7503-2171-6ch12

Thus, for a photon emitted into a solid angle element $\mathrm{d}\Omega$, the number of allowed states in an energy interval $\{\hbar\omega, \hbar(\omega + \mathrm{d}\omega)\}$ can be written as $\rho_{\hbar\omega,\mathrm{d}\Omega}\mathrm{d}\hbar\omega$, where

$$\rho_{\hbar\omega,\mathrm{d}\Omega}\mathrm{d}\hbar\omega = \mathrm{d}N_{k,k+\mathrm{d}k}\frac{\mathrm{d}\Omega}{4\pi}; \tag{B.4}$$

and from

$$\omega = ck, \quad \mathrm{d}\omega = c\mathrm{d}k, \tag{B.5}$$

$$\therefore \rho_{\hbar\omega,\mathrm{d}\Omega} = \cancel{4\pi}\frac{\omega^2}{c^2}\frac{\cancel{\mathrm{d}\omega}}{c}\frac{L^3}{(2\pi)^3}\frac{\mathrm{d}\Omega}{\cancel{4\pi}}\frac{1}{\hbar\cancel{\mathrm{d}\omega}}, \tag{B.6}$$

$$\therefore \rho_{\hbar\omega,\mathrm{d}\Omega} = \frac{V\omega^2\mathrm{d}\Omega}{(2\pi)^3\hbar c^3}. \tag{B.7}$$

CPSIA information can be obtained
at www.ICGtesting.com
Printed in the USA
BVHW010334281020
591959BV00003B/7